Advances in 21st Century Human Settlements

Indexed by SCOPUS

This Series focuses on the entire spectrum of human settlements – from rural to urban, in different regions of the world, with questions such as: What factors cause and guide the process of change in human settlements from rural to urban in character, from hamlets and villages to towns, cities and megacities? Is this process different across time and space, how and why? Is there a future for rural life? Is it possible or not to have industrial development in rural settlements, and how? Why does 'urban shrinkage' occur? Are the rural areas urbanizing or is that urban areas are undergoing 'ruralisation' (in form of underserviced slums)? What are the challenges faced by 'mega urban regions', and how they can be/are being addressed? What drives economic dynamism in human settlements? Is the urban-based economic growth paradigm the only answer to the quest for sustainable development, or is there an urgent need to balance between economic growth on one hand and ecosystem restoration and conservation on the other – for the future sustainability of human habitats? How and what new technology is helping to achieve sustainable development in human settlements? What sort of changes in the current planning, management and governance of human settlements are needed to face the changing environment including the climate and increasing disaster risks? What is the uniqueness of the new 'socio-cultural spaces' that emerge in human settlements, and how they change over time? As rural settlements become urban, are the new 'urban spaces' resulting in the loss of rural life and 'socio-cultural spaces'? What is leading the preservation of rural 'socio-cultural spaces' within the urbanizing world, and how? What is the emerging nature of the rural-urban interface, and what factors influence it? What are the emerging perspectives that help understand the human-environment-culture complex through the study of human settlements and the related ecosystems, and how do they transform our understanding of cultural landscapes and 'waterscapes' in the 21st Century? What else is and/or likely to be new vis-à-vis human settlements – now and in the future? The Series, therefore, welcomes contributions with fresh cognitive perspectives to understand the new and emerging realities of the 21st Century human settlements. Such perspectives will include a multidisciplinary analysis, constituting of the demographic, spatio-economic, environmental, technological, and planning, management and governance lenses.

If you are interested in submitting a proposal for this series, please contact the Series Editor, or the Publishing Editor:

Bharat Dahiya (bharatdahiya@gmail.com) or
Loyola D'Silva (loyola.dsilva@springer.com)

T. M. Vinod Kumar
Editor

Indo-Pacific E-Gateways

Cases from Digital Economic Communities

Springer

Editor
T. M. Vinod Kumar
School of Planning and Architecture
New Delhi, India

Kozhikode, Kerala, India

ISSN 2198-2546 ISSN 2198-2554 (electronic)
Advances in 21st Century Human Settlements
ISBN 978-981-95-4823-1 ISBN 978-981-95-4824-8 (eBook)
https://doi.org/10.1007/978-981-95-4824-8

This Springer imprint is published by the registered company Springer Nature Singapore Pte Ltd.
The registered company address is: 152 Beach Road, #21-01/04 Gateway East, Singapore 189721, Singapore

Contents

The Indo-Pacific E-Gateways in Digital Economic Communities

T. M. Vinod Kumar

Abstract This introductory Chapter is dedicated to the concept and applications of E-Gateways for digital economic communities in the Indo-Pacific region and beyond. This strategic vision aims to foster rapid economic development and cooperation among countries bordering the Indian and Pacific Oceans, as well as beyond. It provides a detailed explanation of E-Gateways, their functioning, and their alignment with global efforts to enhance digital infrastructure and connectivity, as well as international E-Governance, facilitating it through necessary protocols. It explores the role of megacities, international economic corridors, E-Global Governance, and digital public goods in creating and sustaining digital economic communities that are functionally connected to the outside world in a scenario of ever-changing global economics determined by international politics. The E-Gateway is an integral part of the overall entrepreneurial ecosystem, and the digital economic community is crucial for achieving last-mile economic and social connectivity. E-governance practices adopted by individual countries effectively facilitate digital economic communities, reaping economic development benefits within these communities. Thus, the introductory chapter presents case studies of E-Gateways for digital economic communities. The survey of Thrissur in Kerala is an empirical exploration of several digital economic communities to demonstrate the design of the Entrepreneurial economic ecosystem within these communities. Given the strengths, weaknesses, opportunities, and threats that digital economic communities face, an appropriate and timely strategic approach to economic development can be designed and implemented in response to changes in geopolitics and geo-economics as regional economic scenarios emerge. A case study of Estonia within the European Union was undertaken for the E-International Economic global government. As part of E-Governance, entrepreneurs seeking to tap into the European market for their digital economy can apply for an E-Residency and E-Identity card to register a firm in Estonia electronically and conduct business in Europe without being physically present or needing to travel through the European Union. Here, a digital economic community can partner with any entity in Estonia to register a firm and conduct business, including bank transactions and tax filing in Europe or abroad, without residing in Estonia or being

T. M. Vinod Kumar (✉)
School of Planning and Architecture, New Delhi, India
e-mail: tmvinod@gmail.com

T. M. Vinod Kumar (ed.), *Indo-Pacific E-Gateways*, Advances in 21st Century Human Settlements, https://doi.org/10.1007/978-981-95-4824-8_1

an Estonian immigrant or citizen. Such an E-Resident can use all E-Governance services of Estonia from outside, 24 h a day, 7 days a week. Therefore, all existing E-Governance stacks in Estonia are also discussed and elaborated on, concerning the benefits that the digital economic community entrepreneurs have accrued using them. The chapter illustrates how the fourth pillar of the Indo-Pacific Economic Framework of the QUAD, specifically tax and anti-corruption, can be digitally transformed to foster fair competition, accountability, and transparency in the region. Many emerging technologies can be leveraged by E-Gateways, including industrial applications of 5G, Industry 4.0, AI, Blockchain, Quantum computing, Cloud computing, and embedded IoT systems. They are discussed in detail, along with their usage patterns by digital economic communities, highlighting both advantages and disadvantages. Many countries and country groupings have legislated many acts that support the E-Gateways. Some examples from the ASEAN and European Union are presented for illustration. In addition, individual countries participating in E-Gateways can leverage their cultural and industrial potential through digital transformation and integration with the megacity system, the Indo-Pacific Economic Framework and International economic corridors. The chapter concludes with practical suggestions and examples for businesses and governments interested in participating in this initiative, providing a clear roadmap for their involvement. It concludes by highlighting key trends and technologies expected to shape the future of the digital economy, including AI, 5G, blockchain, and the metaverse. It underscores the importance of digital trust and security for the success of the digital economy, reassuring the audience that safety measures are in place.

Keywords E-Gateways · Digital economy · Digital economic core and periphery communities · Smart global economic community · The Indo-Pacific region · E-Gateways functioning · Digital infrastructure · Connectivity · Megacities system · International economic corridors · QUAD and I2U2 projects · Digital public good · Case studies-E-Gateways for digital economic communities Thrissur in Kerala · India · International E-Governance case study-Estonia the European Union · Digital transformation · Integration with the megacity system · The Indo-Pacific economic framework · The fourth pillar · Tax and anti-corruption · Competition · Accountability · and transparency · AI · 5G · Machine learning · Data mining · Blockchain · Wi-Fi 6 · Quantum computing · The metaverse · International unified payment interface (UPI) · DigiLocker · Open network of digital commerce ONDC · E-government Estonia · R-residency · ID card Estonia · Functions and benefits E-governance stack Estonia · Digital market act · The digital service act · Artificial intelligence act · 2030 digital compass · Digital skilling · ASEAN trade facilitation framework · The ASEAN free trade area · Cross-border E-commerce framework · Trade lens · Implementing road map · Geopolitical uncertainty

1 What is the E-Gateway, and How Does It Function?

The book uncovers the captivating history of Indo-Pacific E-Gateways for Digital Economic Communities. A city gate is a fortified gateway set within a city wall. These gates served various functions, including controlled access to and departure from a walled city for people, vehicles, goods, and animals. City gates fulfilled roles in defence, security, health, trade, taxation, and representation, depending on the historical context. Military or municipal authorities often staffed them. City gates were also used to display public information, including announcements, tax schedules, local ordinances, and legal documents. Some gates were heavily fortified and adorned with heraldic shields, sculptures, or inscriptions. In the Middle Ages, towns in the Indo-Pacific were fortified with gates, magnificent feats of art and engineering, and adorned with towering walls and a moat of water bodies guarded by crocodiles. Initially designed for security, these gates gradually disappeared as advancements in warfare technology occurred. Today, we witness the metamorphosis of these gates into gated communities, independent homes, and industrial plots, where we build single, group, or multi-storey apartments or flatted factories within a secure enclosure. We also encounter country gates, such as the India-Bhutan gate, at road entries, airports, and seaports, which serve as gateways (Fig. 1).

Physical gates for cities may or may not have practical uses, especially for megacities like Tokyo and Delhi, which span vast geographical areas. However, in the face of modern threats, such as hypersonic bombs with speeds of Mach 10, glider bombs, and deadly yet inexpensive drones, the traditional concept of gates has become obsolete. Yet, the essence of gates persists in various forms. It is more efficient and practical to transform these physical Gateways into virtual gates on a web page platform, enabling the seamless flow of goods and services and facilitating necessary Government functions.

The term 'E-Gateway' is derived from the use of electronic devices to operate gates in modern communities. This concept has further evolved with the advent of the Internet, which provides a more convenient and widely accessible platform for E-Gateways. The operation of these gates and devices, such as PCs, Macs, iPads, or Smartphones, requires the availability of Internet broadband.

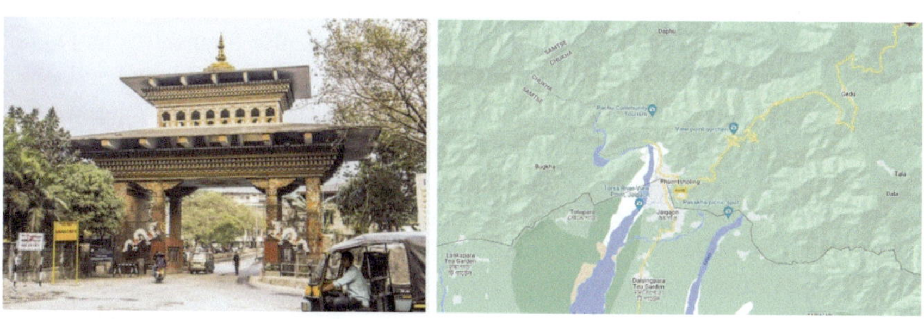

Fig. 1 Gateway to Bhutan from India at Phuentsholing

The "E-Gateway for Digital Economic Community" concept aligns with global efforts to enhance digital infrastructure and connectivity, fostering economic development. For instance, India is working towards digital transformation to augment its economy to $5 trillion or more, emphasising the importance of digital trust and security. The Digital India Mission aims to build a digital ecosystem based on these principles.

Similarly, the European Union's Global Gateway initiative aims to support infrastructure development worldwide, focusing on creating intelligent, clean, and secure connections in the digital, energy, and transport sectors. This initiative complements the EU-ASEAN economic cooperation, prioritising the digital economy and connectivity.

Moreover, the ASEAN Digital Economy Framework Agreement (DEFA) will establish a secure and interconnected ASEAN digital economy, transforming the region into a leading digital community and economic bloc by 2025.

These initiatives underscore the global recognition of digital infrastructure's crucial role in socioeconomic growth and resilience, particularly in the face of challenges such as the COVID-19 pandemic.

The Internet has permeated households at a remarkable pace. Smartphone access has increased significantly due to the availability of affordable smartphones and lower internet usage charges, making it accessible to all income groups. This widespread access to the internet and smartphones is crucial for the operation of E-Gateways, highlighting the importance of these technological advancements in the context of urban development.

There has been an advancement from 2 to 5G, with an increased possibility for E-Gateways. Now, 6G has an enhanced capacity compared to 5G in terms of latency and bandwidth. Now, remote heart surgery, which doctors have already performed in India, where doctors are located on faraway continents, operates with 5G, connecting one distant location to patients, groups of doctors, and support staff on another continent. This has evolved traditional telemedicine into telesurgery, allowing emergency surgery to transcend continental and national boundaries. It can be performed in remote and inaccessible locations for much-needed emergency surgery, provided the necessary preparations are in place.

Today, the world has free access to many public goods, such as a unified payment interface free of usage charges and numerous other accessible interfaces, making it easy to do business if households and individuals organise themselves as Digital Economic Communities. Therefore, the time has come to design a digital economic community worldwide that fosters faster-than-average economic development in specific locations. This would enhance the community's quality of life, as determined by its inhabitants, who primarily utilise their efforts to create emerging economic opportunities and become part of global supply chains for goods and services. It proposes an alternative Digital Economic Community model to the existing capitalist, socialist, and Chinese economic development models, featuring global factories and a Belt and Road initiative to transport goods from these factories worldwide.

Generally, a community of individuals executes economic activities in any country using their awareness, intelligence, entrepreneurship, and capabilities, which is not

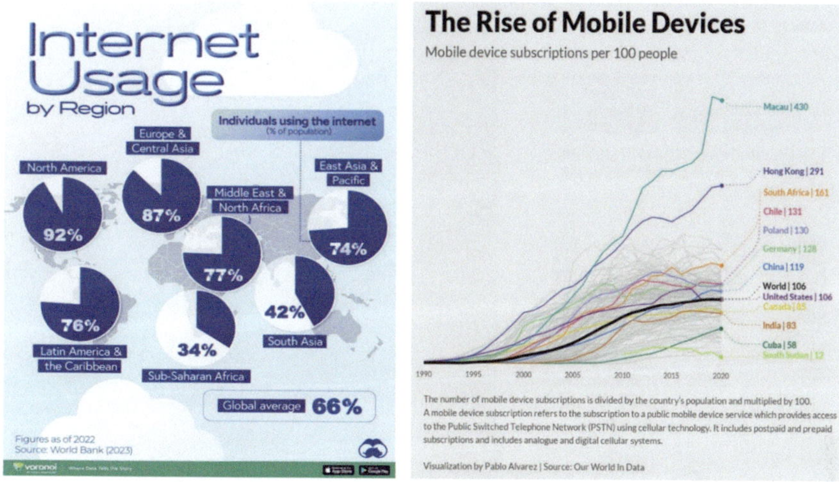

Fig. 2 Internet availability usage in 2022 and the rise of mobile users from 1990 to 2020

a job cut out for Government administrators, as practised in Communist countries in the name of socialism. A group of adjacent households can form an economic community, part of a global supply chain. These households can be non-contiguous, such as a community of software engineers and computer scientists jointly designing electronic networked e-commerce platforms or providing technical support to the digital economic communities that utilise the platform.

Internet usage and device ownership are presented below, illustrating the rapid progress in supporting global digital economic communities (Fig. 2).

Fewer women than men participate in economic activities in many countries in the Indo-Pacific region, including India and other Asian nations. Households, adults, and children often participate jointly in gainful economic activities. Existing laws do not permit children to work as labourers and prevent them from accessing the education to which they are entitled.

It is found that most workers, whether in India, China, or the rest of the Indo-Pacific, join the workforce by participating in the community's economic activities in a majority as self-employed from non-formal sectors of the economy or as casual/temporary labourers, followed by regular company employees in decreasing proportion, with private or public sectors, or government servants, in tiny percentage in the organised workforce in a firm as shown in Fig. 3.

Such existing patterns of work participation can be quickly reorganised into digital economic communities that enhance the household economy and pave the way for improved living according to their aspirations, through suitable organisation, leadership, and design of innovative digital community economies, primarily through self-help rather than government grants (Fig. 4).

The above survey challenges the actions of digital economic communities and Governments. The labour market is oriented towards lower educational levels, with

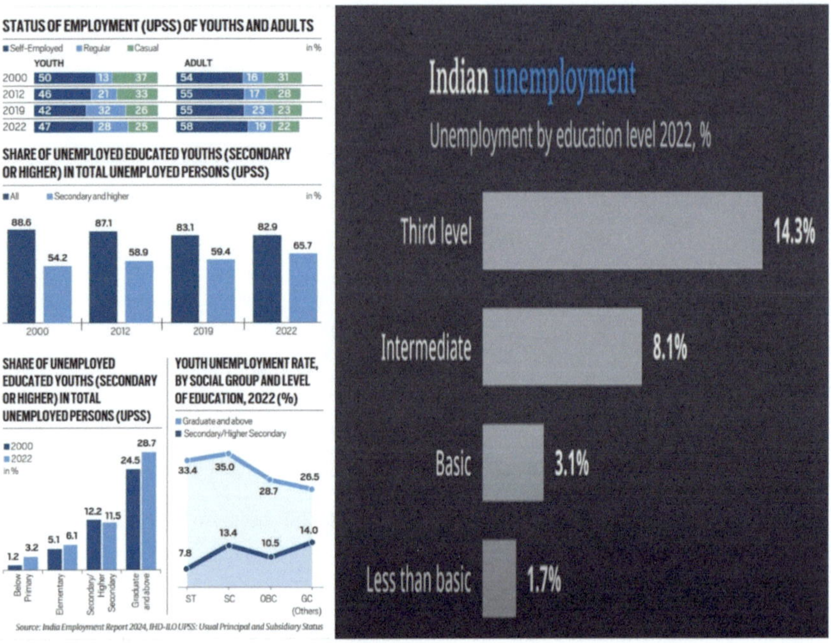

Fig. 3 Status of current employment of youth and adults by education level and unemployment by education level in India

Fig. 4 Comparison of male–female participation in the workforce 2021 of India and selected countries. *Source* ILO and Institute of Human Development, (2024) India Employment Report Youth, Employment, education and skill, PDF

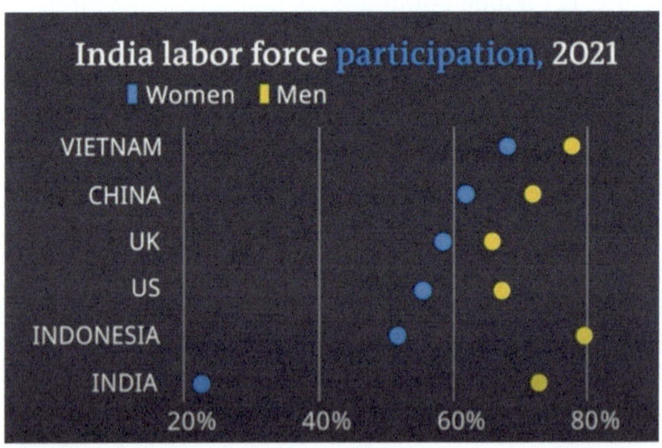

graduates and those with higher education primarily unemployed in a higher proportion. Such economic community products and services generally yield low-value additions. Graduates and postgraduates can unlock the full potential of the digital economic community if they are skilled, trained, and organised to be part of high-value-added supply chains. The aim should be to enable all eligible age groups, including graduates and postgraduates, to continually develop their skills and participate in high-value-added economic activities. Additionally, the community requires self-financing to educate all its members to the graduate level and ensure ongoing skill development and education. Many governments in the Indo-Pacific target literacy percentages as National goals and do not take action to increase the graduate enrolment ratio to 100%. The survey results indicate that economic development is more than just investment alone; the government's administration often governs the investment with a high likelihood of corruption and economic inefficiencies, due to chronic delays that hinder business ventures in the marketplace.

Large fish eat smaller fish in the ocean, and the king cobra, a giant snake, preys on all other snakes on land. This can happen to the Digital economic community. China has world-scale giant factories and excess capacity to produce unbelievably cheap industrial and household products with Government subsidies combined with under the economy of scale, supported by the Belt and Road initiative for logistics that gives "free passage through sovereign debt diplomacy" to their goods in the market of digital economic communities to kill all of them by unfair competition. China ranks second in GDP, with the immense capacity to subsidise its exports and undermine all local digital economic communities by flooding markets with cheaper products at artificially low commodity prices. The digital economic communities are then threatened with the loss of their sustenance, which compels smart digital economic communities to design and activate their E-Gateways.

An e-gateway for the digital economic community is urgently needed in the above circumstances. The next question is how it should function. E-Gateways for digital economic communities have two broad functions: they can sustain a global digital economic enterprise ecosystem in a specific location and facilitate its activation and self-regulation through global E-governance.

1.1 What Constitutes the Digital Economic Global Enterprise Ecosystem Design and Digital Economic Community Global Governance Design?

E-Gateway is the culmination of the Digital Economic Global Enterprise Ecosystem and the Digital Economic Community Global Governance System.

The Digital Economy is based on a study of the digital economic community's economic and financial strengths, weaknesses, opportunities, and threats, identifying the goods and services they can provide globally in the megacity's subregions and the economic corridors that extend beyond these subregions. This process evolves a strategy for the digital economic community based on the specialisation of local land use systems. It also reflects the Digital Economic Community's specialisation in land use configuration. Identify and self-finance the gaps in digital and physical public infrastructure/ goods and services that can gradually extend the economic community's ability to take advantage of more economic/financial opportunities, Feasibility assessment, and local community actions on selected program initiatives of QUAD,

I2U2, and other Bilateral Trades in the Indo-Pacific that can be implemented opens newer international opportunities.

Digital Economic Community Global Governance Design involves Digitally embracing the fourth pillar of tax and anti-corruption in the Indo-Pacific economic framework. "We are committed to promoting fair competition by enacting and enforcing effective and robust Taxes, anti-money laundering, and anti-bribery regimes aligned with existing multilateral obligations, standards, and agreements to curb tax evasion and corruption in the Indo-Pacific region. This involves sharing expertise and seeking ways to support the capacity-building necessary to advance accountable and transparent systems". Advocacy and determining the adoptability of legislations such as the Digital Services Act (DSA), Digital Markets Act (DMA), Data Act, Artificial Intelligence Act (AI Act), and Media Freedom Act at the National or subnational/Federal State Level can be another Governance support activity of the Digital economic communities.

1.2 Are Indo-Pacific E-Gateways for Digital Economic Communities Special and Unique? [1–7]

Yes, it is unique since each country within the Indo-Pacific is sovereign and has autonomy in making decisions. Nothing like NATO for collective defence or the European Union for unified Governance of the Indo-Pacific exists now, and it will not be possible soon. However, collective trade treaties, such as those of ASEAN with others, exist but do not involve significant countries like India in multilateral trade, which generally focuses on bilateral trade and autonomy in the country's defence.

1.3 What Constitutes the Unique QUAD Partnership Based on the Unity in Diversity in the Indo-Pacific Region, and How Does It Implement that in the Design of E-Gateways? [8–10]

The E-Gateway Design is unique for the Indo-Pacific Region and those outside it, and specific to the country where such gateways are located. The following discussion covers all aspects in a step-by-step manner.

The Indo-Pacific is a region that encompasses the two large oceans, the Indian Ocean and the Pacific Ocean. Four crucial countries of the Indo-Pacific—namely, the United States, Japan, Australia, and India—came together to form the QUAD, which aims to lead this initiative in the region. They have been formulating developmental and disaster management, defence-related programs, monitoring programs, and implementing essentially economic development programs, as briefly presented by the Author in his edited book [8]. QUAD has a common goal of achieving peace, tranquillity, and prosperity in the Indo-Pacific region. No countries within the Indo-Pacific are expected to be allies of the QUAD and follow all its programs. They can join some or part of some programs while abstaining from others totally or partially, thereby maintaining the country's autonomy. They have conducted several summits in these four countries in rotation, announcing and monitoring several QUAD programs. Peace and Tranquillity in the Indo-Pacific is the goal of QUAD, and it requires defence preparedness to face numerous threats effectively. The United States has established the Indo-Pacific Defence configuration and periodically revised its strategies. Additionally, some QUAD programs facilitate the joint monitoring of the Indo-Pacific Ocean by regional countries. Defence technologies are shared, as in the case of AUKUS. AUKUS is a nuclear-powered submarine configuration that protects the Indo-Pacific through the participation of Australia, the United Kingdom, and the United States. It aims to ensure peace and tranquillity through deterrence. E-Gateways, therefore, reflect the above unique character of the Indo-Pacific and can be used to implement individual partnerships based on countries' autonomy, emphasising and implementing unity in the diversity of the Indo-Pacific geopolitics and resulting geoeconomics that drive the digital economic communities.

1.4 *Indo-Pacific and Asia–Pacific: What Are the Key Differences?*

All countries in the US, Australia, the UK, and some others, as well as their establishments, refer to the Indo-Pacific, whereas China and Russia focus on the Asia–Pacific. It is essential to know the area of these regions. Below is a diagrammatic representation of the differences between the two areas. They denote two different places. Such discussions are better visualised by maps as given below (Fig. 5).

The beneficiary countries of the Indo-Pacific are listed below, along with some key parameters. Some Indo-Pacific countries are landlocked and rely on the Indo-Pacific for their international trade and movement. Many trade corridors under construction may make more countries dependent on the Indo-Pacific for trade movements (Table 1).

1.5 *What Role Does the Indo-Pacific Megacity System Play for E-Gateways? [12–27]*

Suppose you discuss Gateway trade functions in the commercial or industrial districts of Thimphu, Bhutan, or Kathmandu, Nepal, in the Indo-Pacific Region. In that case, you will be directed to an address in Kolkata Port, where all trade, import, and export functions are readily available for end-users. Again, suppose a Nepalese factory wants to market its product in India under a bilateral trade agreement. In that case, a

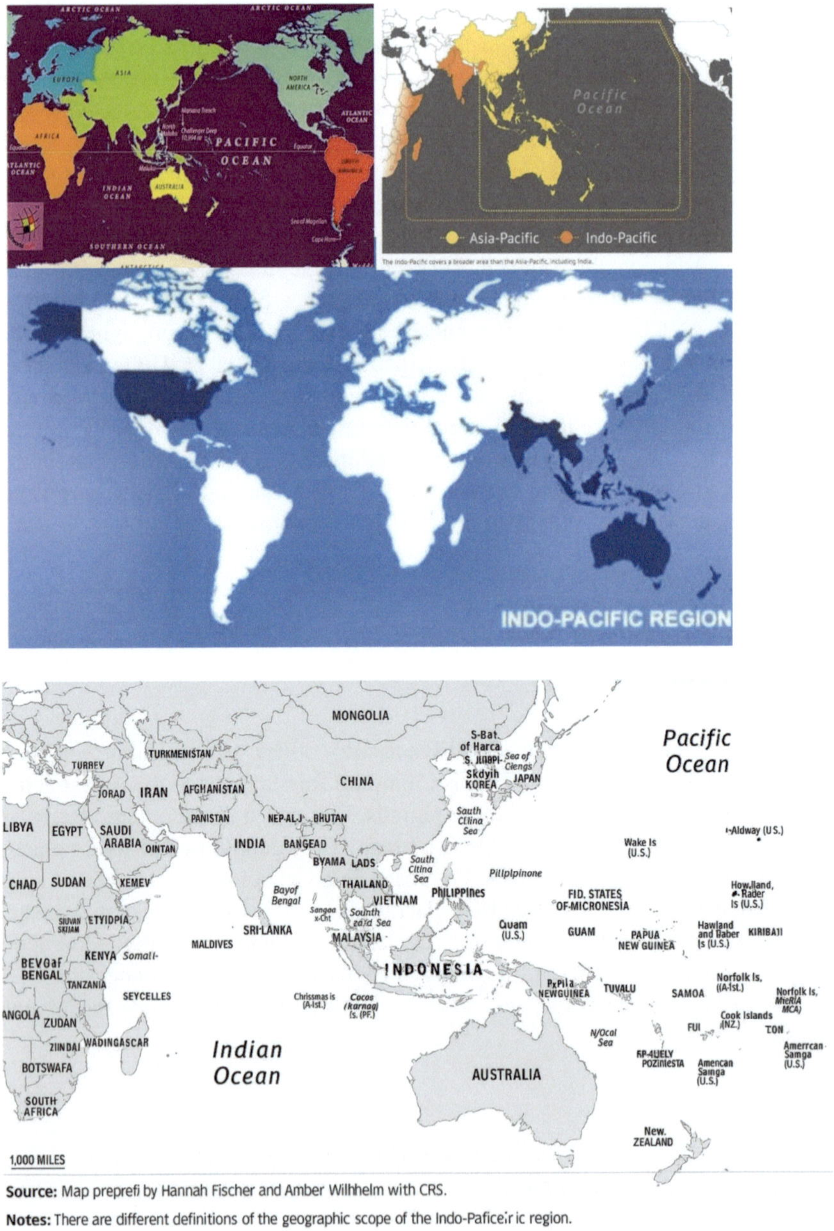

Fig. 5 Indo-Pacific Oceans, Indo-Pacific and Asia Pacific Regions and Indo-Pacific Region with small Islands. *Source* [5, 6]

Table 1 Major countries of the Indo-Pacific region population and GDP

S. No.	Country	Land Area (sq. km)	Population 2017	GDP in Millions US$ (2017) GDP per Capita US$ 2017	GDP per capita US$2017	Capital
1	Australia	7,692,024	23,731,000	1,482,539	49,928	Canberra
2	Bangladesh	147,570	156,594,962	205,715	1359	Dhaka
3	Bhutan	38,394	742,737	6384	2804	Thimphu
4	Brunei	5765	415,717	17,426	26,939	Bandar Ser Begawan
5	Cambodia	181,035	13,205,539	16,899	1270	Phnom Penh
6	Fiji	18,274	909,389	8330	5153	Suva
7	India	3,287,263	1,324,171,354	1,690,000	1939	New Delhi
8	Indonesia	1,904,569	252,164,800	856,066	3570	Jakarta
9	Japan	377,944	126,434,969	4,769,804	38,884	Tokyo
10	Laos	236,800	6,320,000	11,707	2353	Vientiane
11	Malaysia	329,847	30,185,787	336,913	9503	Kuala Lumpur
12	Maldives	298	341,356	4920	8602	Male
13	Myanmar	676,578	50,496,000	65,291	1275	Naypyidaw
14	Nepal	147,181	26,494,503	62,384	730	Kathmandu
15	New Zealand	268,021	4,357,437	201,028	39,427	Wellington
16	Papua New Guinea	462,840	6,732,000	16,096	2183	Port Moresby
17	Philippines	343,448	107,242,000	289,686	2951	Manila
18	Singapore	710	5,183,700	307,085	52,961	Singapore
19	Sri Lanka	65,610	20,277,597	233,637	3835	Colombo
20	Taiwan	36,191	23,119,772	505,452	31,900	Taipei
21	Thailand	513,120	67,764,000	380,491	5908	Bangkok
	Timor Leste					
22	United States	983,520	328,271,859	19,386,800	59,531	Washington
23	Vietnam	331,210	88,069,000	187,848	2186	Hanoi

Source [11]

market specialist will not give you a logistic plan encompassing villages and towns of India for marketing but ask you to move the product only to Kolkata from where the existing support system available will take the product to the sub-region of Kolkata and then to the rest of India with Nepalese or Bhutanese entrepreneurs involved in it.

No wonder we notice agricultural food self-sufficiency in feeding all 1.4 billion Indians today. Despite India being self-sufficient in feeding all citizens with locally produced food, not imported, the BJP confirmed that, in its 2024 election promise, those 80 crores (one crore is ten million) will continue to receive free rations for the next five years, since the COVID-19 days. Digital economic village communities accomplish it with their respective E-Gateways.

It is worth noting that the nation's wealth is mainly outside the rural areas that produce sufficient food products for India. Figures 6 and 7 illustrates this in urban areas of India.

The urbanisation trend in India indicates that the urban population is growing faster than the rural population and is expected to surpass it in urban areas soon. The urban share of GDP is also expected to increase proportionately to 75% by 2031. All these show that urban centres control gateway functions, and the larger the urban centres, the more advanced the gateway functions are handled. The same applies to the GDP share between rural and urban areas. The Indo-Pacific region has 21 megacities, each of which efficiently serves its catchment area, comprising both urban and rural areas, based on accessibility (Table 2; Figs. 8, 9 and 10).

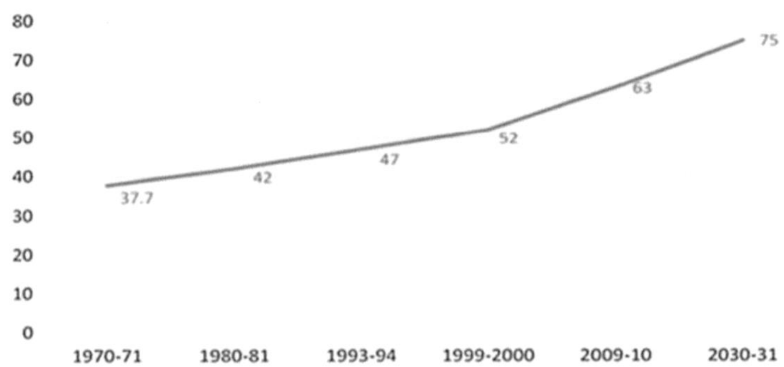

Fig. 6 Urban share of GDP in India 1970–71 to 2030–31. *Source* [10]

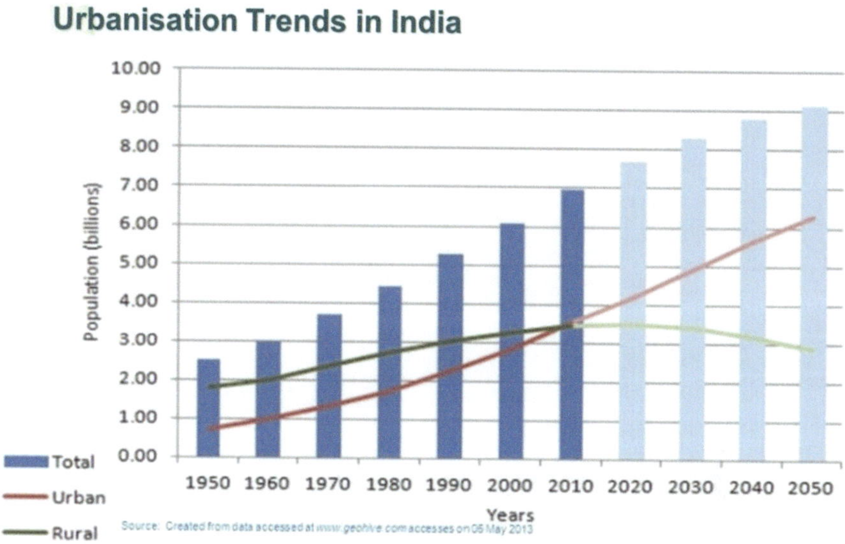

Fig. 7 Urbanisation of India 1950 to 2050

1.6 How Are the E Gateways of Digital Economic Communities of One Megacity Region Connected to Many in the Fastest Ways for Products and Services to Move?

E-Gateways are essential for every digital economic community worldwide, but the Indo-Pacific region is particularly crucial for several reasons. E-Gateway performs various digital functions, including an enterprise ecosystem for industrial production, supply chains, and services, as well as e-government services related to international trade, anti-corruption, money laundering, and taxation. The Indo-Pacific is a vast region encompassing the world's two largest oceans. The existing sea route via the Suez Canal is a circuitous route with heavy congestion, hindering Europe's connectivity to the Indo-Pacific, Russia, the Middle East, and Central Asian countries. Piracy by the Houthis and a possible attack by China as part of its aggressive and salami-slicing strategy are other threats in the Indo-Pacific.

The emerging trade relationship in the Indo-Pacific is outlined below, which includes provisions for the movement of selected goods and people (Fig. 11).

With the advent of sanctions against Russia following the Ukraine-Russia conflict, a group of beneficiary countries has attempted to establish connectivity between St. Petersburg and Mumbai, known as the International North–South Trade Corridor (INSTC). This corridor, which benefits Russia, India, and the countries of Central Asia, and the Chennai to Vladivostok Sea route, which benefits Asian Russia and India, are being constructed. The multi-model connectivity between countries in

Table 2 Megacities and Meta cities in the Indo pacific 2016–2030

Rank	City, country	Population in 2016 (thousands)	City, country2	Population in 2030 (thousands)
1	Tokyo, Japan	38,140	Tokyo, Japan	37,190
2	Delhi, India	26,454	Delhi, India	36,060
3	Shanghai, China	24,484	Shanghai, China	30,751
4	Mumbai (Bombay), India	21,357	Mumbai (Bombay), India	27,797
5	Beijing, China	21,240	Beijing, China	27,706
6	Kinki M.M.A.(Osaka)Japan	20,337	Dhaka, Bangladesh	27,374
7	Dhaka, Bangladesh	18,237	Karachi, Pakistan	24,838
8	Karachi, Pakistan	17,121	Kinki M.M.A.(Osaka)Japan	19,976
9	Kolkata (Calcutta), India	14,980	Kolkata (Calcutta), India	19,092
10	Chongqing, China	13,744	Guangzhou, Guangdong, China	17,574
11	Manila, Philippines	13,131	Guangdong, China	17,574
12	Guangzhou, Guangdong, China	13,070	Chongqing, China	17,380
13	Los Angeles-Long Beach-Santa Ana, USA	12,317	Manila, Philippines	16,756
14	Tianjin, China	11,558	Bengaluru, India	14,762
15	Shenzhen, China	10,828	Tianjin, China	14,655
16	Jakarta, Indonesia	10,483	Chennai (Madras), India	13,921
17	Bengaluru, India	10,456	Jakarta, Indonesia	13,812
18	Chennai (Madras), India	10,163	Los Angeles-Long Beach, Santa Ana, USA	13,257
19			Lahore, Pakistan	13,033
20			Hyderabad, India	12,774
21			Shenzhen, China	12,673

Source United Nations, 2016 "World Cities in 2016 Data Booklet", New York. United Nations, 2016 "World Cities Report 2016", New York

Europe, the Middle East, and India to the Indo-Pacific, which considerably shortens the distance, is being implemented through the market mechanisms of these regions. The I2U2 and QUAD initiatives benefit from establishing connectivity to the Indo-Pacific through the India-Middle East-Europe Economic Corridor (IMEEEC), a significant development in the region's economic connectivity (Fig. 12).

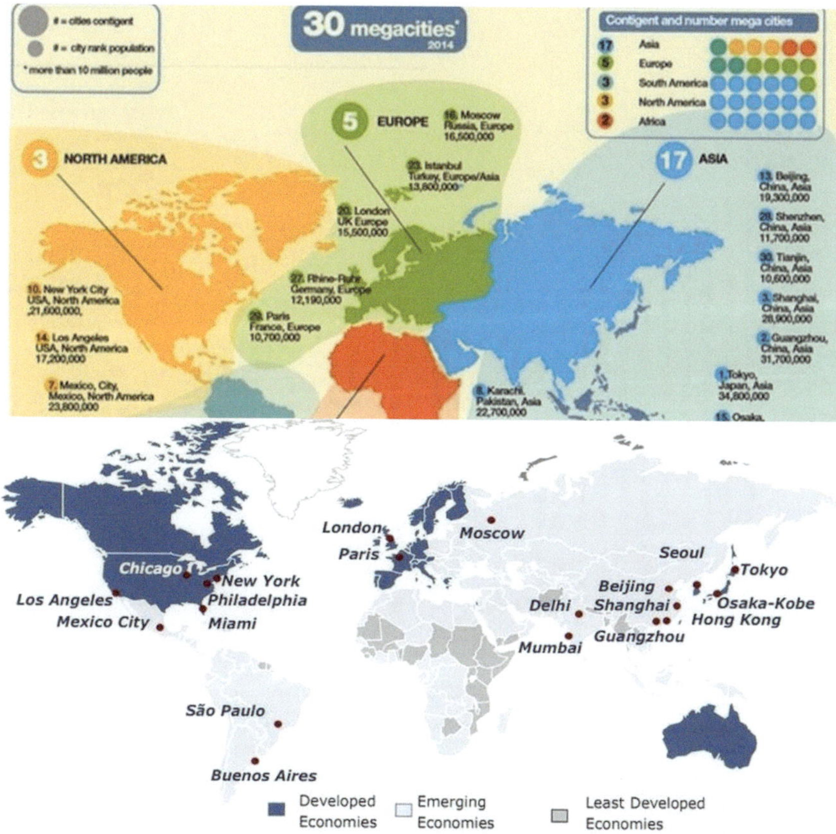

Fig. 8 Megacities in 2014 and megacities in 2025 by country classification

1.7 How Gateways Evolved from Communities of Ancient Towns to Today?

Gateways are transformed into E-Gateways, first as pieces of art, craft, and architecture for security and cross-border Governance, then into visible computer platforms as part of any economic activity, such as payment or logistic gateways, and finally invisible as part of any platform related to any part of the business process.

The need for a gateway has existed since immemorial, even before the megacity system, which facilitated Cross-Border Trade before the Digital Age. Trade is facilitated nowadays by megacity systems, especially in megacity international regions. First, it provides a ready market with a considerable population and purchasing capacity, and then it gives an already existing marketing system from megacity subregions. Then, many trade and economic corridors came first, such as the Spice Routes or the Silk Routes. It traded and exchanged culture, religion, civilisation, science, philosophy, and scholarship between countries. Then came international

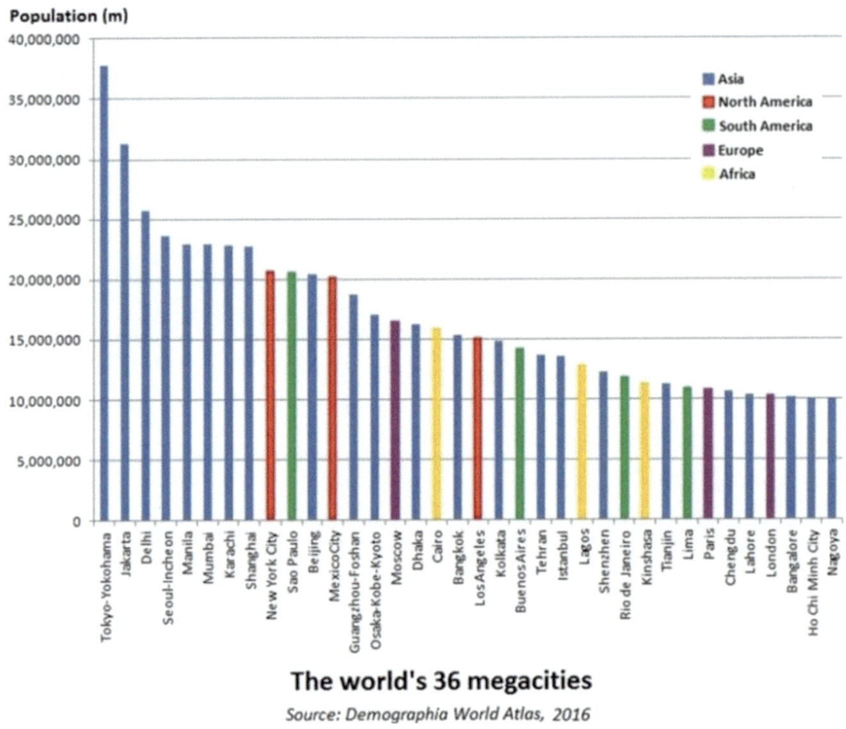

Fig. 9 Population of thirty-six megacities

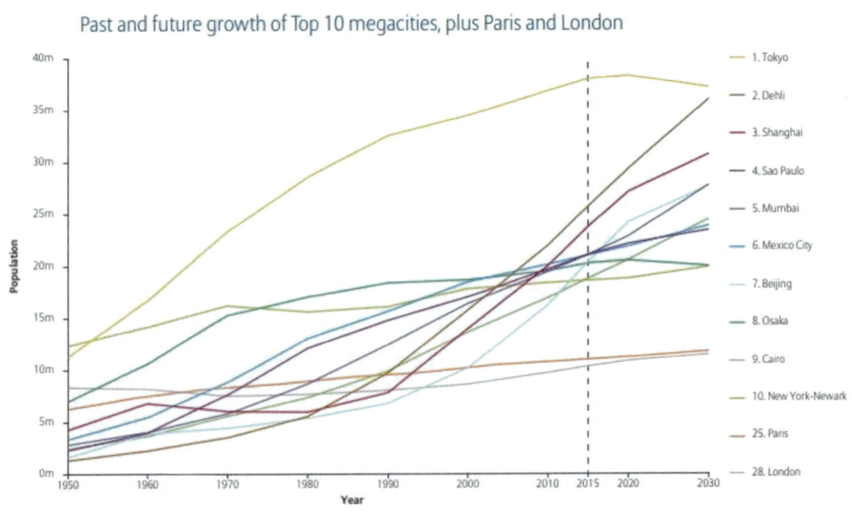

Fig. 10 Population growth patterns of selected megacities. *Source* United Nations Population Division: World Urbanisation Prospects 2014 edition 24–4-2022

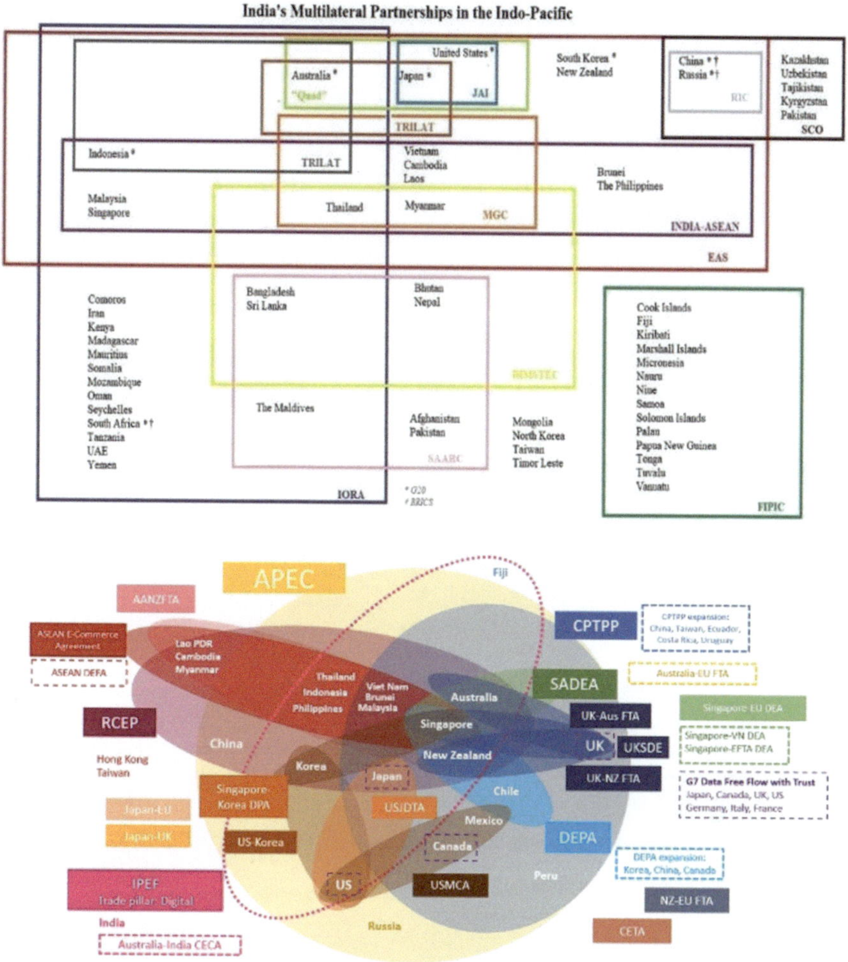

Fig. 11 Emerging trade relations and institutions. *Source* [7]

trade, utilising economic corridors that connected many megacity regions into a single, long trade route, thereby increasing global population as part of the global economic system. It is worth noting that there were no megacity systems during the ancient Silk and Spice Route era. Still, international trade routes with necessary infrastructure shared by traders existed, were significantly different from modern corridors, and are conserved as heritage today.

There are different models of E-Gateways: Public, Private, and Hybrid. The Chinese initiated the Belt and Road project, providing easy loans with sovereign security at higher interest rates than the prevalent ones and constructing corridor infrastructure with Chinese labour, engineers, and materials. There was purely private sector, market-driven corridor infrastructure construction, such as the India-Middle

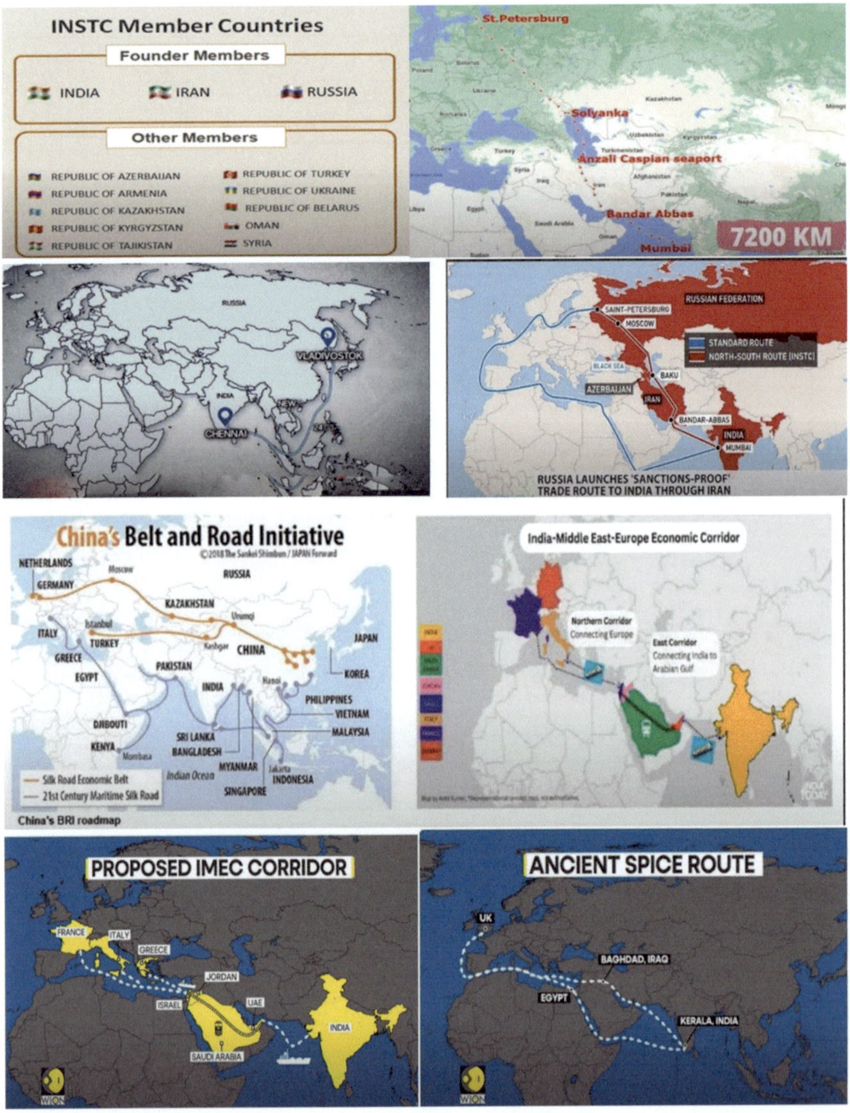

Fig. 12 International North–South Trade Corridor (INSTC), Vladivostok-Chennai corridor, Northern Sea route, Ancient Spice Route. Belt and Road Project, and India-Middle East-Europe Economic Corridor (IMEC)

East-Europe Economic Corridors, or another hybrid where the government and Private sector jointly executed and managed international economic corridors.

E-Gateways were considerably easier to embrace and efficient. Their function is to develop an efficient business ecosystem with megacity regions and long economic

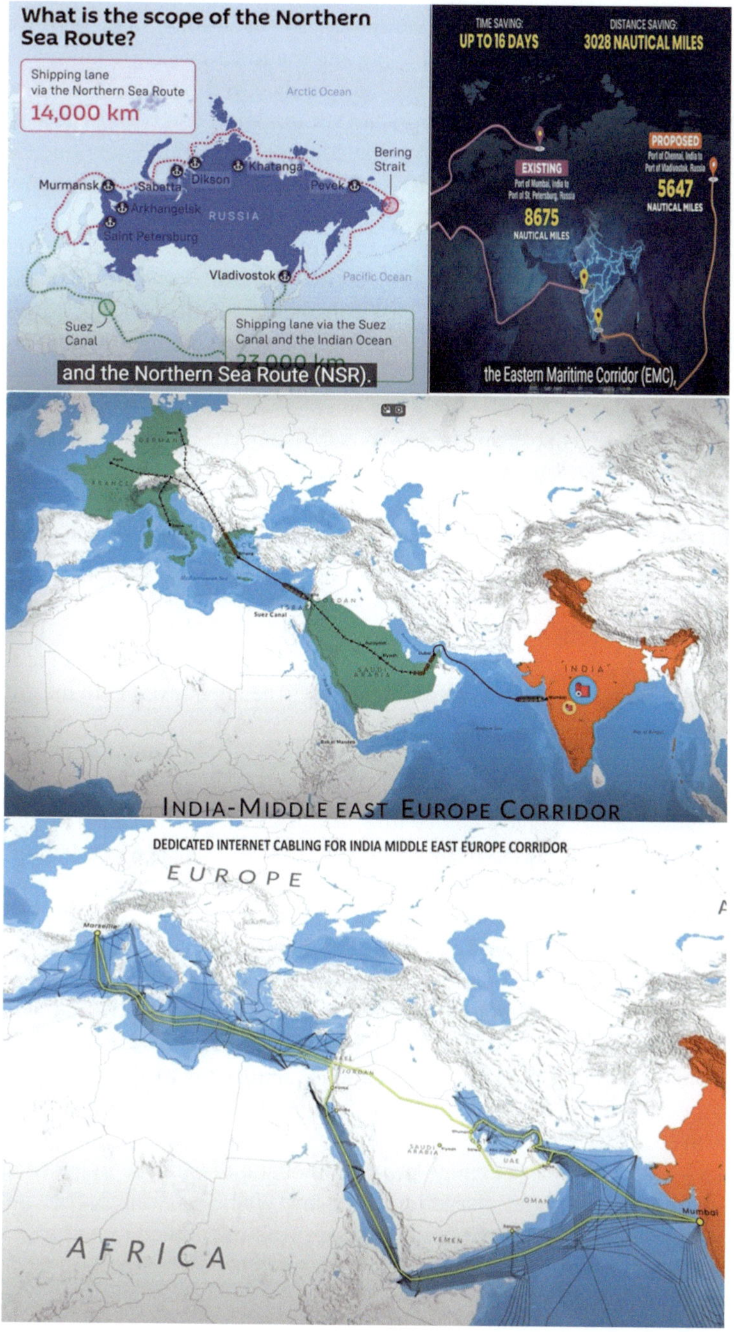

Fig. 12 (continued)

corridors passing through many megacity regions. In addition, they perform international governance functions by obeying international and national laws related to economic activities. This highlights transparency and protection of local business systems from any unacceptable practice of dumping goods and services to kill local economic activities at low prices. The concept of 'Digital Economic Communities' refers to the integration of digital technologies into the economic activities of megacity subregions, ensuring they are organised and efficient.

E-Gateway protocols can create an acceptable regulatory environment and universal standardisation, ensuring high quality of products and services and security in all business transactions. It is easier to succeed if all economic activities of megacity subregions are organised as Digital Economic Communities and provided with E-Gateway access to all these communities and households.

Digital technologies are also used for interactive connectivity across megacity regions to conduct business and communicate using videos, if required, for inter-country business interaction. These may include using 8 k resolution videos and images with augmented reality, real-time money transfers in different country currencies, and simplifying country customs administration for digital governance related to international trade.

Robotic interaction is available in one distant megacity using electronic means. To illustrate one example in surgery, New York Hospital can participate collaboratively in robotic heart surgery in Delhi Safdarjung Hospital without being physically present there, and such a facility exists if one uses the latency of 5G in the emerging technologies.

A cohesive factory floor can be integrated between two distant megacities in two countries and still have integrated robotically operated smart co-production that AI and communications channels can control independently of humans. We have also seen the space technologies used in operations over the planets Mars and the Moon for landing and moving hardware from Earth. These could be done by a group of countries instead of one country, such as multi-country space stations and multi country production and delivery of services globally.

The Digital Economic Community is integral to the international and regional megacity system. For example, Indo-Pacific keen to uplift their multi-national communities to a desired standard of living commensurate with their capability to use digital technologies/digital public goods and the ability for hard work and willingness for households and entrepreneurs to operate internationally, and then the targeted GDP for an enhanced living of the community is realised using emerging geopolitical opportunities generated as a change in accommodation/resilience in geoeconomics as part of the supply chain disruption for many emerging war/conflict points in the regional megacity system whose benefit are well distributed by purpose and by design among all households since they know each other and since they are face-to-face communities within walking distance. With emerging 5G and above communication and AI-driven technologies, spatially, co-production and sharing services transcending national boundaries are entirely acceptable to digital economic communities since response time for such co-production is beyond the human nervous system. To become an individual digital household, they require 5G-based mobile and broadband internet connections with appropriate technology based on affordability and convenience. Some state government projects, like K-Phone in Kerala, India, connect the lowest-income households with fibre optic internet connections free of cost.

1.8 *Objectives and Scope of the Book*

The book explores E-Gateways of digital economic communities in the Indo-Pacific region and those outside the area who wish to develop digital economic communities, focusing on the Indo-Pacific. The first chapter examines the concept of E-Gateways in all geographic settings. A few case studies follow this; the other chapters are on case studies, and the final chapter is on the conclusions.

2 How Did the Rise and Evolution of the Digital Economy, the Emergence of Global Digital Economic Communities, and Their Management of E-Gateways Materialise? [28–32]

The digital economy refers to the economic activities that emerge from connecting individuals, businesses, devices, data, and operations through digital technology. It encompasses online connections and transactions across multiple sectors and technologies, such as the Internet, mobile technology, big data, and information and communications technology. It also refers to the social bonding of the face-to-face community for community-based local economic development and the economic survival of digital economic communities in times of war, pandemic, natural calamities, and ever-changing geopolitics. The digital economy differs from the traditional economy because it relies on digital technology, online transactions, and its transformative effect on conventional industries. Digital innovations such as the Internet of Things (IoT), artificial intelligence (AI), virtual reality, blockchain, and autonomous vehicles create a digital economy.

A community comprises families or households that come together to achieve common goals of peace, tranquillity, and prosperity. Economic communities can be spatial and aspatial. It is spatial if the community's households live adjacent to each other in a neighbourhood. The community becomes aspatial when using a shared computer platform to enable industrial productions in multiple countries coordinated by Artificial Intelligence, automation and blockchains for economic development or jointly develop a platform for international transactions on commerce, logistics and industries, such as an Open Network of Digital Commerce/logistics/industries cutting across boundaries of nations and sectors of economies. Digital Economic

Communities intend to create a community with higher-than-average incomes, high-paying employment, and enhanced living standards by using digital technologies in community living and working. We call these communities global if they can transact internationally. This is achieved in two ways. First, the community identifies as part of the Indo-Pacific Megacity System. It intends to use all required and existing megacity system capacities for the community's Digital and Global Economic Development. Secondly, the community wants to become a customer of digital economic services provided by many economic corridors, able to move products faster, with ease, and with less bureaucracy, where all paperwork is performed digitally compared to manually.

The Indo-Pacific region has shown rapid economic development recently, with China first, followed by India and many ASEAN countries exhibiting rapid growth. The GDP growth from 1990 to 2022 in China and India, as well as the projection of the relative rise in GDP by PPP 2016–2050 of leading emerging markets and G7 countries, and their global share of GDP are given below. Their share of the global GDP trend is also shown. Further, the richest and emerging markets, with no reference to population size, are also shown for comparison. We present a time series of data on countries' rankings by GDP here (Fig. 13).

The first question is, why are these changes? Why can the G7 countries not sustain their ranking? Why do emerging markets seem to dominate more than the G7 countries? What should governments, both G7 and emerging markets, adopt in such circumstances? The answer is a development approach based on geopolitics, to a certain extent, but only partially, based on the country's development model and leadership.

When we study history, we see the rise and fall of many countries. We had many empires in the past. We have studied the rise and fall of the Roman Empire. Once upon a time, we used to say all roads lead to Rome. Nowadays, with Trillions of dollars of China invested in Belt and Road initiatives, we can say all roads are leading to China all over the world. We have also seen the once worldwide colonial power of the United Kingdom disappearing, and the UK is no longer a superpower or the leading economy of the world. The UK had a GDP of the ninth rank in 2016 and will become tenth in 2050. India is steady in the third rank based on PPP and will soon be in the real GDP. The only rational answer for changes in GDP is to follow the changes of time series data on critical variables, and based on that, each country adjusts their stance, which is most beneficial in the short and intermediate range.

We all know we have superpowers with higher economic, defence, and nuclear capabilities, such as the USA, China, and Russia. Some countries with inferior defence capabilities compared to superpowers and GDP are called middle powers, like India and some BRIC and G7 countries. Why are not all countries allies of one or other of the superpowers, and why are some just partners? The sovereign government decides what level of autonomy in economic development and self-defence the country wants based on the parameters the country presents itself in the short and intermediate range. Smaller countries seem more accepting of being allies than prominent countries with vast geographic areas and populations. These countries like to relate with other countries in partnership with a superpower or regional

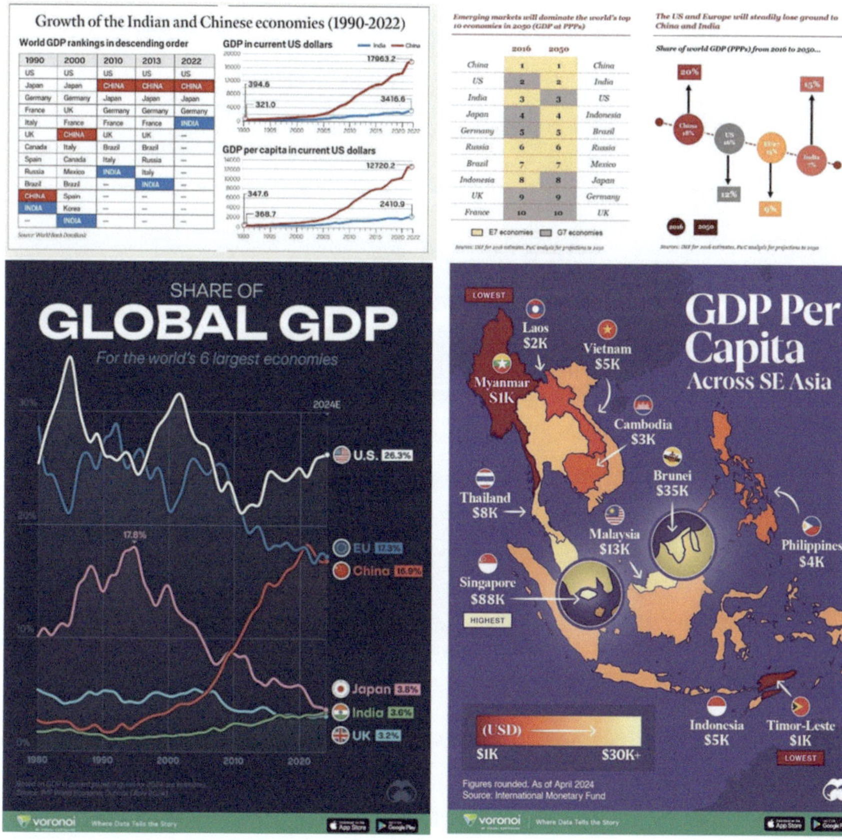

Fig. 13 GDP of G7 and emerging markets and ASEAN countries

hegemon who may immediately help them in case of aggression or calamities while maintaining the country's autonomy in decision-making. While Superpowers can project defence forces around the universe in collaboration with allies, the regional hegemon can reach the region during war and natural disasters. An excellent example of a superpower is the USA and its involvement in World Wars 1 and 2, later in Iran, Iraq, Palestine, Libya, Afghanistan, Vietnam, and now in Ukraine. An example of a regional hegemon is India or China, with the aspiration to grow as a superpower with global reach. The regional hegemon shared its vaccines regionally or even internationally. It helped countries during the tsunami and other natural calamities or military aggression, such as India helping with defence forces during China's aggression with Bhutan or the Bangladesh liberation war (Fig. 14).

The sustenance of global economic development because of ongoing inter-country economic transactions can never be assumed or taken seriously. There can be unexpected twists and turns to anticipate concerning sudden and unanticipated changes in geopolitics triggered by unrelated events that have nothing to do with economics but

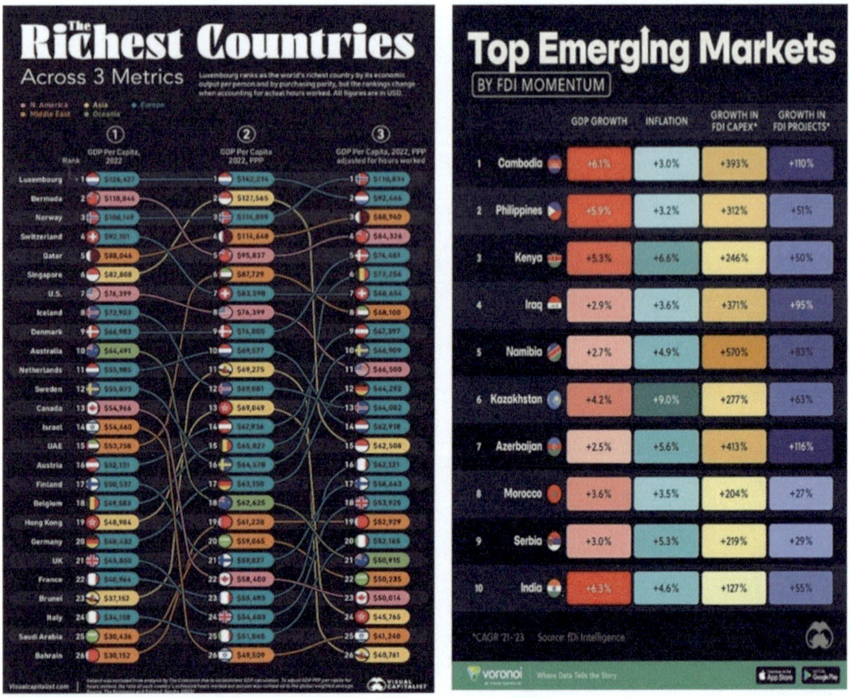

Fig. 14 Richest countries and emerging markets by three parameters of GDP

change the current geoeconomics. Therefore, as the situation emerges, the community shall embrace and adjust the two aspects of E-Gateway, namely the redesign of the ecosystem for entrepreneurs and the global governance system that the country can embrace, given the current world events, to achieve the desired goals and objectives from time to time.

E-gateway, in tune with geopolitics, geoeconomics, and the resulting entrepreneurial ecosystem, provide a pathway to identify international collaboration for producing and distributing goods and services in ever-changing international circumstances. This is a brilliant opportunity to investigate pathways using artificial intelligence's generative large language models to formulate suitable strategic interventions.

The second pathway of the E-gateway is for producing standardised goods and services at the most competitive price in conformity with all national and international laws regarding trade, commerce, and industrial production and compliance with goods and services tax and income tax internationally, with the least tolerance for corruption and money laundering.

Another aspect of E-Gateways is the E-governance of the digital economy at the community, national, and subnational levels. Such governance needs to be totally and 100% digital E-Governance for efficiency and transparency, and to avoid the usual bureaucratic delays, the hallmark of post-colonial nations in the Indo-Pacific.

Digital innovations such as the Internet of Things (IoT), artificial intelligence (AI), virtual reality, blockchain and autonomous vehicles all play a part in creating a digital economy. In its earliest days, the digital economy was sometimes called the new or web economy because it relied on Internet connectivity. Therefore, one definition is economic value derived from the internet. However, economists and business leaders assert that the digital economy is more advanced and complex than the internet, such as the transition from the third industrial revolution to the fourth industrial revolution. The third industrial revolution refers to the changes in the late twentieth century with the transition from analogue electronic and mechanical devices, such as robotics, to simple digital technologies. The fourth industrial revolution builds on the digital revolution of the third industrial revolution as digital technologies continue to bridge the physical and cyber worlds. The digital economy continues to evolve and expand rapidly, with emerging technologies and innovations shaping its trajectory worldwide and not confined to certain countries.

The following are some notable examples of the digital economy's evolution:

Digital trade and e-commerce. The surge of e-commerce, such as Amazon, Alibaba, Flipkart and eBay, has transformed online buying and selling, reshaped retail, and created new technologies and business models. Some are open-source and shared as an Open Network of Digital Commerce; others are not, such as Amazon's model.

Social media. The emergence of social networking platforms such as Facebook, X (old Twitter), Instagram, and LinkedIn has changed how people communicate, connect, and promote their products and services, including brazen and unethical self-promotion.

They increased remote work adoption. The pandemic caused a change in workplace culture as more people accepted remote work and began using apps such as Zoom, Slack, and Microsoft Teams to promote online collaboration. Due to this trend, the digital economy has evolved, which has reshaped how businesses function and manage their workforce.

Omni-channel approach to sales. Many retailers reach and serve customers through multiple channels, such as online sales and mobile apps. This lets them identify buyers, whether they're shopping via the internet or in person. They can collect and analyse customers' browsing and sales data to better understand their interests. They can use that data to reach customers via social media, enabling better service and, ultimately, higher sales and increased brand loyalty.

AI and automation have significantly shaped the digital economy. Virtual assistants, chatbots, and recommendation algorithms powered by AI improve consumer experiences and provide more personalised services.

Digital payments and cryptocurrencies. Digital payment systems such as the Unified Payment Interface of India, PayPal, Venmo and mobile wallets have changed how people conduct financial transactions.

Digital entertainment. The entertainment industry has undergone significant changes due to the rise of streaming services such as Netflix, Spotify, and YouTube. These platforms have revolutionised media consumption by providing instant access to various content based on their browsing history.

Telemedicine and Telesurgery. The COVID-19 pandemic accelerated the spread of telemedicine and made remote medical care possible through digital platforms. Today, telehealth is a crucial component in providing total healthcare. The low latency of 5G is now making it possible to move from telemedicine to telesurgery with multi-centre participation in surgery using robots.

The sharing economy. The sharing economy has transformed how people share resources such as cars, lodging and services, as exemplified by the Uber, Airbnb and TaskRabbit platforms. Peer-to-peer sharing has reshaped traditional industries and made new business opportunities possible.

"The digital economy highlights the opportunity and need for organisations and individuals to use technologies to execute tasks better, faster, and often differently than before. The related concept of digital transformation encompasses opportunities for existing entities to do better, do more, do things differently, and do new things. Numerous entrepreneurs seized on the technologies that fuel the digital economy to create new companies and business models that couldn't have existed or existed at the size and scale they do today in past generations. Examples include the ride-sharing platforms Uber and Lyft, the home rental platform Airbnb, and content-on-demand services such as Jio, Google, Netflix, and Spotify. The digital economy is expanding rapidly with new technologies that improve connectivity, enable automation, advance data analysis, and create new business prospects".Standard technologies that are accelerating the digital economy include the following:

AI. AI technologies, including generative AI, machine learning, and natural language processing, facilitate automation, data analysis, and decision-making for organisations across various industries. With the help of AI-powered systems, businesses can analyse large amounts of data, improve customer experiences, automate activities, and increase operational efficiency.

5G. 5G technology enables rapid downloads, low latency, and a wide range of device connections. 5G offers many advantages, including facilitating smooth data transfers, enhancing mobile and TV experiences, and fostering the development of innovative applications and services.

Wi-Fi 6. Compared to earlier Wi-Fi standards, Wi-Fi 6, also known as 802.11ax, provides faster data transfer rates, decreased latency and increased network efficiency. It also accommodates the increasing number of connected devices and the demand for high-bandwidth applications, making connections faster and more dependable, especially in congested areas. Now, Wi-Fi 7 is being marketed as further advancement.

Augmented and virtual reality technologies revolutionise gaming, education, healthcare, and training by developing immersive experiences and simulations.

Blockchain. Blockchain technology enables decentralised and secure recording and verification of transactions. It eliminates the need for intermediaries and secures digital transactions' transparency, immutability, and trustworthiness. This technology is transforming Industries, including finance, supply chain management and healthcare.

IoT is a system of networked sensors and devices used for data collection and exchange. By enabling the fusion of physical items with the digital world, this technology creates new possibilities for automation, real-time monitoring, and data-driven insights. Smart homes, smart cities, smart agriculture, and industrial automation are some areas where IoT applications improve efficiency, productivity, and convenience.

Quantum computing. While still in its early stages, quantum computing can tackle complex problems at previously unheard-of speeds. It has applications in cryptography, materials science, and optimisation.

Advantageous

The digital economy provides numerous benefits, contributing to its rapid expansion and positive effect on various industries. Businesses can improve their productivity and efficiency by automating their operations and processes using digital technology.

Reduced costs. Cloud computing and digital frameworks eliminate the need for substantial physical infrastructure and capital expenditures, enabling organisations to scale up and down as needed.

Extended reach. Businesses can foster a global economy and presence through online platforms and technologies, thus expanding their customer bases and market opportunities.

Access to more data. The digital economy produces large amounts of data that can be analysed for insights, trends, and data-driven decision-making. Businesses can access this data to better understand customer behaviour, customise experiences, and increase operational effectiveness.

Greater convenience. Consumers can purchase digital goods and services from the convenience of their homes. E-commerce and mobile commerce let customers buy products and services whenever and wherever they want.

We have improved customer experience. Businesses can deliver faster and more responsive customer service through digital channels and chatbots.

Personalisation. Data analytics and AI allow businesses to customise products, services, and marketing campaigns, ultimately improving customer satisfaction.

Disadvantages

Privacy and security concerns. The digital economy is significantly dependent on acquiring and storing personal data, which can create data privacy and security issues. Events such as data breaches, cyber-attacks and unauthorised access to private records can lead to financial losses, identity theft and various adverse outcomes.

Waves of disruption. The digital economy has created new companies and new ways of interacting. However, many companies and industries that didn't or couldn't capitalise on the technologies to change their operations have faced declining sales, falling market share and even complete collapse. For example, Blockbuster and other content rental shops that needed to adopt streaming technologies quickly enough shuttered their operations. The taxi industry is another example, as it needs help to compete with customers who find Uber and Lyft easier to use.

Job displacement. Automation and digitalisation can displace jobs, rendering some roles obsolete. Individuals might need to acquire new skills for ongoing employability, which can cause temporary unemployment and economic disruption.

Monopoly. The digitalisation of the economy has resulted in a small number of large providers, such as Apple, Amazon, and Google, gaining substantial power, resulting in monopolistic conditions in specific sectors.

Digital divide. The existence of a digital divide, which refers to the disparity between those who have access to technology and those who don't, is a prominent disadvantage of the digital economy. This division can result in inequalities concerning access to information, education, employment prospects and economic advancement.

Environmental footprint. The digital economy's energy use in data centres and electronic device production has ecological consequences, with rising demand for digital services leading to more significant carbon emissions, e-waste, and a more extensive environmental footprint.

The following key trends and technologies are expected to shape the future of the digital economy:

AI and machine learning. AI, including machine learning, deep learning, and neural networks, is advancing and will play a growing role in the digital economy. AI is essential for extracting insights from big data, automating complex tasks, making predictions, and managing autonomous systems and robots.

Transformation of traditional sectors. The digital economy is also changing conventional industries such as agriculture. For instance, smartphone apps can help farmers get real-time updates on crop quality, soil conditions, and irrigation.

Digital connectivity. Robust broadband connectivity and infrastructure are crucial to the success of the digital economy. Technologies such as 5G are predicted to play a vital role in enhancing digital connectivity. They will result in quicker and more reliable communication while supporting the growth of the digital economy.

The metaverse. Immersive technologies like the metaverse can create new consumer experiences and open, innovative business applications. These digital environments have the power to revolutionise entire sectors, and in the future, a parallel universe with a distinct financial and economic system might also exist.

Healthcare transformation. Integrating telemedicine and digital health tools and applications is expected to improve healthcare delivery and accessibility.

Cybersecurity advancements. The rapid adoption of a digital economy is evolving cybersecurity measures to address increasingly sophisticated cyber threats, including AI-powered attacks. By applying machine learning algorithms, AI-powered cybersecurity systems can detect anomalous behaviour, identify potential vulnerabilities, and proactively lower risk factors.

2.1 *What Happens When the Digital Economic Communities Use the Extraordinary Capabilities of Megacity Systems? [33–35]*

Emerging communication technology, such as the Internet-based, 5G industrial applications in production and services, AI, and Iot and its applications in the community for action-oriented international and regional development rooted in the Megacity system of the Indo-Pacific, is the goal of the Design of a Digital Economic Community in the Indo-Pacific. One visualises that the Digital Economic Community can be easily organised, planned, designed, and implemented by one person with immense ability to collaborate with the local community, industries, and service providers outside, with the help of all commercial and industrial firms, institutions, and establishments. Local associations jointly implement emerging technology locally, such as 5G in industry-4 and medical applications, for example, the embedded and integrated IoT systems for joint production in many locations, Edtech, FinTech, UPI (unified Payment Interface), ONDC (Open Network of Digital Interface), and the fourth industrial revolution. This can be based on our knowledge base produced in the past by authors and study teams. The community shall be self-reliant, and the marginal cost of production and delivery of high-value-adding products and services shall be based on the near-zero marginal cost of production possible in a circular economy that uses renewable energy. Services can be more cost-effective using electronic means, fewer intermediaries, and renewable-based power.

Alongside economic growth, the designer digital community aspires to an enhanced quality of life hitherto experienced, appropriate to their unique and uncommon socio-cultural characteristics, and not a poor copy or imitation of another nation's way of life in a developed country, for example, the American dream. Access to high-quality and enhanced living and high-value-adding production and services shall be made accessible by cost-effective intelligent design.

The current thinking in many democratic countries of the Indo-Pacific, like India, is that Governments at the National and Federal levels are not constituted to work as businesspeople or entrepreneurs taking calculated and managed risks and running public sector undertakings such as hotel chains, tourism resorts, public utilities and services and industries which Governments are found to be not capable by the experience of countries like India. In the complex world, their attention should be only on governance as given in their constitution for law and order, delivery of speedy justice, defence against enemy countries and terrorists, protecting human rights and territorial integrity of the nation for intelligent and digital communities to perform with no hindrance to working globally. Digital and global economic communities are exclusive avenues of livelihood for citizens to exercise the right to work and live.

Income distribution, equity, anti-terrorism measures, anti-corruption drives, defence of the Nation, foreign policy, independent scientific monetary management, legislative reviews and revisions, ensuring judicial independence, and affirmative actions are concerns of Governments, to mention a few examples and not of citizens who can only cooperate and obey the Government in these realms. On the other hand, the citizen's job can be taken care of by creating a governing space that increases universal access to work for high-value-added products and services within households, bringing more women into high-value-adding household economic activities. The marginal cost required for enhanced living is taken care of by finding ways and means to keep it within easy access to the community of citizens with near-zero marginal cost. All these are achievable by proclaiming that the digital economic community is family in the Indo-Pacific culture.

The community's spatial design is collaborative work to restructure the community spatially to accommodate community workspaces through cashless urban land management. This would reduce the community's dependence on local, federal, and central government freebies, an unhealthy characteristic of the current Indian democracy. The society's network and services shall be for emerging technology. A detailed technically, economically, and financially feasible design that community consensus accepts is the first step in restructuring community spaces and networks.

The new urbanism principle in community spatial design is acceptable. The walkability shall be increased to reduce car use, and co-creating digitally enabled community spaces for all age groups and genders to interact securely and activate can be designed to augment community communications and use cashless, well-known urban land management approaches. A self-governing and efficient management mechanism that listens to community voice and opinion in real-time as issues emerge and gives everyone in the community an equally active role in high-value-adding economic activity and creating enhanced life may be designed in digital economic communities.

Megacity systems have built over time a reliable system for all economic activities supported by

Physical and communication infrastructure and Digital Public Goods. Digital economic communities can integrate with Megacity systems more cost-effectively than conventional economic linkages with middlemen. The digital economic community's digital interaction with money and communication capability can be incorporated into the Megacity system more quickly than in traditional neighbourhoods.

2.2 How Different Are the Core and Peripheral Digital Economic Communities of the Indo-Pacific?

The Core Digital Economic Communities constitute all island countries of the Indo-pacific within and those that envelop the confluence of two oceans, the Indian Ocean, and all countries bordering the Pacific Ocean in adjoining land masses, which also include North and eventually Southern Pacific, Indian Ocean countries bordering up to East African countries and goes all the way to up to central Asia a part of International North–South Economic Corridor of all countries which are part all sub-regions of Megacity system of 21 megacities discussed.

Peripheral Digital Economic Communities are those outside the Core Indo-Pacific Countries with robust and long-range sustainable economic, political, and military interests to safeguard their assets and business interests in Indo-Pacific in the Indo-Pacific region. They may have National policies to promote long-range economic engagement with the Indo-Pacific. These policies enhance living and working with more inclusive participation, striving for higher income for all using digital means. Since we are giving equal importance to the core and periphery, it takes work to draw a boundary as the boundary of the Indo-Pacific region. Peripheral and core digital communities can change the border in a community, emphasising digitisation for economic development and transcending distances that separate them.

There is no unified and one accepted approach to the theory building of a Digital Economic Community. Many quantitative, qualitative, normative, and value-based approaches can be loaded with democracy, autocracy, socialism, communism, and capitalism. We adopt a theory of accurate historical interpretation against the usual distortion of history ideologically motivated by a particularly national stake. The history of nations shows that the nations under study have definite natures and characters like human beings. Since history substantiates this repetitive behaviour, there is no need for calibration or validation of the model other than the historic behaviour pattern of countries; with these, presenting the theory of the digital economic community is realistic.

2.3 How Did Digital Economic Communities and E-Gateways for Supply Chains Emerge? [36–46]

Anthropologists, Sociologists and Political Scientists were the first groups of scientists to understand Communities. Anthropologists and Sociologists talked about the ability of groups of people based on their income cohort, class, region they are from, religion they practice daily, castes, and shared goals to come together as a community. They also spoke about urban anonymity in urban communities, where you do not know your neighbours in a metropolitan area. This became an existential threat in recent years in the era of urban terrorism when a terrorist lived and prepared bombs to detonate in the local market to kill some in the community who are present there.

Yet, the neighbour did not know the terrorist neighbour. Sociologists also discussed how these communities were formed online based on shared economic interests, religious affiliation on platforms, and conducted activities, including terror. So, WhatsApp groups, LinkedIn, Facebook, and Telegram view current developments, react, communicate, and are even empowered to act rightfully or wrongly. All such activities fall into the Digital Economic Communities.

Political scientists study the affiliation of the group mentioned above, those inclined to vote for a particular political party, and those who do not, and use such information to predict which party will win the elections. They also discuss the misinformation spread in specific TV news channels and Newspapers whose sole objective is to create opinions that support their views (owners of the news channel/paper) on current events. The owner's affiliation determines the views of the newspapers and TV channels. These communications channels do not value factual and objective information based on verifiable data as academicians do.

Economists grouped the population based on income, formed low, middle, and high-income groups and nations, and studied their consumption patterns. They were also communities in our broad definitions in a communication age where ideologies like Marxism and capitalism or a particular religion such as Hinduism, Islam, Christianity, and Jews could unite in groups and politically compete.

Spatial planners believe in community and discuss the new urbanism movement at length, influencing spatial planning and designs for reinventing communities. Walkable communities were ideal for the Greeks in the past to assess the opinions of people they met during the walk and use them to debate in their assembly for community decision-making and actions.

Political scientists and economists talk about competition. So, we have competition between China and the West, mainly between China, the US, its European allies, and Russia. We do not see such a West Vs China competition among ASEAN countries in the Indo-Pacific. Even India has trade relations in tune with $ 135 billion with China, even when Indian soldiers stand eyeball to eyeball on the Himalayan borders, and any misadventure can create a big war between the two nuclear powers. Above all, the Indian government defines its relationship with China as abnormal but participates jointly in the BRICS and SCO summits where China is present. This competition gives way to supply chain disruptions and identifies new supply chains. Before the Ukraine-Russia war, pipelines from Russia were used to meet European energy needs. This supply chain is politically and physically broken, and energy flows from India from the oil supply they received from Russia in India and is marketed to Europe after refining with value addition, and to Europe from the US and the Middle East. The same is true for India and Pakistan. Ever since there have been several terrorist attacks from Pakistan, there has been practically no relationship, including trade, between the two countries. Now, Indian goods go to the UAE, then from the UAE to Pakistan, and a new supply chain route with higher costs to Pakistanis is being implemented.

Competition is the pet subject in modern democracy and capitalism, which competes with socialism and Marxism. Western-educated elites in democracies like India accepted these concepts, although it is contrary to the Hindu view of life.

Hindus believe in *Vasudeva Kudumbakam*, which says the world is a family. It was accepted by the G20 summit in 2023 and led to many G20 deliberations in hundreds of towns in India under the Indian presidency, under this logo and theme. How can competition on the floors of parliament and in marketplaces based on many reasons not necessarily lower price and high quality be acceptable to Hindus, the majority in India who firmly believe the world is a big family in which father and son or husband and wife or brothers and sisters do not compete but live together? If the son has more achievements than the father, there is no competition between father and son; the father accepts it as a pleasant, acceptable defeat in competition. The Indians believe more in consensus in political debates for constructive law-making than competition based on specific attributes and phobias. In addition to India, Indonesia and many small countries and still smaller island communities in the Indo-Pacific practice consensus after debate as a cultural norm and not competitive behaviour in parliament, followed by a walkout without reaching a consensus, as we see in the Indian parliament deliberations.

During emergency days in India, the word secularism, which is again a Western concept, was added by the majority Government, the Congress, to the Indian constitution. How can anti-secularism exist in predominantly Hindu-dominated India when they say the world is a family? The only benefit of secularism is to create a vote bank of Muslims for Congress, or a Marxist party based on disinformation, who are made to oppose Hindus and divide the Hindu fabric of unity of many religions of one big family, the Vasudeva Kudumbakam. Despite the advocacy of secularism, the secularist parties do not share their political power with Muslims, as evident in the time series data of members of parliament and assemblies of every election and Government formation in states and Nations. Traditionally, Hindus and Muslims even practice jointly many Hindu rituals in temple festivals, even in Ayodhya, Sabarimala and in many parts of India. The concept of secularism may be more valid in the West, the land of crusades between Christians and Muslims, than in India.

In contrast to the use and abuse of communities these days in politics, Digital Economic Communities aim to create an enterprising and law-abiding economic community with potential international reach for economic interactions and showcase the extensive use of the power of digital technology to achieve economic interaction internationally. They aim to enhance living beyond what is promised by the ruling party in power and targeted by the community through their effort and local economic development for high-value-added products and services provision and global transactions enabled by digital means. There cannot be any opposition to such goals and objectives, and it is entirely agreeable to the central concept of Vasudeva Kudumbakam.

2.4 *Case Study E-Gateways Thrissur*

Case Study E-Gateway Enterprise Ecosystem Design and Activation Thrissur [47–56]

There are two aspects of E-Gateways.

The first aspect **is E-Gateway Enterprise Ecosystem Design** and Activation. The second aspect is the **E-Gateway for Global Governance**.

The Enterprise ecosystem case study is from Thrissur, part of the Kannur-Kochi Megacity of Kerala, Bharat, and the Global Governance case study is from Estonia in Europe (Fig. 15).

Thrissur, located in the central region of Kerala state, is the cultural capital of Kerala and is a place of immense cultural richness. Its name, 'Thrissivaperur', translates to the abode of Lord Siva, a testament to its deep-rooted religious heritage. Thrissur is renowned for its art academies like Kerala Kala Mandalam and vibrant festivals like Thrissur Pooram. Its history is equally fascinating, having been under

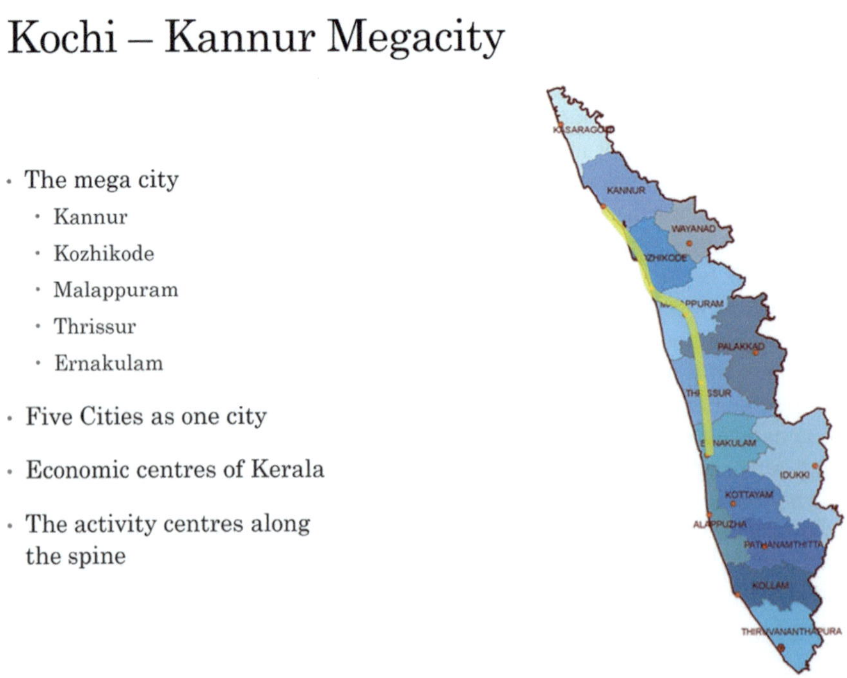

Fig. 15 Kochi-Kannur megacity. *Source* Muhammed Zulfikar Ahmed, Ria Sethia, Thrissur District Trend Analysis, an unpublished class project by course coordinator Prof. T.M. Vinod Kumar for Thrissur. (2019) Smart Global Economic Community in the Megacity Kannur-Kochi of Kerala, October 2019, National Institute of Technology, Calicut

the rule of Chera, Perupadapappu Swarupam, and Zamorin before becoming a part of Aikya Kerala.

Two National Highways, 66 and NH 544 (connecting N-S corridor highway), 9 State highways—21, 22, 49, 51, 61, 62, 69, 75 and 76, and MDRS such as M G road. Railway connectivity to Kerala and Chennai with four stations. The city is served by Cochin International Airport, about 55 kms away: the municipal corporation (Thrissur MC), 6 Municipalities and 92 panchayats under the Thrissur district. TMC is the third largest urban agglomeration (18 lakhs/1.8 million). The mayor heads the Corporation. The Thrissur Urban Development Authority and Town and Country Planning Department (TCPD) are the agencies that prepare development plans for the city. 6 taluks and 13 constituencies.

The district has a Tropical humid climate. On average, there are 124 rainy days in a year. The average daily maximum temperature in March and April is about 31 degrees Celsius in the coastal regions and 36 degrees Celsius in the interior. Relative humidity is generally over 70%.

Demographic Trend Analysis

The Thrissur district's population was 3,121,200 as of the 2011 census; the last census since the 2021 census was abandoned because of the COVID-19 pandemic.

The population growth follows a similar trend to that of the Kerala district, which increased from 2,128,877 in the 1970s to 31 21,200 in 2011. It is on the lower side.

Thus, the population has grown 1.5 times in 50 years (Fig. 16).

Over the years, the district has seen a decline in its growth rate due to the reduced birth rate,

- increased marriage age and nuclear families with reduced household size.

A similar trend and reasons to those of Kerala state itself have led to an incremental increase in the district population. The population density, along with the Municipal Corporation, has been increasing over the decade, even though municipalities still have the highest population density. The urban agglomerations along the municipal corporation, such as Madakanthara, Kolazhy, and Mulankunnathkaavu, have seen an influx of people due to their proximity to the Municipal Corporation.

Regarding the urbanisation rate, the district has been registering an increase in urban areas. The urbanisation rate is primarily driven by urban population growth

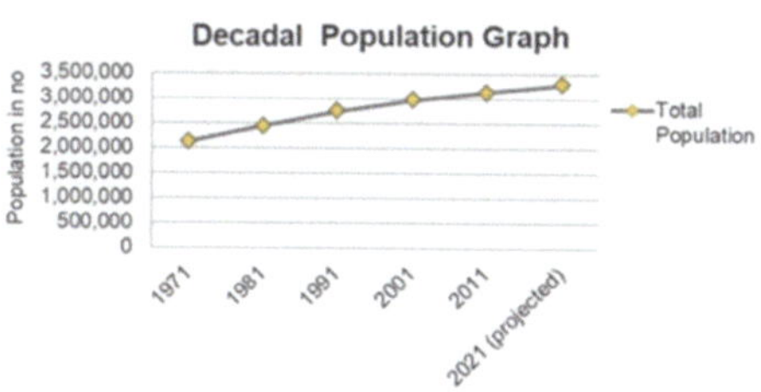

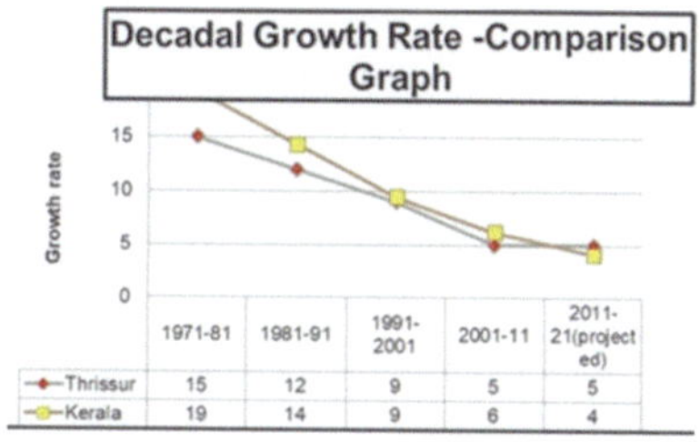

Fig. 16 Decadal population growth Thrissur District and Kerala

and the formation of census towns, resulting from increased population density, as well as people transitioning from agricultural to secondary and tertiary sector jobs. In 2011, 107 new cities were formed in Kerala, with Thrissur taluk accounting for 36 new census towns and Chavakkad for 13. All the reasons cited have contributed to the urban share increasing to 67%, and most of these merge into a single city in the district.

It has a healthy sex ratio of 1089 and a population of 16.8 lakh females as opposed to 14 lakh males. The state and the district are steadfast in their outlook on anti-gender discrimination.

Thrissur's literacy rate is 97%, above the state average of 93.91% and the national level of 73%. The availability of social and health-related infrastructure is the main reason for this.

The age-wise population also shows that 65% are of working age. While the young population below working age accounts for 22%, the older population above working age comprises 13%. This implies a higher dependency load for the district in the coming years, as well as rapid growth in the senior citizen population (Fig. 17).

In the case of built-up areas,

The built-up area in and around urban areas of the district has increased from 2006 to 2011.

- The significant reasons contributing to urban sprawl from the city to outside are better land availability and lower land prices. Also, the migration of people from rural areas led to increased built-up areas.

The district's HDI can be compared to that of developed nations. Over the years, investments in education, health, and enrollment rates have been one of the primary reasons for higher HDI. This is mainly due to outmigration to the Middle East and the subsequent remittances, which boosted the housing and infrastructural improvement of the state and the district, and higher expenditures by the household for health and education (Fig. 18).

Economic Trend Analysis

The district's working population's share ranges, on average, from 30 to 35%. A projected decline in the working population growth rate by 2021 indicates a decreasing workforce participation rate. The female workforce declined from 32% in 2001 to 25% in 2011, indicating the need and potential to involve females in the workforce (Fig. 19).

The share of primary workers in the economy is decreasing, whereas the number of marginal workers increased from 1981 to 2011. The number of workers in the primary sector has declined, but that in the tertiary sector has grown from 1981 to 2011, indicating a shift in the sectoral economy and more job opportunities.

From 2016, the district's GDP growth rate has been less than the state's GDP growth rate, indicating a drop in the economy at the district level from 2016 (Fig. 20).

The tertiary sector contributed 63% to the district GSVA in 2017–18, making it a major economic driving force. However, the district's share of state GDP suddenly dropped due to a lack of and failures in the secondary and tertiary sectors. Per capita

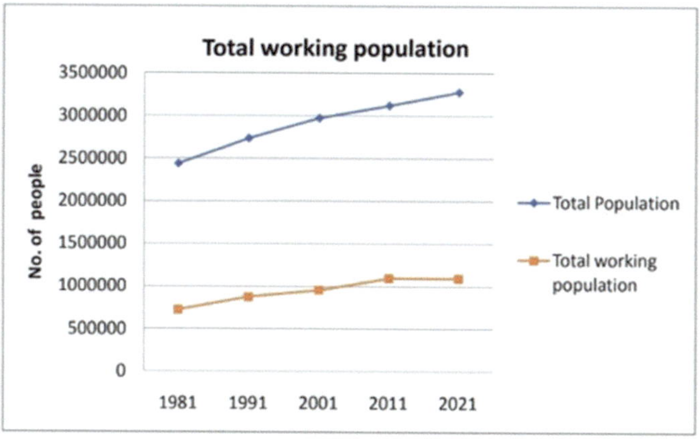

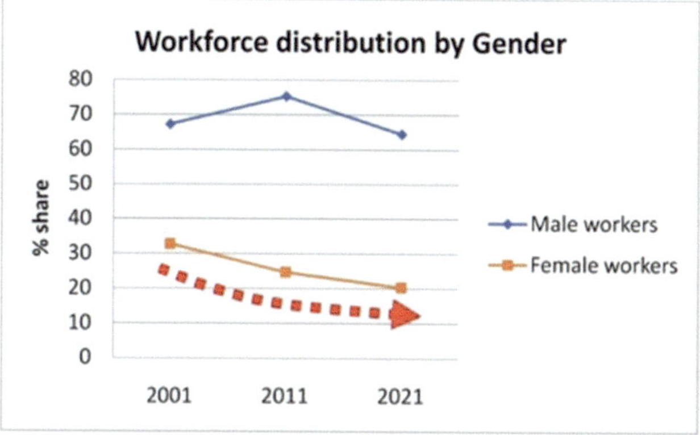

Fig. 17 Work force Thrissur District

Fig. 18 HDI India, Kerala, and Thrissur

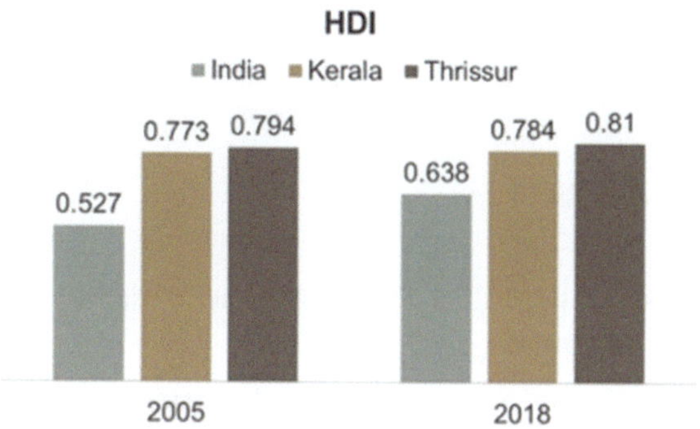

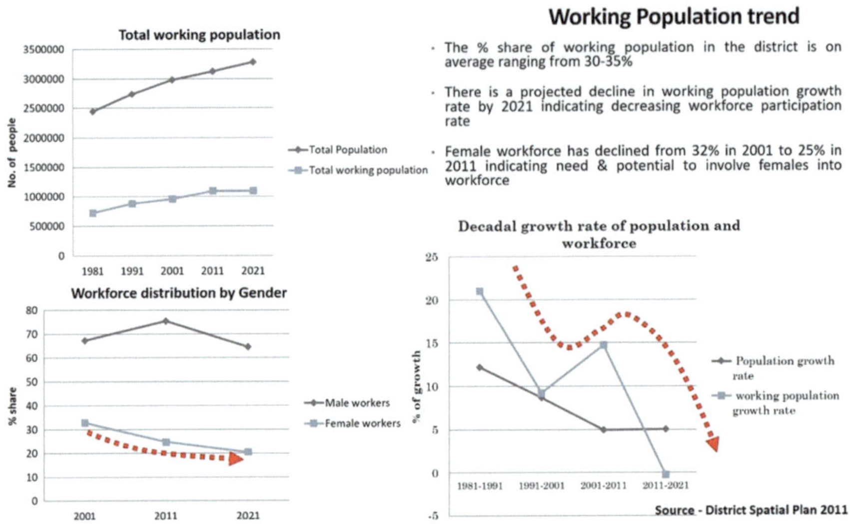

Fig. 19 Work population trend

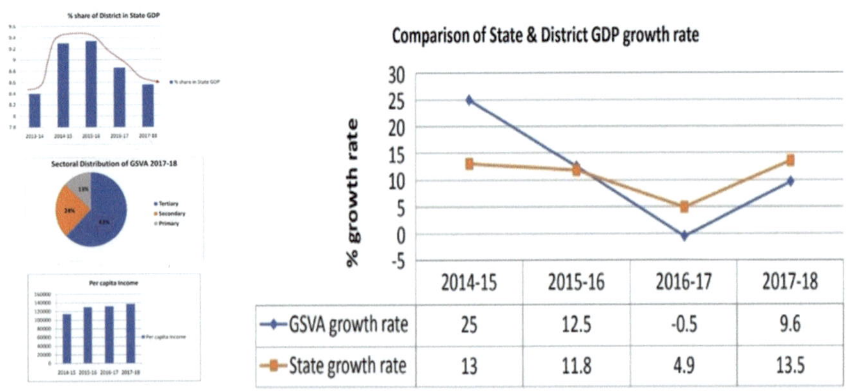

Fig. 20 GDP Thrissur

income in the district didn't show much variation after 2016, indicating a slowdown in the economy.

Industry and Service Sector in the district—trends, prospects, and issues

2017–18 registers the maximum number of industrial setups and employs approximately 17,000 people, the highest in the district's history. The growth of employment and industrial units in the past two years indicates potential secondary sector activities for the district's growth. The total registered industrial units are 14,596 as of 2018 (Fig. 21).

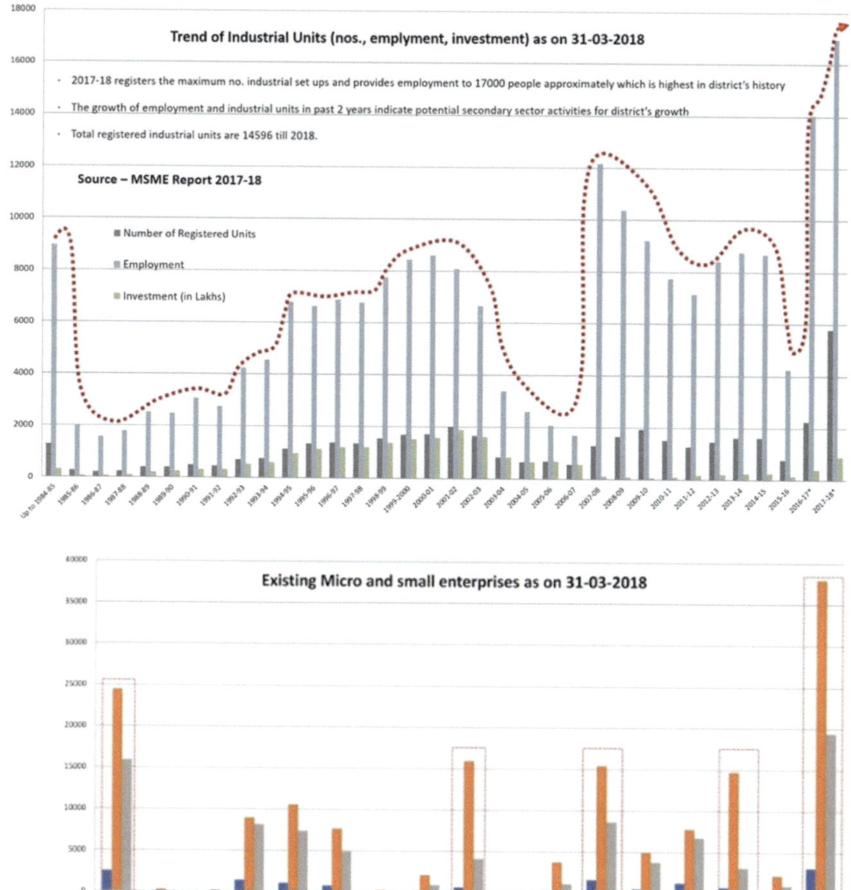

	Agro & Food based	Soda Water	Cotton Textiles	Readymade Garments & Embroidary	Wood/Wooden based furniture	Paper & Paper Products	Leather based	Chemical/Chemical based	Rubber/Plastic	Mineral Based.	Metal based (Steel Fab)	Engineering Unit.	Electrical Machinery & Transport Equipements.	Repairing and Service	Gem Designing.	Artisan Unit.	Others.
No of Units	2548	35	19	1455	1088	770	29	201	668	2	242	1627	501	1265	761	275	3110
Investment (Lakhs Rs.)	24458	268	29	8955	10579	7677	257	2146	15930	28	3791	15402	4977	7788	14712	2196	37998
Employment	15883	84	228	8168	7384	4921	212	909	4140	16	1108	8617	3824	6744	3100	971	19365

Fig. 21 Trend industrial units Thrissur

The district's central micro and small enterprises are agro and food-based, rubber/plastics, engineering unit, gem designing, and IT.

Major Clusters identified in the Thrissur District

1. Thrissur Traditional Furniture Cluster, Chevoor Pvt. Ltd.
2. Terra Tile Consortium Pvt. Ltd.
3. Plastic Cluster
4. Diamond Cluster Development Society

As per the 2012 MSME report, clusters present in the past but not effectively working today are (Fig. 22).

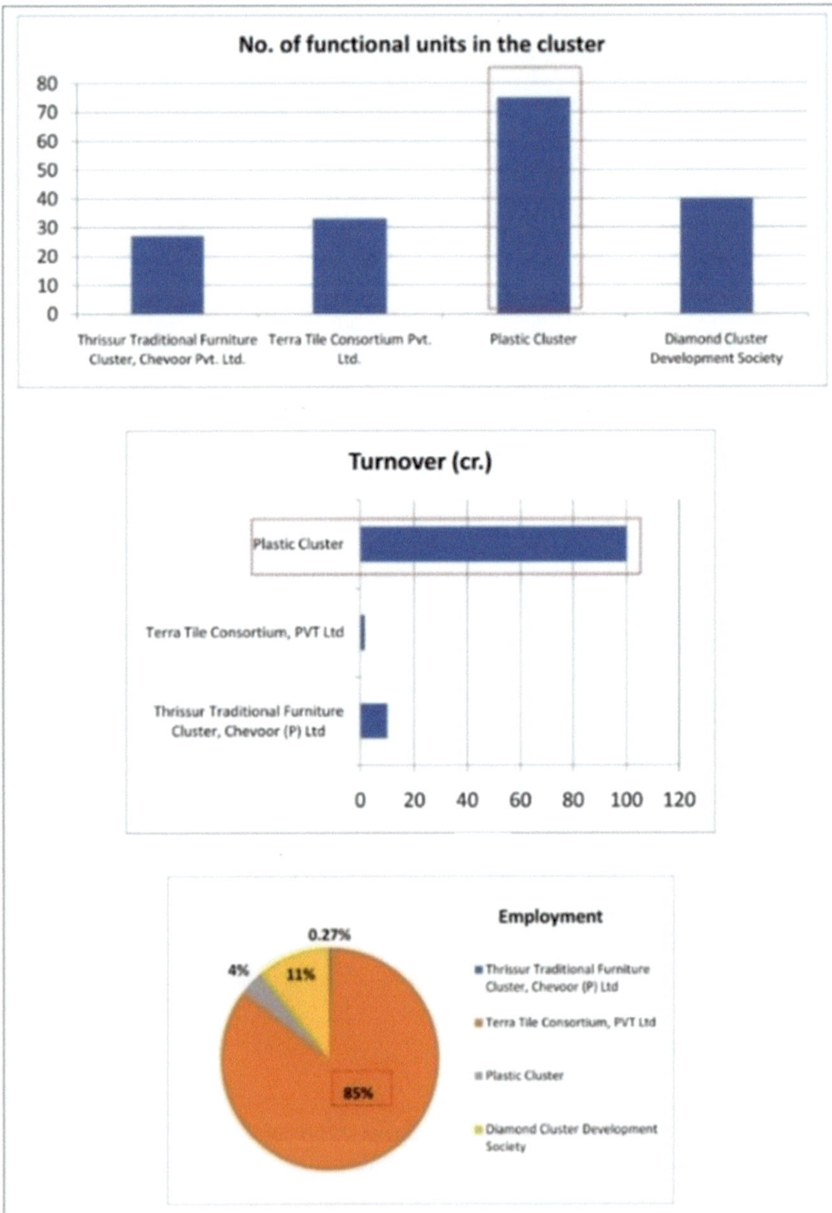

Fig. 21 (continued)

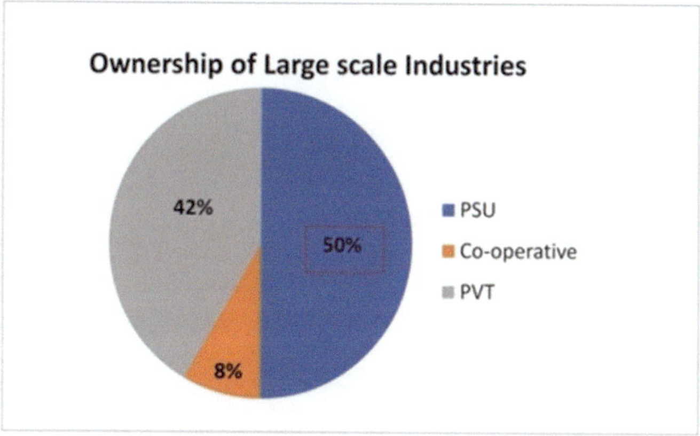

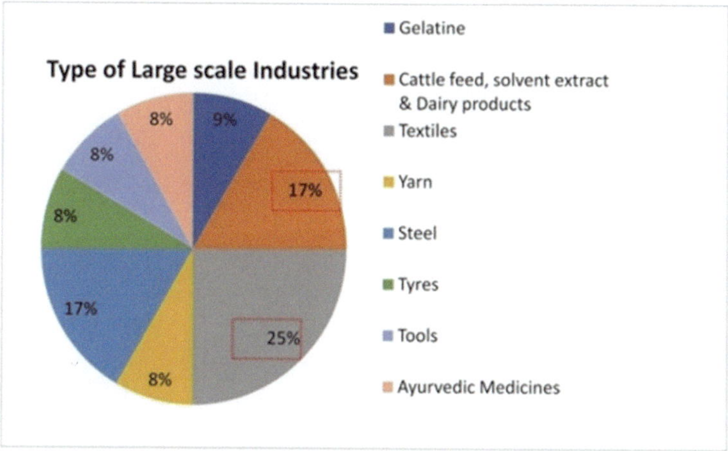

Fig. 21 (continued)

Kuthumpally Handloom Cluster
Ayurvedic Cluster
Proposed clusters as per the 2018 NSNE report
ThalikkulamVasthragramam—garment sector
Cherpu handicraft sector
Agro-food sector cluster
The Major exportable items are

- Ayurvedic Products, Handicraft items, Textiles
- Food Products, Dry Fish, Rubber Products
- Coconut Oil, Soap, Spice Oil, Food Processing Machinery
- Concrete Product machinery, etc.
- The major Service enterprises

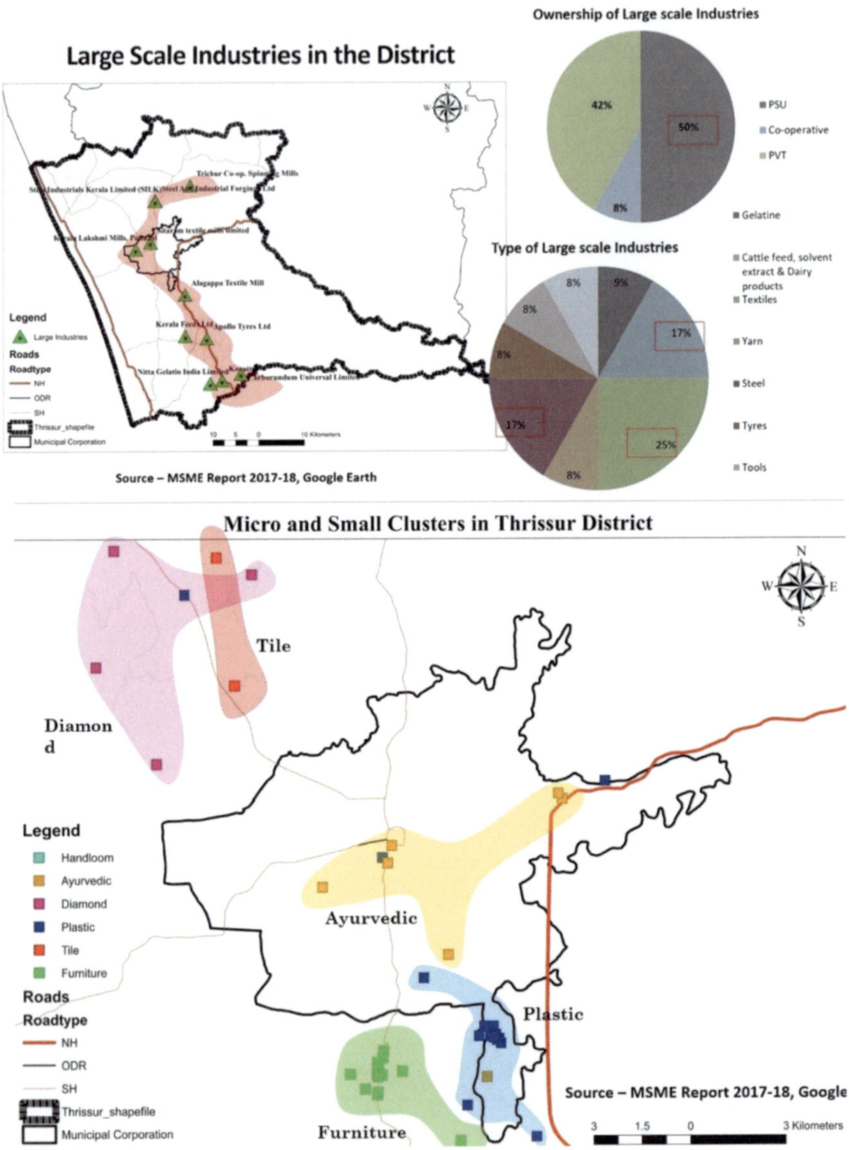

Fig. 22 Industrial clusters

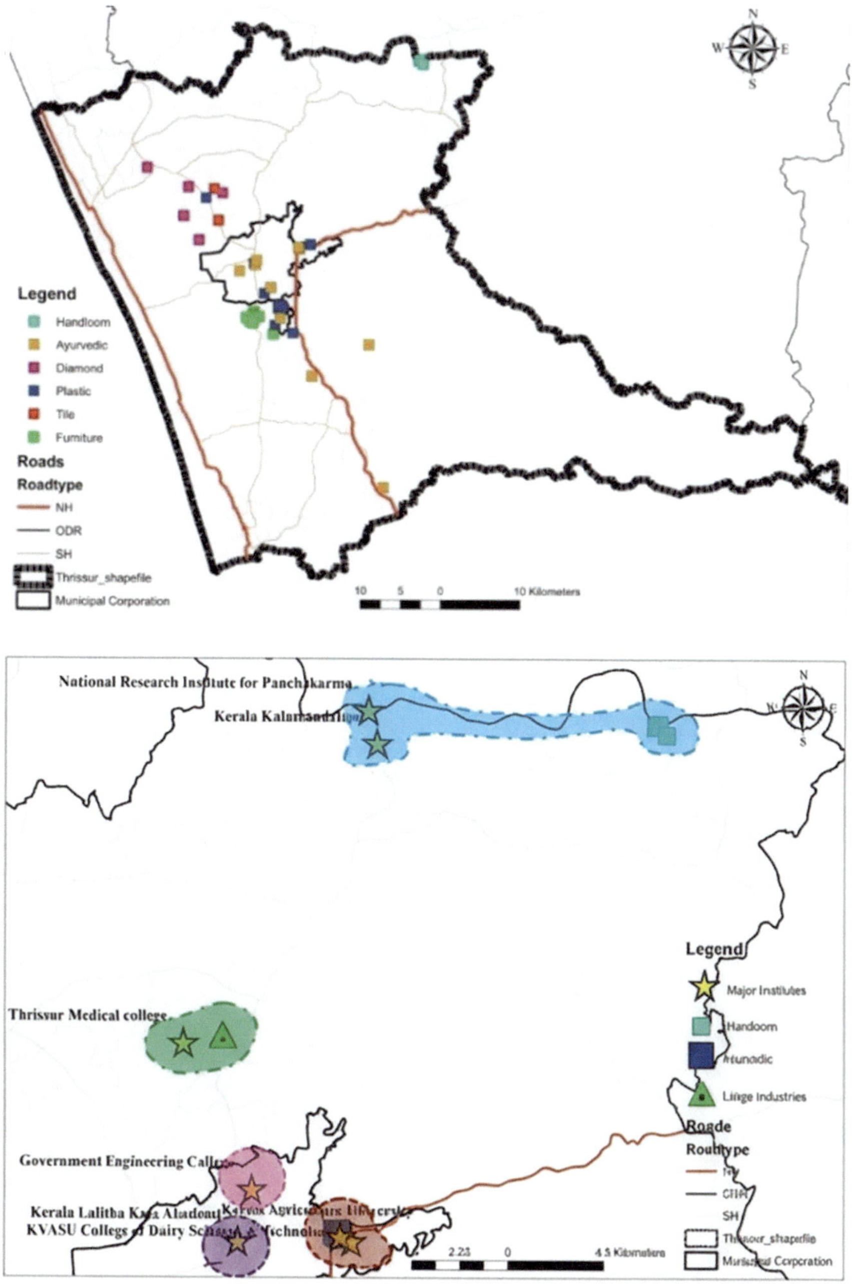

Fig. 22 (continued)

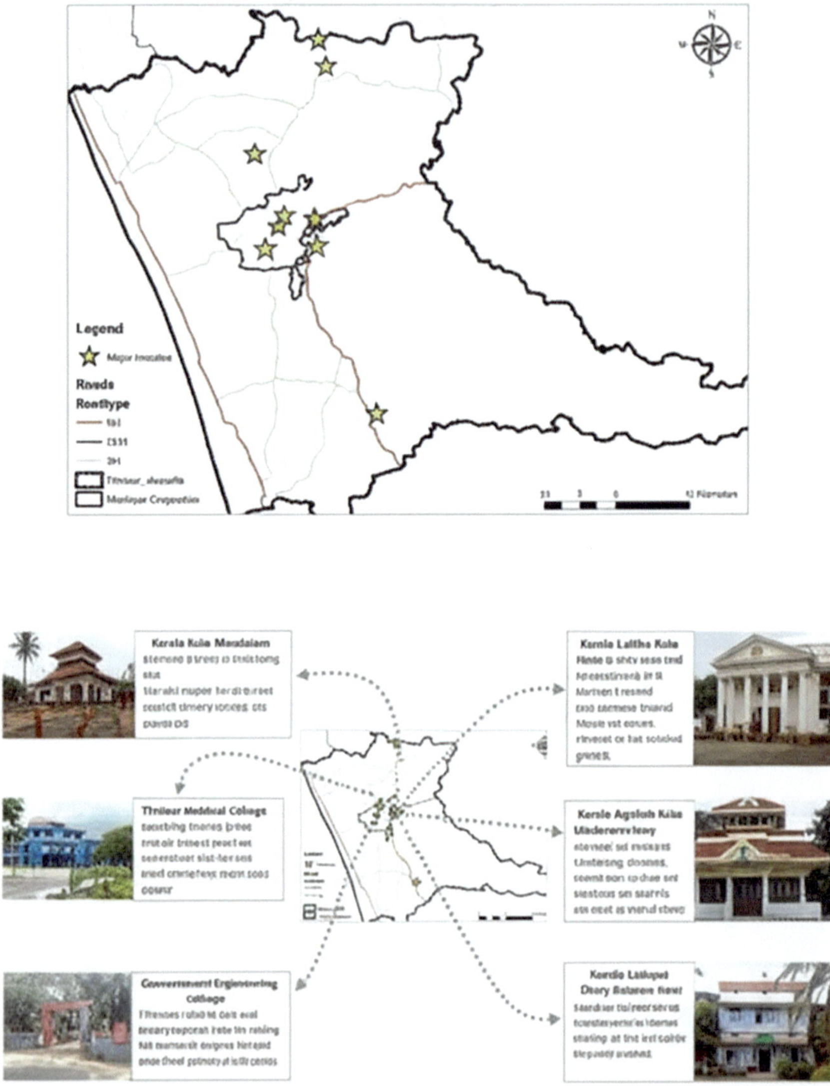

Fig. 22 (continued)

- Automobile Repairing and Service
- IT-enabled services, Tire retreading
- Gold chain cutting
- Health services, Hotels and restaurants
- meat processing, food grain milling/processing
- Tourism and Ayurvedic health care packages, etc.
- Potential Areas for the Service Industry -

- ITES, Transport, Studio, Healthcare services
- Tailoring, Flour Mills, Power laundry.

Issues

- Lack of availability of sufficient Industrial Land and high land prices.
- Lesser linkage with technical institutions in advanced industrial technologies.
- Low productivity in local units makes the products less price-competitive.
- No proper utilisation of manpower, especially among women.
- Procedural delay for issuing clearances, licenses, power connection, etc.
- Lack of proper marketing support.
- Lack of Entrepreneurship development programme and technology training.
- Lack of willing Human Resources and IT Support.

SWOT Analysis

Strengths

- Increasing workforce in secondary and tertiary sectors over the past decades
- The highest number of industrial setups has been established in the last two years, and employment in the secondary sector is increasing (Fig. 23).

Weaknesses

- Declining share of the district in state GDP since 2015, and almost stagnant per capita income

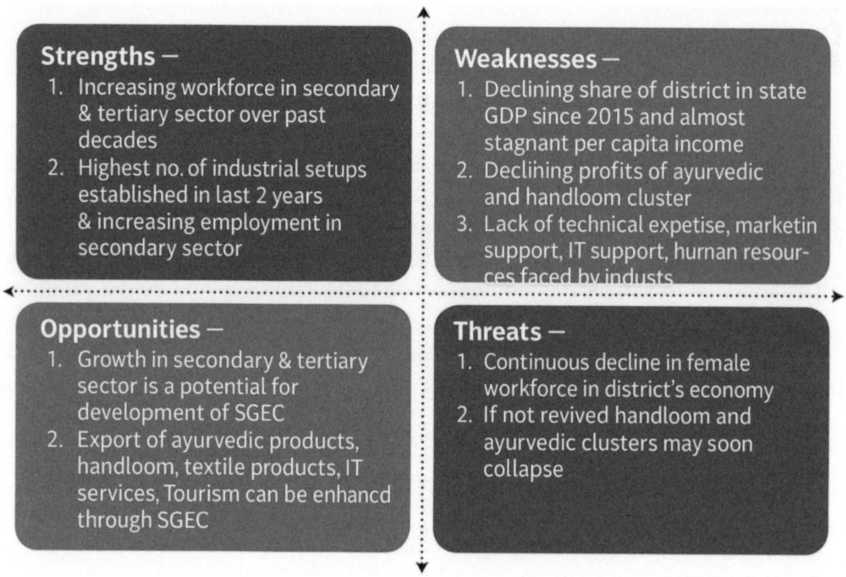

Fig. 23 SWOT

- Declining profits of the ayurvedic and handloom cluster
- Industries face a Lack of technical expertise, marketing support, IT support, and human resources.

Opportunities

- Growth in the secondary and tertiary sectors has potential for developing SGEC.
- Export of ayurvedic products, handloom, textile products, IT services, and Tourism can be enhanced through SGEC.

Threats

- Continuous decline in the female workforce in the district's economy
- If not revived, handloom and ayurvedic clusters may soon collapse.

Proposed SGEC in Thrissur District

The identification of communities is to be proposed as SGEC (Figs. 24 and 25).

Products/Services

- Handloom Products like silk sarees, cotton sarees, dhotis, etc.
- Ayurvedic Products like Herbal medicines, etc.
- Ayurvedic services like massage, treatments, etc.

SGEC Kerala Agriculture University—Agro Ayurvedic community

The cluster consists of two significant institutes, namely Kerala Agricultural University (KAU)-KVASU College of Dairy Sciences and Technology (Fig. 26).

This also involves major Ayurveda factories in the region. The idea is to promote the cultivation of herbs and other economic products in the cultivable areas of the community cluster and then promote feasible and economic agriculture.

The idea is to create the raw agricultural materials required for the Ayurveda factories and global market with the help and guidance of agro and dairy institutes. The idea is to make agriculture feasible for the disinterested farming community, which shifts to various other secondary jobs.

The overall idea of services from each community is detailed in the tables below:

The community can be trained to produce more Ayurveda and herbal products, such as beauty products and supplements.

Herbal powders, Ayurvedic Supplements, and Skin creams are some of the products that can be produced locally (Fig. 27).

The cluster consists of

- Kerala Lalitha Kala Academy
- Community around.

The idea is to correlate the institute–industry and create a community based on a knowledge-based economy from the institute. The cluster concept is as follows:

The services and interaction of the Institute-Community is as follows -

The expected outcomes of the collaboration are:

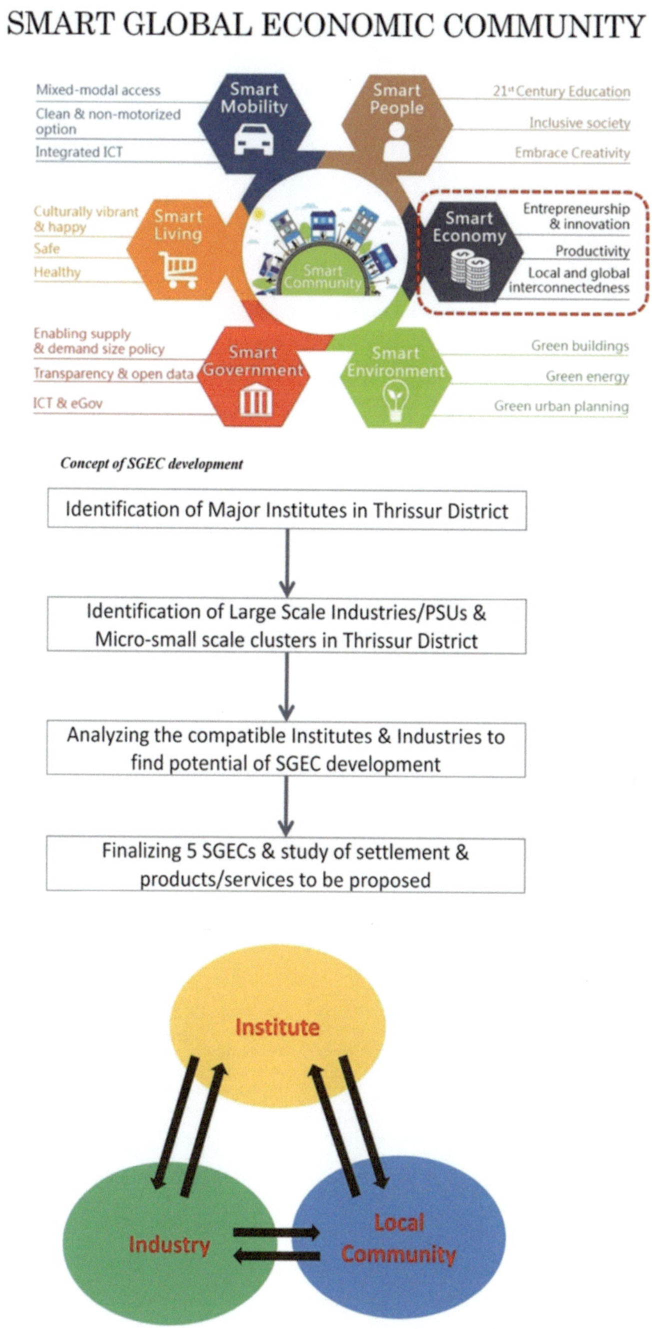

Fig. 24 Thrissur smart global economic community (SGEC) concept and development process

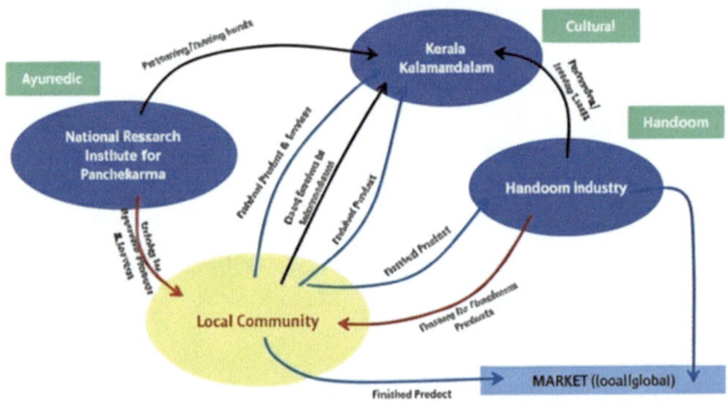

Institute-Industry	Institute-Community	Industry-Community	Indnste-Institute
Kalamandalam-Handloom	**Kalamandalam-Local people &Panchakarma Institute-Local people**	**Handloom-Local people**	**Kalamandalam-Panchakarma Institute**
Handloom products made based an demand & market anslysis, type of product and quaniity	Kalamandalam can enable services to employ local people after tharough training Panchakarma Institute can provide training & involve community to produce ayurvedic products and services.	Handloom clusters to utilize petenttial involvement of community to manufacture handloom products for local market and export	Partnening of both institutes to enable ayurcedic services and products for local market and particularly glabal market

Products/Services -

- Handloom Products like silk sarces, cotton sarces, dhotis, etc.
- Ayurvedic Products like Herbal medicines, etc.
- Community servicees massaee metments, etc.

Fig. 25 SGEC Development Kalamandalam Institute

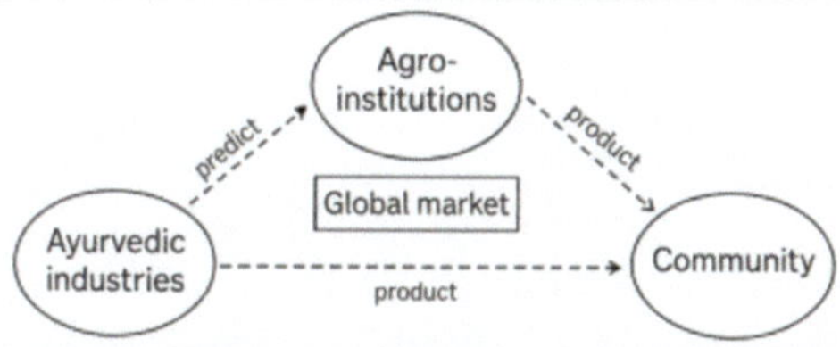

The overall idea of services from each community is being detailed in the tables as below:

Institute-Industry	Institute-Community	Industry-Community
Agro-Dairy Institutes and Ayurvedic Institutes	**Agro-Dairy Institutes and Community**	**Ayurvedic Institutions-Community**
- The institutes-industry collaboration caun work out various ayurvedic products which are of 'local and global demand	- Provision of seeds and expertise for cultivation - Quality control - Market Demand Guidance	- Provision of raw materials such as a herbs and other household produces

Fig. 26 SGEC Kerala Agriculture University

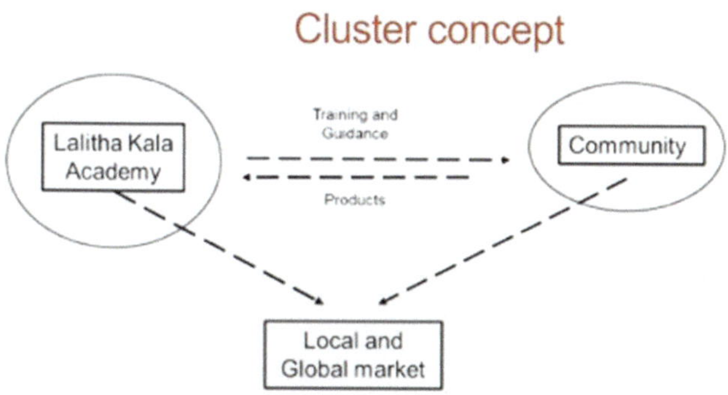

The services and interaction of the Institute- Community is as follows -

Institute-Community
Lalitha Kala Academy-Community
-The institute and interested people in community can work for manufacturing products and goods which will be of artistic and economic values -The academy can become an enabler and a trainer -The institute becomes the facilitator for local and global market

Fig. 27 SGEC—Kerala Lalitha Kala Academy

- The community can be guided to produce products such as sculptures and conduct exhibitions that could be marketed.
- The academy can facilitate a market for the community arts and be a part of marketing

SGEC—GEC Thrissur

See Fig. 28.

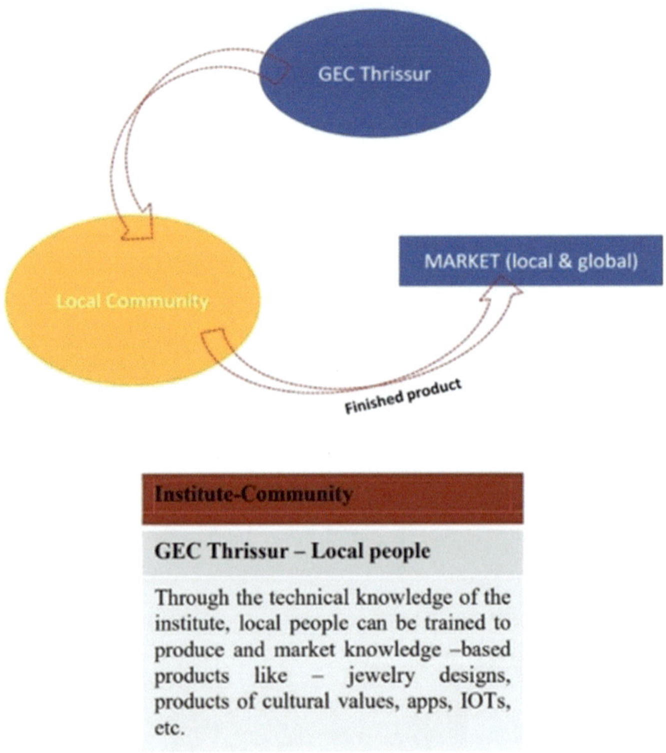

Fig. 28 SGEC–GEC Thrissur

2.5 *Case Study E-Gateway Enterprise Ecosystem Design and Activation*

Projects and Programs of QUAD and I2U2 in the Indo-Pacific and Role of Digital Economic Communities

The Indo-Pacific region is dynamic and diverse, offering many opportunities and challenges for the Digital Economic Communities. The program initiatives of QUAD, I2U2, and other regional, bilateral trades aim to enhance the region's security, stability, and prosperity and address the regional stakeholders' common issues and interests. However, these initiatives are driven by the major powers in the Indo-Pacific region, and the local communities own and manage local economic development, which plays a vital role in shaping the region's economic future. Therefore, it is essential to assess the feasibility and impact of these initiatives from the perspective of the local digital economic communities and identify the actions they can take to participate in and benefit from them. This is a brief overview of some of the selected program initiatives of QUAD, I2U2, and other regional, bilateral trades and how they can be implemented and adapted by the local economic communities at the local level. Not all projects are included here.

Projects of QUAD [57–60]

The Quadrilateral Dialogue, or the QUAD, is a strategic forum for all needs of the Indo-Pacific between the United States, Australia, India, and Japan. The Quad has undertaken several initiatives to tackle 21st-century challenges. Here are some of the critical projects, and for details of equally important other programs, the editor's book on the Indo-Pacific System of Megacities and emerging Architecture refers to:

1. **COVID-19, Cervical Cancer Vaccine and Global Health**: The QUAD leaders launched the QUAD Vaccine Partnership to help enhance equitable access to safe and effective vaccines in the Indo-Pacific and the world. They have pledged to donate more than 1.2 billion vaccine doses globally. India's cervical cancer vaccine is offered to the Indo-Pacific region for use.
2. **Climate Change**: The QUAD has formed working groups on climate change.
3. **Cultivating Next-Generation Talent**: The QUAD is working on cultivating next-generation talent in all its countries.
4. **Technological Innovation and Supply-Chain Resilience**: The QUAD is working on spreading innovative technologies and supply chain resilience throughout the Indo-Pacific region.

These projects aim to deepen ties and advance practical cooperation among the QUAD countries. The QUAD is not a formal alliance, but the group has intensified its security and economic ties as tensions with China rise.

Projects of I2U2 [61–63]

The I2U2 Group, which stands for India, Israel, the United Arab Emirates, and the United States, has undertaken several projects. This unique grouping of countries focuses on joint investments and new initiatives in various sectors. Some of them are given below.

Here are some of the critical projects.

1. **Food Security**: The UAE aims to invest $2 billion to develop a series of integrated food parks across India. These parks will incorporate state-of-the-art climate-smart technologies to reduce food waste and spoilage, conserve fresh water, and employ renewable energy sources. The goal is to help tackle South Asia and the Middle East food insecurity.
2. **Clean Energy**: The I2U2 Group is advancing a hybrid renewable energy project in India's Gujarat State. This project comprises wind and solar capacity complemented by a battery energy storage system.
3. **Infrastructure Modernisation**: The group aims to mobilise private sector capital and expertise to modernise infrastructure,
4. **Low-Carbon Development**: The group aims to mobilise private sector capital and expertise for low-carbon development.
5. **Public Health Improvement**: Improving public health is another key focus area for the I2U2 group.

However, the specifics of how a Digital Economic Community in a particular country and state could participate in QUAD and I2U2 projects depend on these groups' interests, policies and agreements and the specific interests and capabilities of the Digital Economic Community. It's also important to note that while the QUAD and I2U2 are open to cooperation with Digital Economic Communities, they do not have a military component to their cooperation like NATO. It is best to contact the respective organisations, consult with international relations and digital economy experts, or use AI as a substitute to help progress in this direction through Digital Economic Communities. Please note that this is a general analysis, and the possibilities may vary based on the specific circumstances and policies of the parties involved, including the Government.

2.6 Case Study E-Gateway Enterprise Ecosystem Design and Activation Digital Transformation for Tax and Anti-Corruption in the Indo-Pacific Economic Framework [64–84]

The Indo-Pacific Economic Framework is a strategic vision aiming to promote a free and open Indo-Pacific region that fosters economic growth, security, and prosperity. The framework is based on four pillars: market access and trade facilitation, infrastructure and connectivity, energy and digital economy, and tax and anti-corruption.

The fourth pillar of tax and anti-corruption focuses on enhancing fair competition by establishing and enforcing effective and robust taxes, anti-money laundering, and anti-bribery regimes that align with existing multilateral obligations, standards, and agreements to curb tax evasion and corruption in the Indo-Pacific region. This pillar also entails sharing expertise and seeking ways to support the capacity-building necessary to advance accountable and transparent systems.

This part provides an overview of digitally transforming the fourth pillar of the Indo-Pacific Economic Framework, tax and anti-corruption. It offers practical suggestions and examples for businesses and governments wanting to participate in this initiative.

Why Digitally Transform the Fourth Pillar of Tax and Anti-Corruption?

Digital transformation uses digital technologies and solutions to create new or modify existing processes, products, services, and business models to meet changing customer and stakeholder expectations, improve performance, and increase value. Digitally transforming the fourth pillar of tax and anti-corruption can bring many benefits for businesses and governments in the Indo-Pacific region, such as:

- Enhancing tax compliance and revenue collection by reducing tax gaps, simplifying tax administration, and improving taxpayer services,
- Strengthening anti-money laundering and anti-bribery measures by increasing transparency, accountability, and traceability of financial transactions and activities.
- Boosting economic competitiveness and innovation by creating a level playing field, reducing costs and risks, and fostering stakeholder trust and confidence.
- Supporting sustainable development and social welfare by contributing to public goods and services financing and addressing the negative impacts of tax evasion and corruption on the environment, human rights, and democracy.

How to Digitally Transform the Fourth Pillar of Tax and Anti-Corruption?

Digitally transforming the fourth pillar of tax and anti-corruption involves adopting and implementing digital technologies and solutions to enhance the effectiveness and efficiency of tax and anti-corruption regimes and facilitate cooperation and coordination among relevant actors and institutions.

Some of the key steps and actions that can be taken are:

- Adopting international standards and best practices for tax and anti-corruption, such as the OECD/G20 Base Erosion and Profit Shifting (BEPS) Project, the OECD/G20 Inclusive Framework on BEPS, the OECD Global Forum on Transparency and Exchange of Information for Tax Purposes, the Financial Action Task Force (FATF) Recommendations, the United Nations Convention against Corruption (UNCAC), and the Extractive Industries Transparency Initiative (EITI).
- Implementing digital tax and anti-corruption tools and platforms, such as electronic invoicing, electronic filing and payment, online taxpayer services, digital identity verification, blockchain-based registries and ledgers, artificial intelligence and big data analytics, and e-government and e-procurement systems,

- Establishing and participating in regional and global networks and mechanisms for tax and anti-corruption cooperation and information exchange, such as the Asia–Pacific Economic Cooperation (APEC) Taxation Administration Forum, the Asia/Pacific Group on Money Laundering (APG), the Asia–Pacific Forum on Financial Inclusion, the Global Forum on Asset Recovery (GFAR), and the Stolen Asset Recovery Initiative (Star).
- Providing and receiving technical assistance and capacity-building support for tax and anti-corruption, such as the OECD/Korea Policy Centre, the IMF Regional Technical Assistance Centres, the World Bank Group, the Asian Development Bank, the United States Agency for International Development (USAID), the Australian Department of Foreign Affairs and Trade (DFAT), and the Japan International Cooperation Agency.

The fourth pillar of tax and anti-corruption of the Indo-Pacific Economic Framework is a critical component of the vision for a free and open Indo-Pacific region that fosters economic growth, security, and prosperity for all. By digitally transforming this pillar, businesses and governments in the Indo-Pacific region can enhance their tax compliance and revenue collection, strengthen their anti-money laundering and anti-bribery measures, boost their economic competitiveness and innovation, and support their sustainable development and social welfare. This part has provided an overview of how to digitally transform the fourth pillar of tax and anti-corruption of the Indo-Pacific Economic Framework and offered some practical suggestions and examples for businesses and governments that want to participate in this initiative.

2.7 Case Study E-Gateway Enterprise Ecosystem Design and Activation

The Unified Payment Interface (UPI) [85–91]

The unified Payment Interface allows anyone to send money to another person whose mobile number is linked to their Bank account or who knows their bank name, branch name, account name, bank account number, and bank IFSC code. The figure below shows real-time money transfers in many countries under different names and their transaction levels in 2020 (Fig. 29).

One needs to enter any of the above in one's mobile and use a third-party application or any app from any reliable entity or bank, which does not need to be these two banks where money transfer happens, to make it more secure. There is no one-time Password (OTP). The one-time Password generated just by entering one's password for one's bank will do, and there is no rejection of payment if money is available in one's bank account, which can be checked easily by the same payment application, even if it is a third party. There is also a robust remedial channel if one has sent money to someone one never intended to by mistake. Security architecture prevents

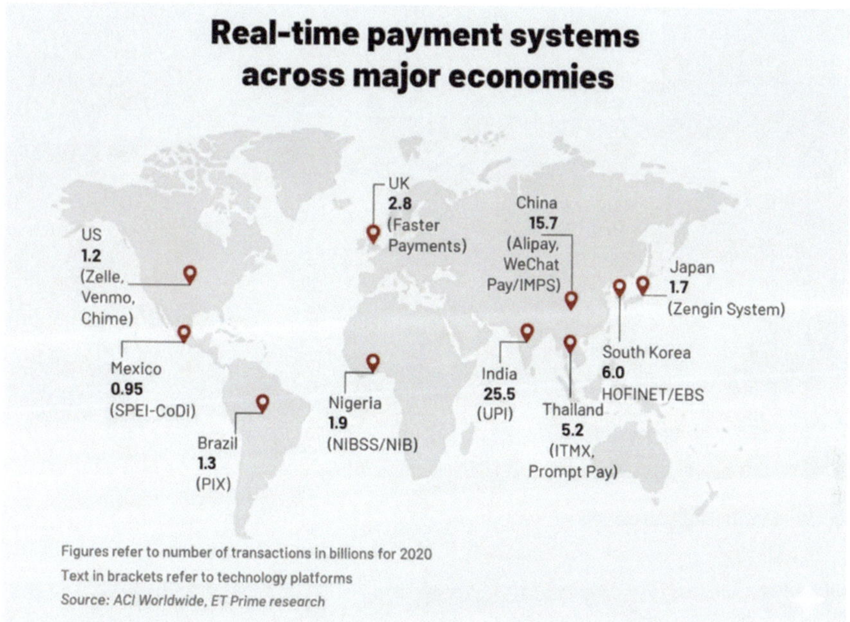

Fig. 29 Mobile money transfer systems of the world

the transferring entities from accessing data-related bank accounts through virtualisation. UPI's popularity and widespread use are due to the transaction speed and security embedded in UPI transactions and the easy way to use it with a mobile device. So, UPI has gone global in a few countries, whose numbers are increasing weekly. UPI also executes international transactions. So, an Individual from such a land, for example, Singapore, UAE, USA, Japan, France, Australia, or some other European country, can have a bank account in the country where they live and send money to any country that has accepted UPI or similar. So, UPI has gone global for the limited number of countries with large Indian migrants to start with, and it is now a way of money transfer for Indians, who constitute the most significant migrant nationality with the largest remittance in dollars home to India from abroad in comparison to any other country in the world.

The credit card transaction is shown (Fig. 30).

The use of UPI transformed it as follows (Fig. 31).

There is an ecosystem that nurtures UPI, resulting in the growth of users (Fig. 32).

The internal process of UPI transactions is machine-controlled, security-enabled, and fast to the satisfaction of users and providers. The process flow chart is given (Fig. 33).

The UPI architecture that enables the process is given (Fig. 34).

As per the usage growth data below, the increasing number of UPI users and transactions in billions of dollars and values have also changed dramatically in India (Fig. 35).

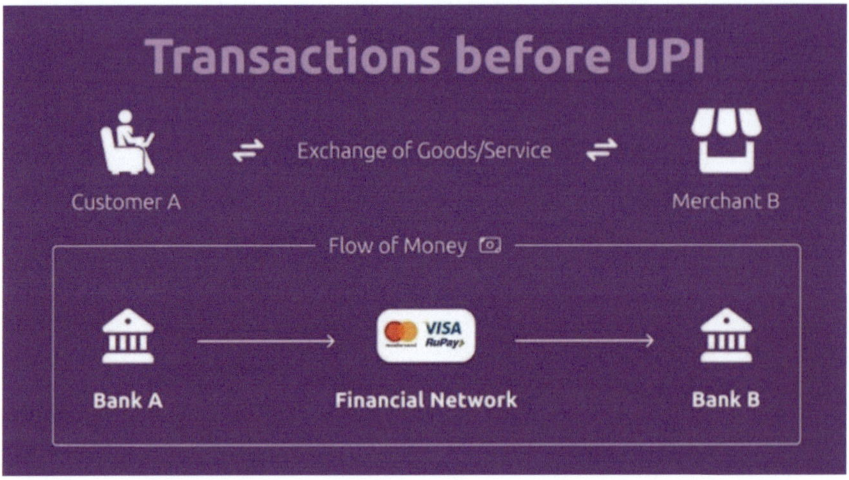

Fig. 30 Credit card transaction

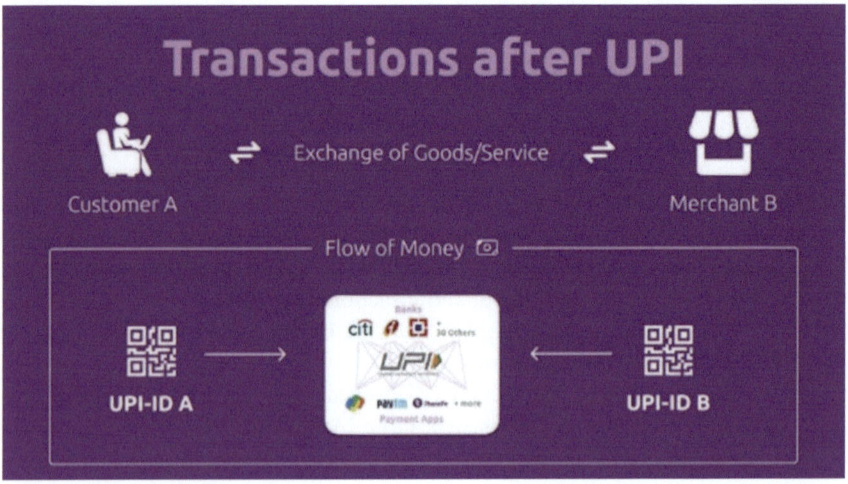

Fig. 31 UPI transaction

Payment service players grew, and some of them were Unicorns like Phonepe with an integrating architecture (Fig. 36).

There was impressive growth in UPI transactions in 2022 in India (Fig. 37).

UPI picked up different use patterns during the past COVID-19 days (Fig. 38).

The market share of UPI apps changed, with some growing faster than others (Fig. 39).

Healthy competition between such service providers created their markets with a status for FY 2022 in India. Today, UPI has become a readily accepted money

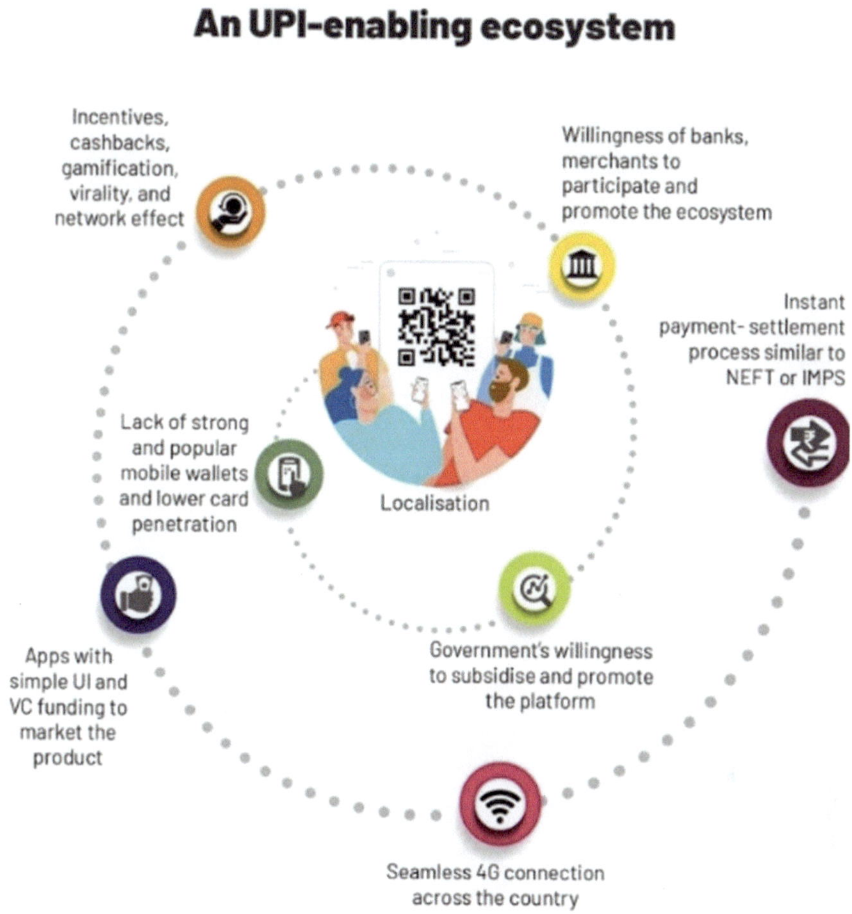

Fig. 32 UPI enabling ecosystem

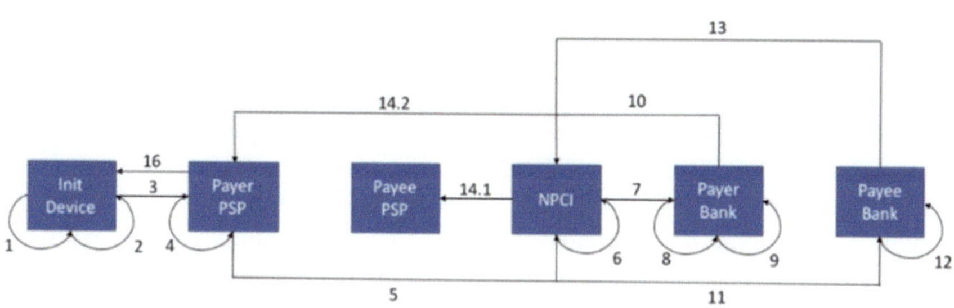

Fig. 33 Successful UPI process. *Source* NPCI

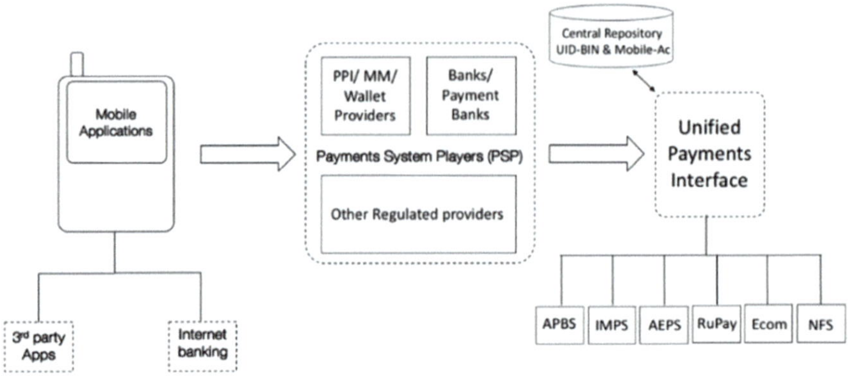

Fig. 34 UPI architecture. *Source* NPCI

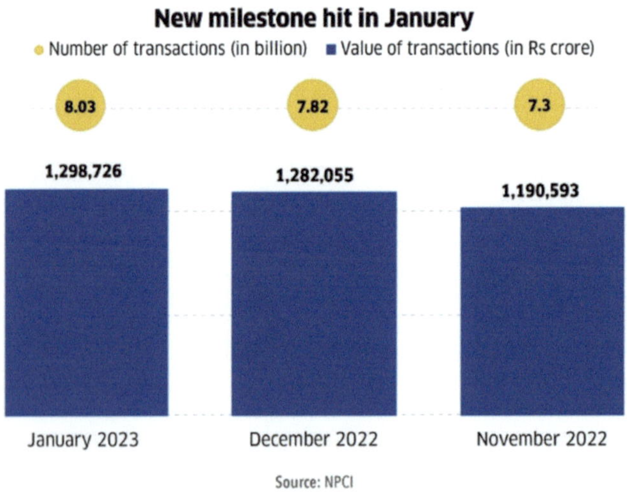

Fig. 35 UPI growth in transactions 2022–23

transfer internet mechanism everywhere, free of cost or tax on fees. It is easy to perform for everybody, including any child. It also spread everywhere from the core Indo-Pacific region to the periphery of the Indo-Pacific region.

Table 3 is a partial list of countries with which the international UPI partners, where UPI transactions are possible with India, and this number shows an increase.

The SCO summit on 5 Jul 2023 adopted UPI and some other digital public goods, so the G20 summit in September 2023 also adopted international UPI transactions and digital public goods.

The interface integration of any UPI application can easily take users to the Open Network of Digital Commerce (ONDC) by integrating ONDC or its sister companies for tech support, logistics, or small and medium industries' supply chains with UPI.

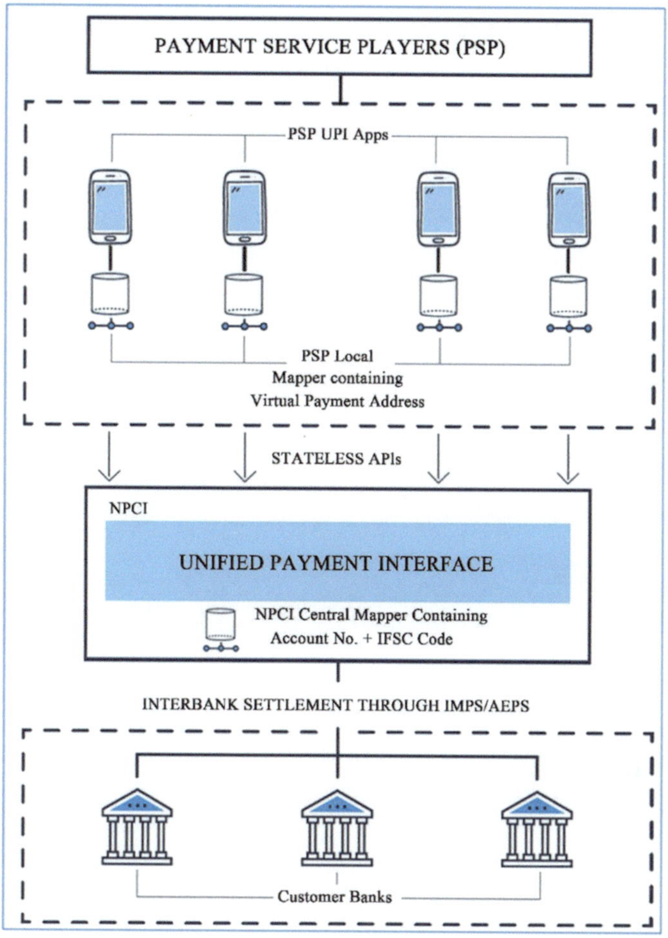

Fig. 36 UPI service players. *Source* NPCI

ONDC has completed the pilot stage with four cities in India, all its phase 1 objectives, and is moving to one hundred cities in the second phase. The people working on this project report their astonishment in wide and straightforward acceptance of all concerned with not much effort since ONDC includes all left out by the ecosystem of individual e-commerce platforms such as Amazon, Flipkart, or Zomato and like, and include Amazon and like and make it a total commerce space in any geography with no merchant cost as user fee unlike credit and debit cards fee of Visa and MasterCard and likes and all e-Commerce platforms. There is something familiar between the COVID-19 pandemic virus and UPI or ONDC; it spreads fast globally. They will soon reach the pandemic level with worldwide infection, UPI acceptance, or possibly ONDC, which is still under development. The growth and acceptance of UPI and

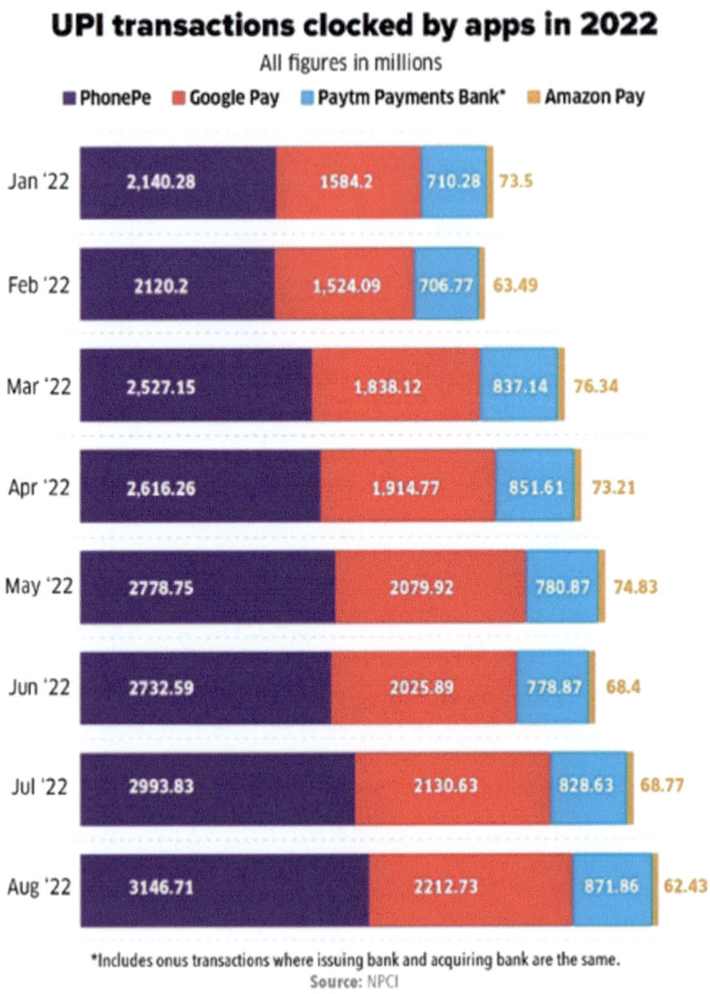

Fig. 37 UPI performance in 2022 by service providers in India

ONDC are nothing short of exponential, just like the COVID-19 pandemic COVID 19.

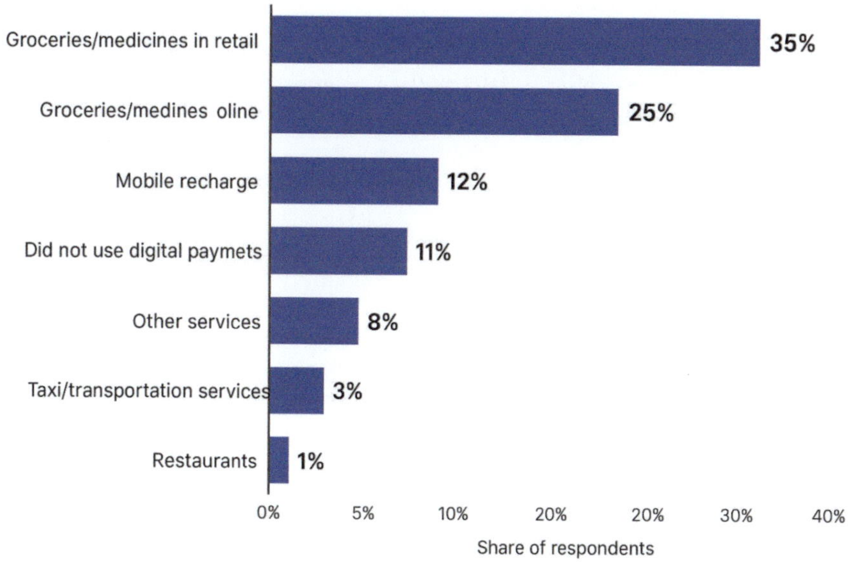

Fig. 38 India impact of COVID-19 on digital payments by type of payment in April 2020. *Source* Sample survey NPCI

2.8 Case Study E-Gateway Enterprise Ecosystem Design and Activation

Digi Locker [88–92]

Digi Locker is a key initiative under Digital India, the Indian Government's flagship program to transform India into a digitally empowered society and knowledge economy. Digi Locker attempts to realise Digital India's vision of providing citizens with a secure document access platform on a public cloud. Digi Locker is a cloud-based platform through which a user can get several government documents issued and verified digitally by the concerned department. Indian citizens get free cloud storage of up to 1 GB for every account. The government launched it to keep official documents like driving licenses, Aadhaar cards, PAN cards, car registration certificates, COVID-19 Vaccination certificates for local and international travel, educational certificates, and birth and death certificates in Digital form (Fig. 40).

Signing up for Digi Locker is easy—all you need is your mobile or Aadhaar number. One Aadhaar number will be authenticated by sending an OTP (one-time password) and setting your security PIN for 2-factor authentication, which will create a Digi Locker account.

One can download the PAN card required for any financial transaction within a ceiling amount for significantly high-value purchases, such as a car and property, in PDF format from Digi Locker. Digi Locker can store and access digital documents and share them online anytime, anywhere.

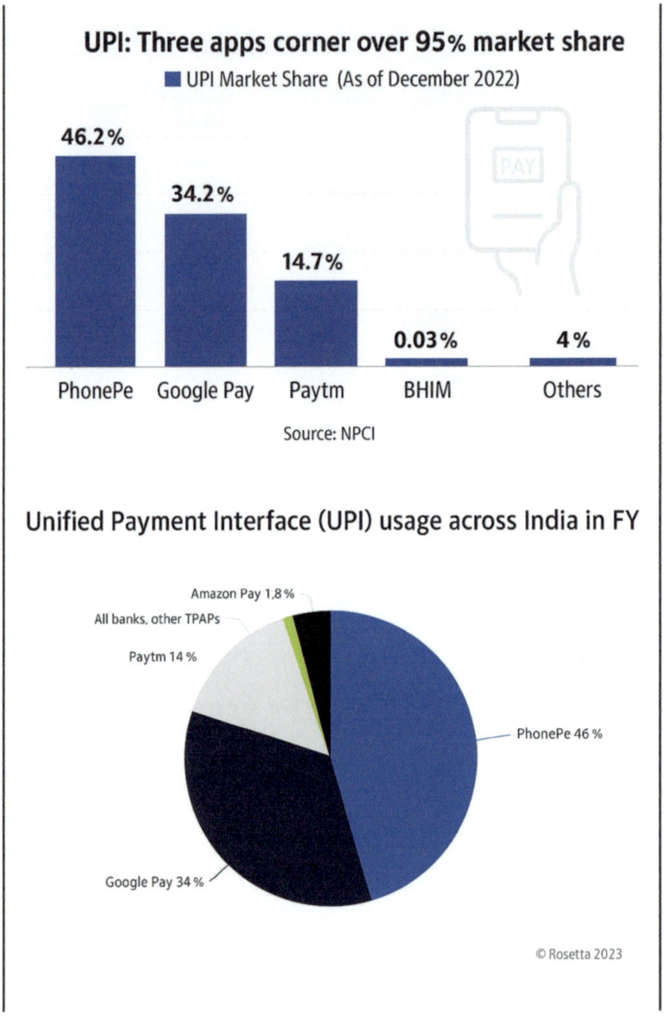

Fig. 39 UPI app share in India August 2022

Government regulators can search for or survey individuals for tax-related, criminal, or civil-related issues. Having verified documents in the Digi locker benefits both parties involved in surveys and avoids hardship.

They intended to ease the Know Your Customer (KYC) process digitally. The Reserve Bank of India (RBI) has now announced that documents stored in the government-backed Digi Locker mobile application can be used to complete the process.

If one stores individual identity proof and other documents in Digi Locker, one can only sometimes carry them physically. One can use the stored documents as proof of valid identification. Traffic control personnel accept these documents on the road while travelling on Indian Railways trains or at airports.

Table 3 Global adoption of UPI agreements with NCPI

Country	Announcement date	Partner
Bhutan	July 2021	Royal Monetary Authority of Bhutan
Singapore	September 2021	Monetary Authority of Singapore-PayNow
Malaysia	September 2021	Merchant trade, Asia
Malaysia, Thailand, Philippines, Vietnam, Cambodia, Hong Kong, Taiwan, South Korea, Japan	September 2021	Liquid Group
South East Asia and East Asia	September 2021	Liquid Group
UAE	November 2021	Network International
UK	January 2022	Lulu Financial, Mashreq Bank and Network International
Nepal	February 2022	TerraPay and PayeXpert
UAE	April, August 2022	Lulu Financial NeoPay, Mananm Infotech
France	June 2022	Lyra Worldline
Europe, Netherlands, Belgium, Luxembourg, Switzerland	October 2022	Worldline
Oman	October 2022	Central Bank of Oman
Saudi Arabia, Baharain	To be announced	To be announced

Source NCPI

Fig. 40 Document wallet 2022. *Source* GOI

Digital documents issued through Digi Locker are valid and at par with the original when used electronically. The concerned authorities should accept it. However, the government of India is working towards awareness and acceptance of such documents by various organisations (Fig. 41).

Following an order issued by the Union Ministry of Transport and Highways, documents saved in the m-Pari Vahan app and Digi Locker will now be considered valid in all states. The government has now made this decision legal.

Now, the Digi Locker is accepted for all passport services. Physical proof of documents and Digi Locker mode are being accepted as authentic documents legally on par with originals. Digital Document Exchange with the citizen's consent results in Faster Service Delivery. Here. Government benefits, employment reduction, financial inclusion accelerated, education, and health sector benefits. The users of Digi Locker are not charged any fee for providing access to content, documents/certificates accessed by one through Digi Locker services or use of Digi Locker API.

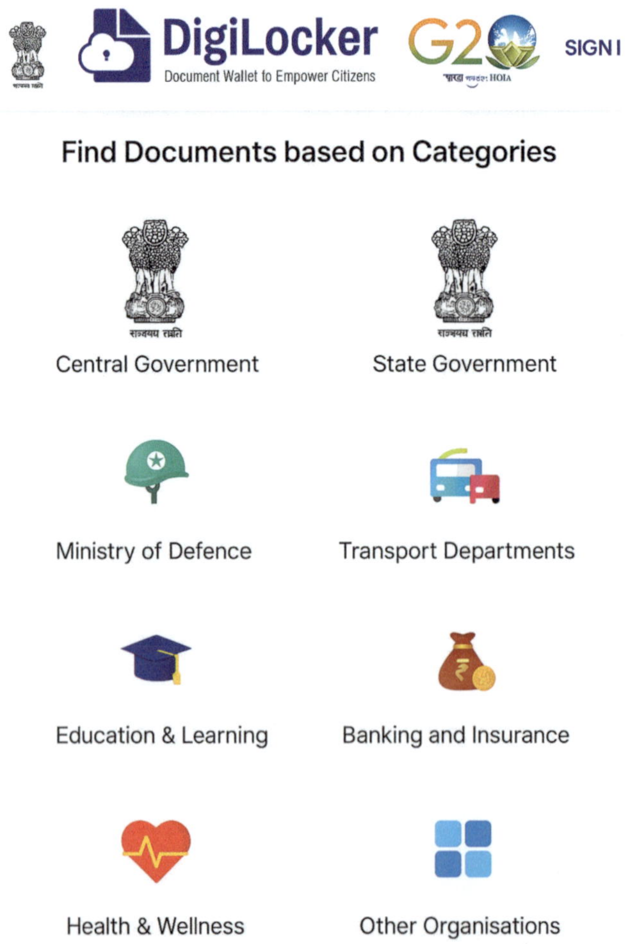

Fig. 41 Digit locker verified documents. *Source* GOI

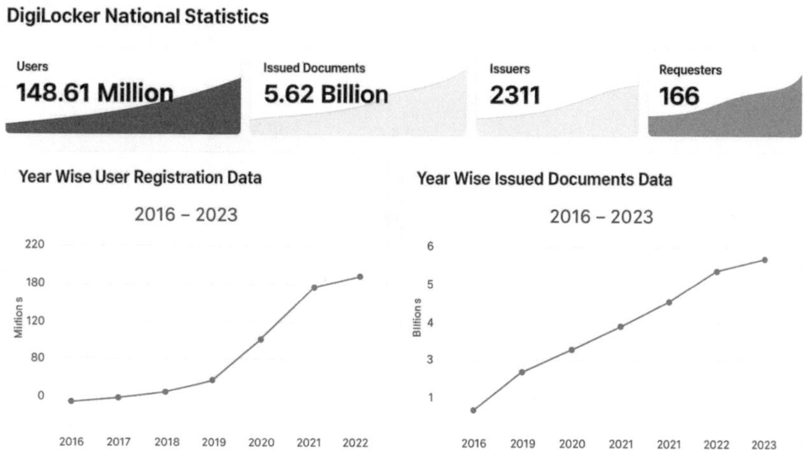

Fig. 42 Digi locker usage growth. *Source* GOI

The popularity of Digi Locker in India is related to its use (Fig. 42).

2.9 *Case Study E-Gateway Enterprise Ecosystem Design and Activation*

Open Network of Digital Commerce (ONDC) [93–95]

E-commerce organises buyers and sellers on an electronic platform economy. The marketplace then expanded for customers from service areas by walking, cycling, or automobile to nations and even many nations because one Indian can legally order a product from the Amazon US site using a credit card after paying all taxes and duties and enhanced charges of money exchange by Amazon with a limitation of only certain products. The popularity of this system in India, based on its growth, is given below.

The ONDC ecosystem goes far beyond the e-commerce platform for total integration with the Digital Economic Community (Fig. 43).

The end-user population in many Indo-Pacific countries also increased, as shown by the growth in retail e-commerce (Fig. 44).

It gave impetus to many economic activities of recent origin, like courier services, which is electronically monitored in its movements, banking using credit and debit cards, UPI and Packaging industries, logistical services, technical support services, and so many others. This platform economy was exclusive to individual platforms of e-commerce firms, as shown below.

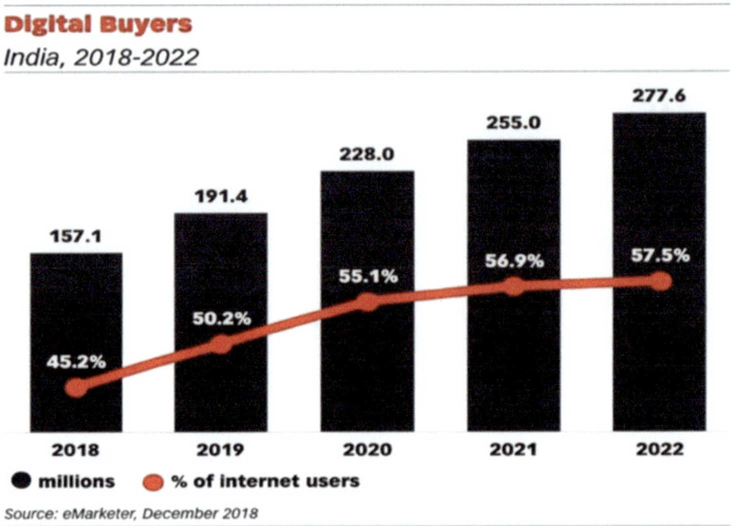

Fig. 43 Digital buyers and percentage of internet users 2018–2022

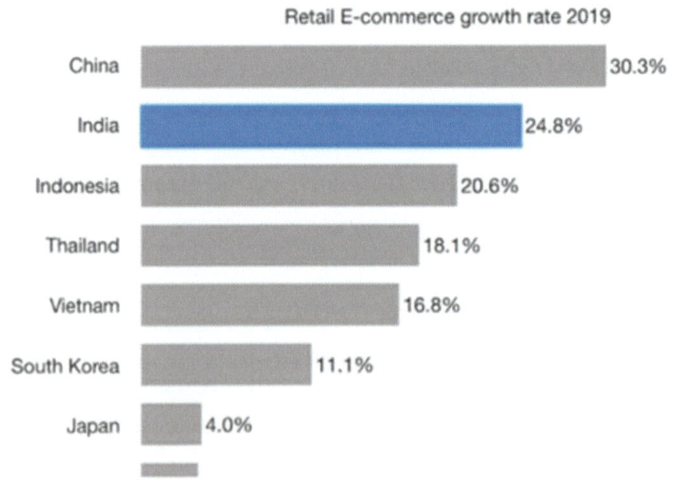

Fig. 44 Retail e-commerce growth rate 2019

ONDC will usher in an open banking ecosystem in India. It will allow bank customers to access financial services from across the spectrum without being limited by their bank's offerings. It will also democratise credit access for last-mile retailers or consumers typically underserved by the formal sector (Fig. 45).

In the Open Network of Digital Commerce (ONDC), an attempt is made to expand the platform concepts to network concepts that combine all platforms and

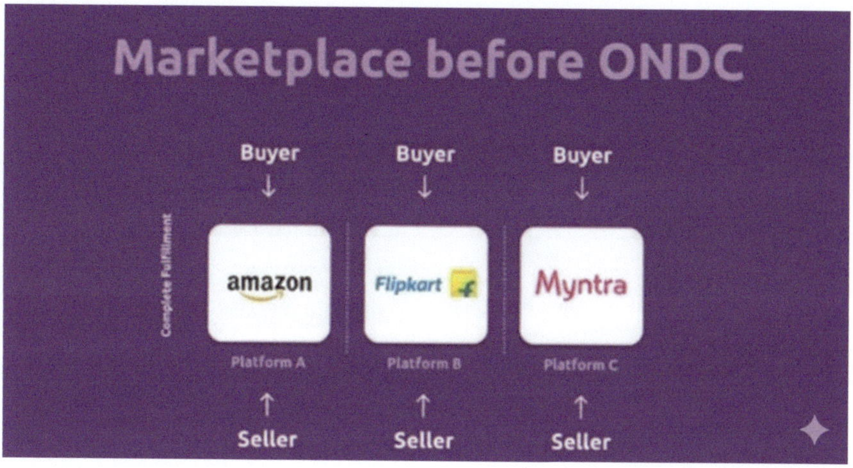

Fig. 45 E-commerce marketplace before ONDC

those outside. This functions like any e-commerce platform, as shown below for end users. The only difference is that it is a much-expanded network, more competitive with lower prices, and money flows shall be equally significant, unlike individual platforms, which are expanding with profitability (Fig. 46).

This open digital commerce network seems to follow the same protocols of email as shown below for its configuration and is in no way so unique (Fig. 47).

The ONDC encompassed many apps and economic activities, as shown (Fig. 48).

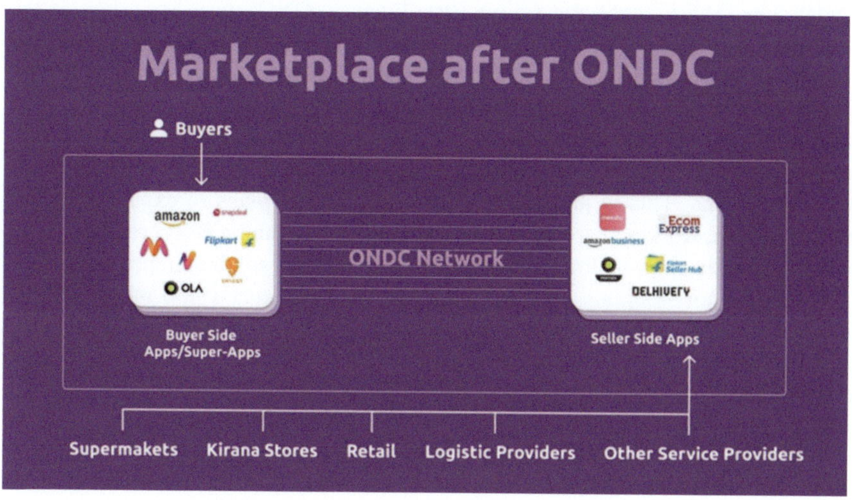

Fig. 46 E-commerce marketplace after ONDC

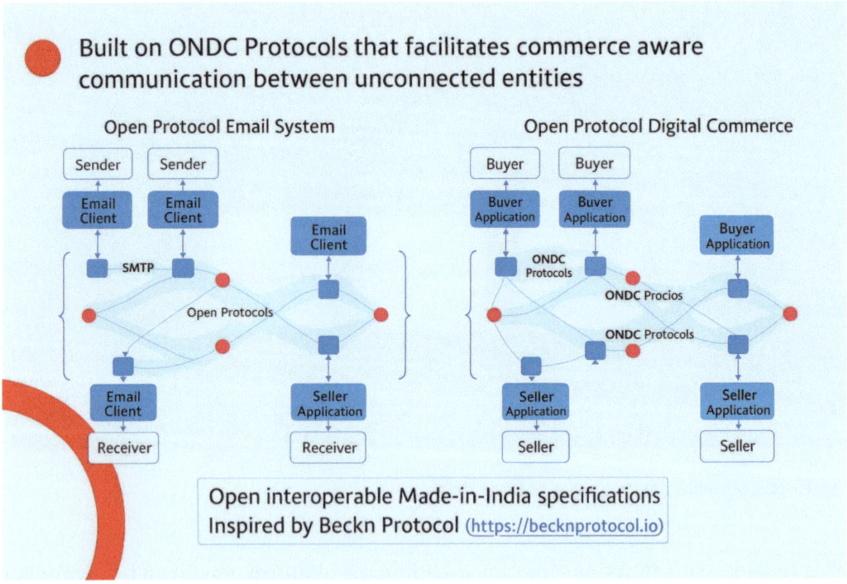

Fig. 47 Similarity of E-mail and ONDC protocols

This resulted in economic activities across many platforms, as illustrated by the following diagrams.

The innovative use of the account aggregator framework further streamlines the loan process, making it more efficient than ever. Through partnerships with leading lenders like DMI Finance and Aditya Birla Finance, ONDC tests apps such as Tata Neu and Paisa Bazaar in India. By harnessing the power of public digital goods like Aadhaar, GST, and UPI, ONDC can deliver small-ticket loans in minutes (Fig. 49).

ONDC's strategic focus is on last-mile retailers and consumers, a segment often overlooked by traditional banks and NBFCs. By formalising their data and institutionalising credit disbursals, ONDC is set to revolutionise the accessibility of unsecured loans. This is just the beginning, as ONDC plans to expand its offerings to include working capital loans, purchase finance, and a B2B credit protocol, potentially reshaping the financial services landscape for small entrepreneurs.

ONDC is not stopping at e-commerce. It plans to expand into other financial services, including insurance and investments. It has already conducted a successful pilot transaction for marine insurance with Insurance Dekho and Bajaj Allianz. It aims to create an open banking ecosystem in India, offering customers various financial services (Fig. 50).

What was more appealing was the observed cost reduction and price reductions for all actors in the ONDC, as shown below, which has the potential to attract more customers (Fig. 51).

Fig. 48 Ecosystem stakeholders of ONDC

Sellers and Buyer's Reputation rankings strictly based on their efficiency in serving customers arising out of a history of successful, delayed and failed transactions are attached to individual sellers or buyers or service providers in ONDC and not to a platform when one embraces ONDC against an individual platform-oriented e-commerce platform which is not transferable from one platform to another. If

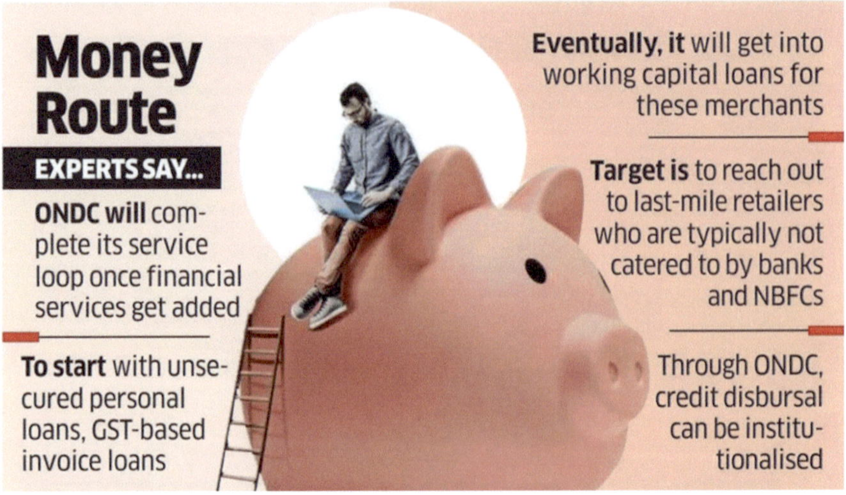

Fig. 49 ONDC approach for credit needs of small entrepreneurs for small-ticket digital loans. *Source* Economic Times 18–06-2024

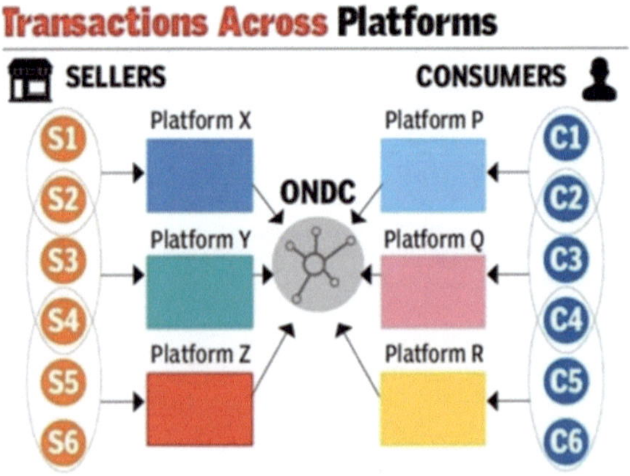

Fig. 50 Transaction of multiple digital platforms in ONDC

many platforms are part of ONDC, reputation ranking stays with the individual and is transferable among all these platforms. This is from one platform to another. An Amazon platform user's reputation ranking cannot be transferred to Flipkart even if the user is the same, unlike ONDC. This resulted in a transition from platform to network marketing in ONDC, diagrammatically represented below. This creates an open network-centric e-marketplace that is much more extensive than individual platforms, which can be international if ONDC becomes as popular as UPI (Fig. 52).

ONDC, like UPI, has the potential for worldwide adoption since it costs less and benefits all, including the monopolies of well-established e-commerce companies.

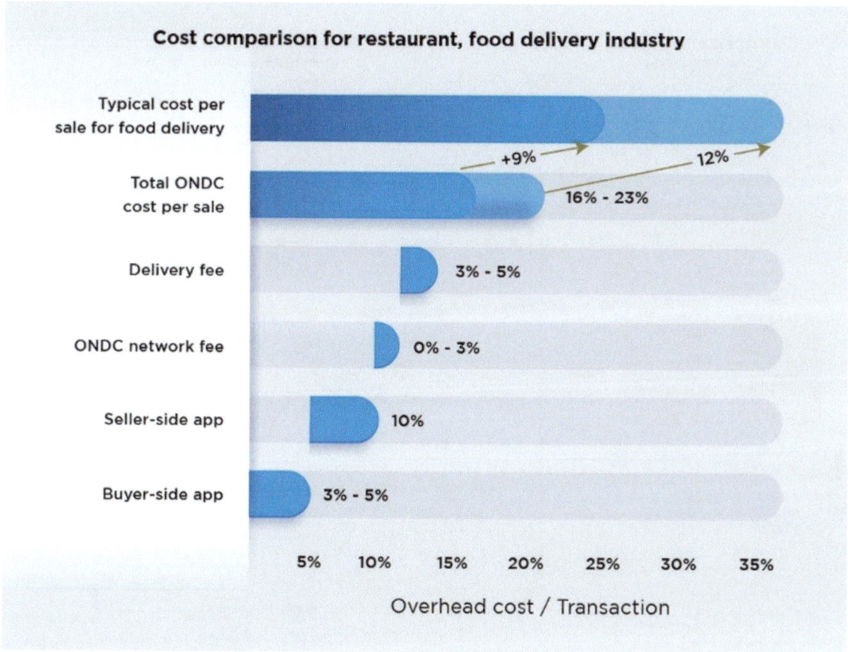

Fig. 51 Cost advantage of ONDC compared to e-commerce

Primarily, wherever the UPI has spread worldwide, the growth of ONDC will also be exponential.

Such a complex ONDC system has a few issues that can be solved systematically with a well-organised governing system (Fig. 53).

This issue resolution created a Governance System for ONDC, as shown (Fig. 54).

Interoperability, decentralisation, and unbundling are essential to ONDC (Fig. 55).

Finally, the building blocks of ONDC are presented in (Fig. 56).

ONDC was developed based on three levels of field testing, starting with phase 1 of five cities and phase 2 of 100 cities and further spreading in alpha and beta stages, as shown in (Fig. 57).

How will ONDC work?

1. The user usually has 2–3 famous applications on their phone, such as Paytm, Amazon, and Flipkart.
2. One logs in to buy a packet of bread. When one searches for bread on Paytm, it shows two options.
    a. Based on local Kirana stores, Rs 40 but no delivery
    b. Breadlinkit Rs120 on delivery
3. One chooses a local Kirana shop and must now choose a service to fulfil the delivery. One goes with Dunzo, which offers a delivery service at Rs 30.

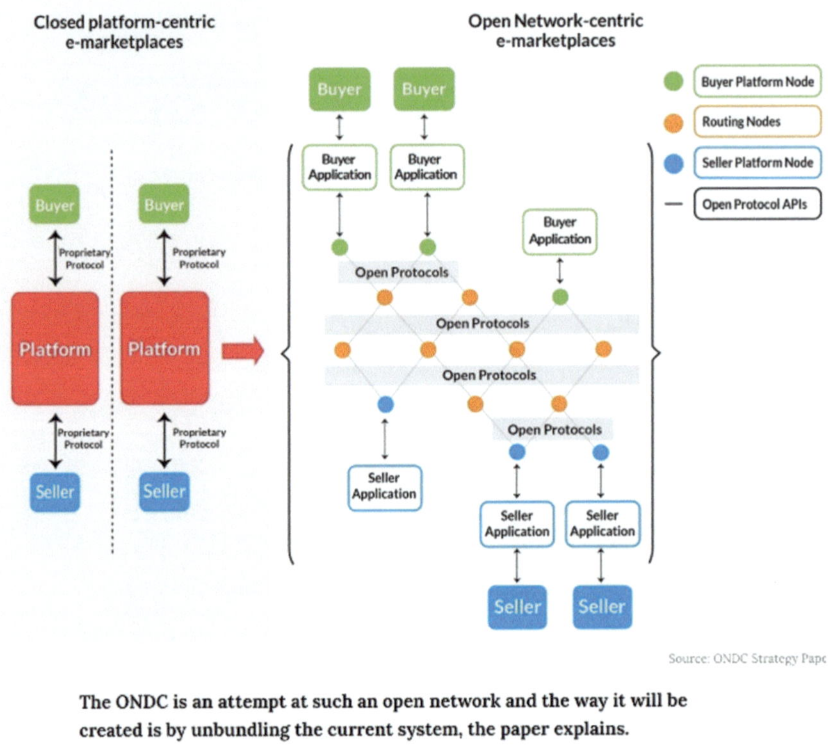

Fig. 52 Digital marketplace comparison single platform, e-commerce versus ONDC e-commerce

4. The total value is Rs 70—one pay using UPI.
5. Now, via the interactions on the ONDC network, Denzo's delivery guy reaches the Kirana store, picks up the ordered bread, and delivers it to her.
6. So here, PayTm brings the user, Kirana store provides the product, and Dezo delivers the product. The value created is distributed accordingly.

What Problems does ONDC aim to solve?
ONDC has very ambitious goals:

1. ONDC expands the market by onboard more retail stores online
   a. Increase gross merchandise value (GMV) from Rs 4.5 lakh crores ($58 billion) to Rs 7.5 lakh crores ($96 billion)
   b. Increase the total retailers from 20 to 90 lakhs
   c. Increase buyers from 9 to 20 crores
2. The second goal is to remove dependence on prominent players like Amazon and Flipkart, bringing the bulk of the users. This will allow retailers to go online on their terms without adhering to marketplace rules set by big players and still reach a huge volume of users.

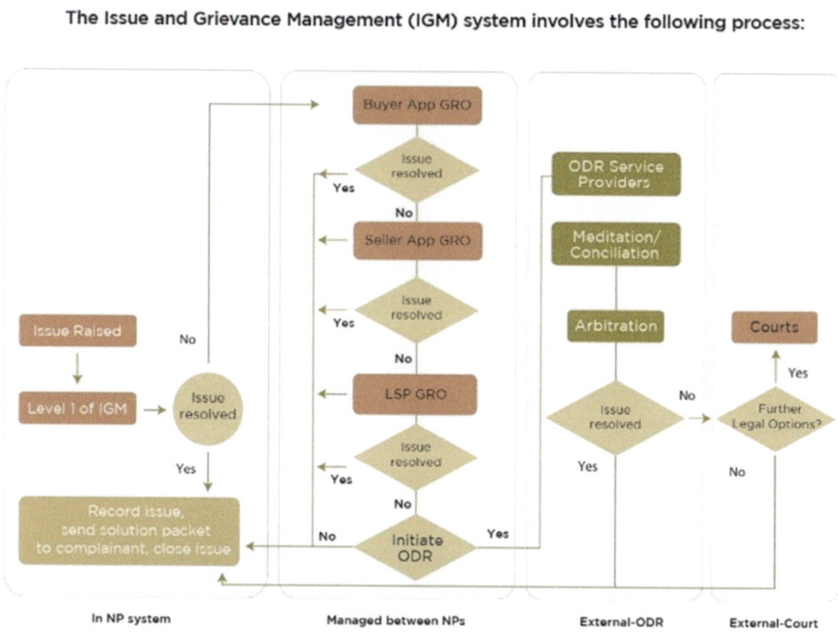

Fig. 53 Governance system for ONDC

3. Rake the power away from aggregators like Amazon, who use their power to control the marketplace and even replicate popular products to sell under their retail brand.
4. Help niche retailers get discovered quickly online without worrying about setting up/delivering logistics systems and attracting customers.
5. Create a solution on top of the preexisting network. Like a pioneer, WhatsApp Pay was built on top of UPI.

Will ONDC be the next UPI?

The UPI was an evident success, with transactions going from 2.38 lakhs in 2016 to 10 lakh crores in 2021. The ONDC uses the same system that makes UPI successful. It is UPI on steroids. The ONDC is a complex system with complex issues.

In the e-marketplace, big players like Amazon, Flipkart, Zomato, Paytm, and Google maximise their profits. How the big players will change still needs to be determined.

It still needs to learn the technical solution ONDC uses to discover sellers and provide consumer logistics services. Still, all players feel it is a just and equitable marketplace to compete.

It is now being added to the beta tests for all emerging issues in India.

Network's Issue and Grievance Management (IGM) system clearly lays out a tiered redressal system.

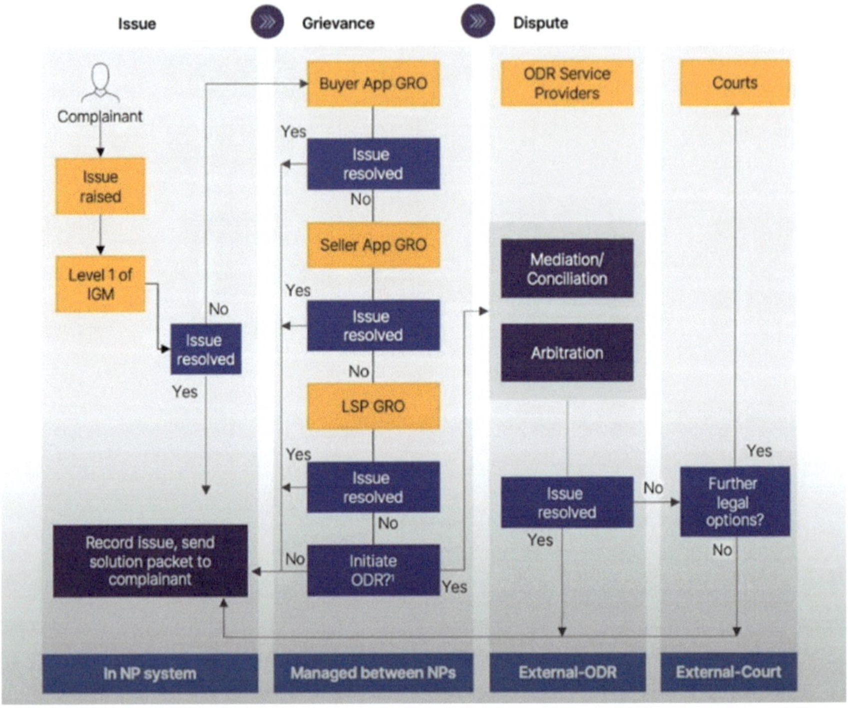

Fig. 54 Tiered Redressal system of ONDC

Let us visualise the end-user experience of UPI and ONDC with the objective of the project's enhanced way of life in the digital economic community and its high-value addition to the digital economic community, which is extraordinarily high and much above the national average.

This is purely hypothetical and in no way an actual incident.

The intelligent couple, Mr. and Mrs. X, are Indian migrant temporary legal migrants in Singapore, a leading economy in ASEAN and a role model for many ASEAN countries.

They married as batchmates and had two diverse undergraduate degrees from one of the National Institutes of Technology. Soon after graduating from BTech, both were selected for a campus interview at NITC and temporarily migrated to Singapore.

They want their first holiday in Singapore, a road trip with a hired self-driven car they do not own, since car ownership is discouraged and very costly to own in Singapore for a long time. Their proposed holiday trip is from Singapore to Penang in Malaysia, where they leisurely stop at numerous motels on seacoasts at night for

ONDC is based on the principles of interoperability, decentralisation and unbundling.

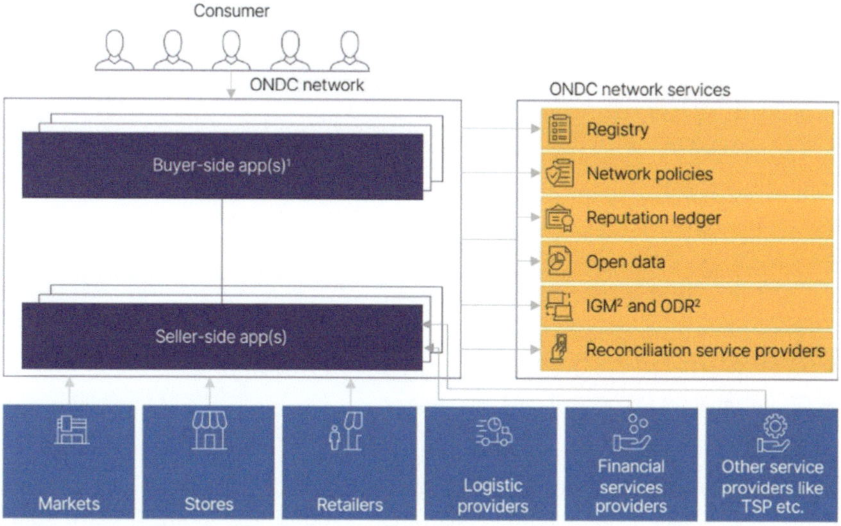

[1] Buyer side apps interact with seller apps via Gateways.
[2] IGM - Issue Grievance Management and ODR - Online Dispute Resolution.
Source: ONDC strategy paper 2022

Fig. 55 Interoperability, decentralisation, and unbundling in ONDC

The building blocks of ONDC

The following figure shows the various components of ONDC:

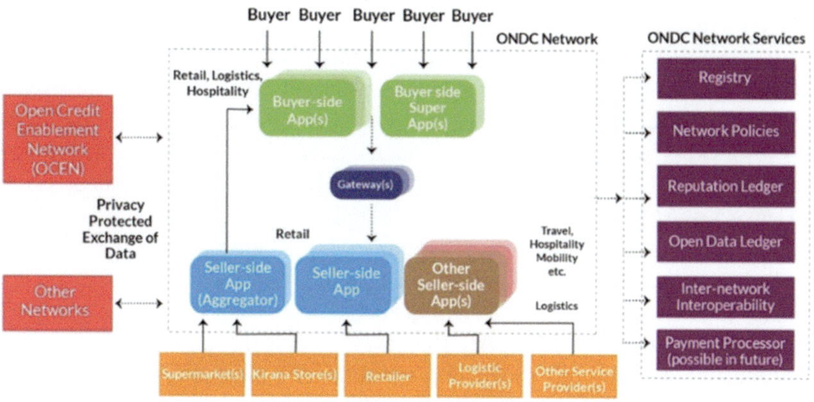

Fig. 56 ONDC-building blocks

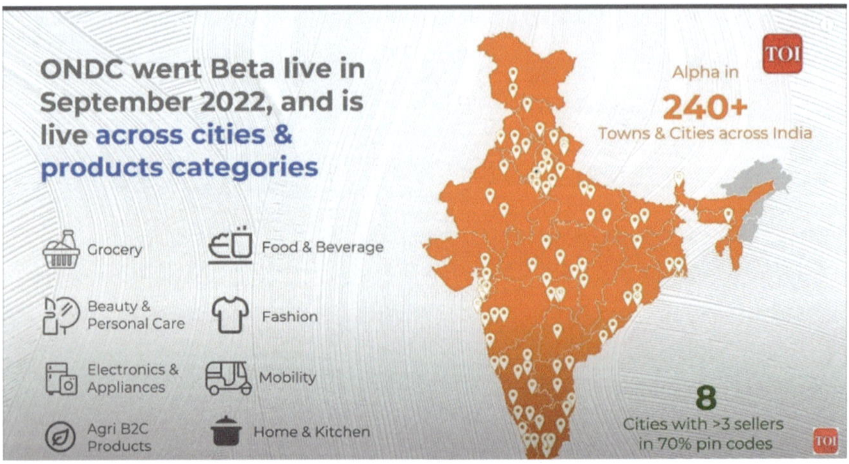

Fig. 57 Footprints of ONDC alpha and beta testing

dinner and a good bath in the sea. They selected breakfast and lunch locations, and evening tea breaks in the public garden after marking it in Google Maps, which guides their road trip, and after making sure that it is open and computing when they reach based on the design speed of the highways and distance from Google Maps.

They also selected hawkers for food to be delivered by them in such locations at such and such time and ordered using their Phonepe app. They had a hawker experience in Bhukit Street in Singapore from a whole night to the early hours of daybreak and liked the hawker food served there. Based on accessibility, they can source food from the streets of Indonesia, Malaysia, and Singapore. They selected these countries' exceptional Indigenous Street food, available in its best form in the streets of these countries.

The Phonepe, used for all activities, is an Indian unicorn that was first registered and is functioning in Singapore. It decided this year to shift to India despite the considerable fee involved in such a shift for reregistering in India. It was executed, resulting in higher profitability envisaged after that. Phonepe is part of ONDC India.

The hawkers' website, connected through the internet's network architecture, had two options. The buyers can select the delivery logistics with companies involved and leave it to AI to choose for them. AI will find the optimal solution for the delivery logistics based on experience and algorithms. Mrs. X left it to AI to do that job.

The hawker and logistics company joined ONDC because it is worth doing it with the only technological support system of image processing to help daily or periodically catalogue their food for sale. The ONDC builds this and is easily accessible. They must photograph this food using a mobile, and then automatically, it gets its label names from AI in many languages, and then apply the price they offer today, with the provision that they need to change the price to what changes they have already made because of the fluctuating price of the inputs.

Mr. and Mrs. X felt refreshed and had a significantly enhanced life throughout their trip back and forth with different venues they selected from Google Maps with good ambience and essentials.

Convenience and sourced their holiday requirements from 3 countries with user-friendly regulations.

2.9.1 Case Study E-Gateway: Global E-Governance

Case Study Estonia [96–104]

E-government is practised in many countries, both developed and developing. The only difference is that one country may have e-government for a greater percentage of the government's routine than another. E-Governance often includes digital public goods such as electronic money transfers, for example, using India's Unified Payment Interface (UPI) or something similar.

An E-Gateway Global E-Governance case study was undertaken in Estonia, a European Union country. In Estonia, E-Governance is citizen-friendly and progresses with innovations that replace conventional governance. Yet this path is long, and almost all countries have miles to go. The best way to accelerate this process is to share the free E-Government stacks with all countries, and each becomes an E-Government advocate just like India, which is willing to share India stack like, for example, ONDC, UPI, as proclaimed in the recent G20 Delhi declaration and declaration of SCO summit and BRIC deliberations. This covers all leading economies of the world.

One problem faced by the author was whether there was anything particular in E-Gateway Global E-Governance that the author needed to include, or all of Estonia's E-Governance stack.

For example, Estonia's E-Residency, presented below, falls in the E-Gateways Global Governance domain.

The UN and several international treaties accept most E-Government stacks globally. For example, there are dual taxation treaties between two pairs of countries in which individuals with income from these pairs of countries need to file income tax returns only in one resident country, which is accepted by the countries concerned.

Given so many provisions, the author concludes that E-Gateway Global Governance is the total E-Government of the Estonia case study in this presentation.

Estonia has had an e-residency provision since 2014, the first in the world. All registered e-residents of Estonia receive an intelligent ID card, which provides digital identification and enables them to enjoy all Estonian government E-Government services available at remote locations outside Estonia. This includes the digital signing of documents.

Hence, an enterprising E-Resident of Estonia from anywhere in the Indo-Pacific can identify themselves as an Estonian enterprise if it benefits them to open to the European market and benefit from all the advantages of being part of the European Union without migrating to and residing in Estonia (Fig. 58).

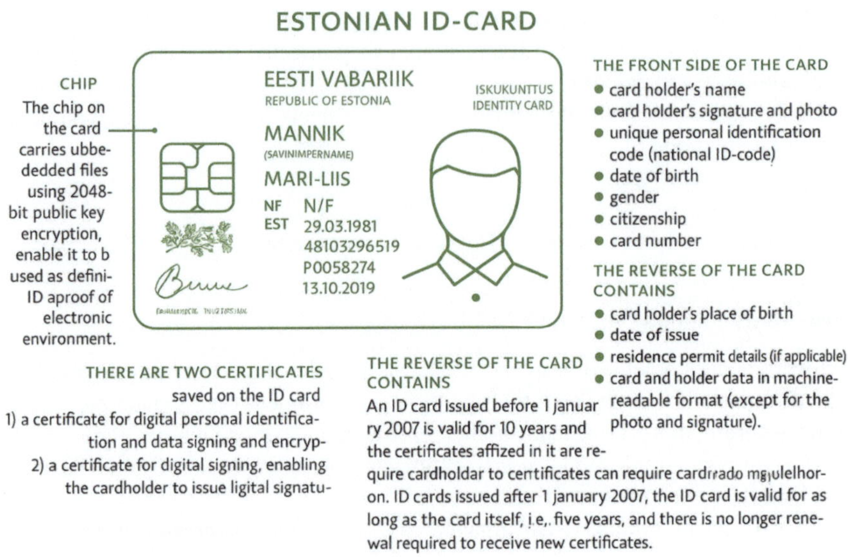

Fig. 58 ID card Estonia

The E-Resident ID card and services are built on state-of-the-art technological solutions, containing two security certificates for authentication and digital signing. However, it is essential to note that e-residency does not confer citizenship, tax residency, residence or right of entry to Estonia or the European Union. The issued e-Resident smart ID card is not a physical identification or a travel document.

Estonia's first e-resident was Edward Lucas, senior editor at The Economist, who called the resident's card an "Estonian Express" that offers a degree of security, convenience, and privacy, unlike any other national ID scheme. In a short period, thousands of foreign entrepreneurs, professionals, and media figures have followed Lucas's example to become Estonian e-residents, among them Shinzō Abe, the then Prime Minister of Japan.

We present three web pages of E-Residency [105–109] below to demonstrate how easy it is for readers to become E-Residents for your E-Gateway to Enterprise Ecology design system (Fig. 59).

All E-Government solutions available in any country for E-Gateway Global Governance, which may or may not include Digital Public Goods for the Digital Economy, are interconnected. For example, filing income tax electronically always ends with payment gateways, where all possible electronic payment schemes are given for making payments after filing.

So, without classifying these electronic services into two compartments, we present all of Estonia's E-Government services in one table below. The first column is the E-Government service; the second column is the functions these E-Government services perform, and the third column is the citizens' benefits.

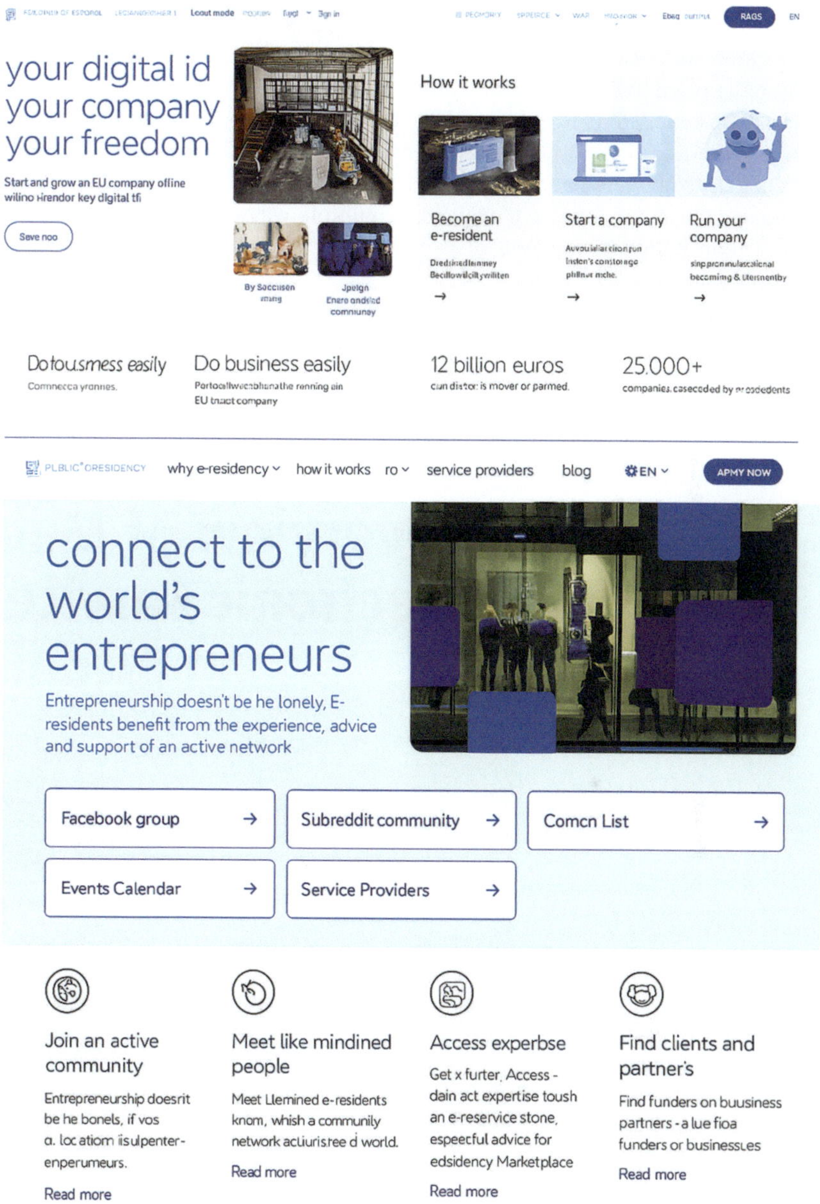

Fig. 59 E-residency web pages for enrolment

Participating in Government services generally does not add value to transactions, so less time spent is better. The time saved can be utilised for enhanced living or engaging in more productive value-adding activities.

The main benefit of E-Residency is that the user can access all Estonian e-services from anywhere in the world. He or she can also declare taxes online, although e-residency does not automatically establish tax residency.

E-Governance services have evolved in multiple ways. The figure below represents the chronological progress of E-governance in Estonia, and the table below gives further details (Fig. 60).

All of these electronic services have a many-fold effect on Estonia. First, it firmly establishes E-Gateway Global Governance and is ready to beneficially access all megacity systems and connect the India-Middle East-Europe Economic corridor. Estonia is an information society. Estonia is one of the most advanced e-societies in the world. It is an incredible success story of a partnership between a forward-thinking

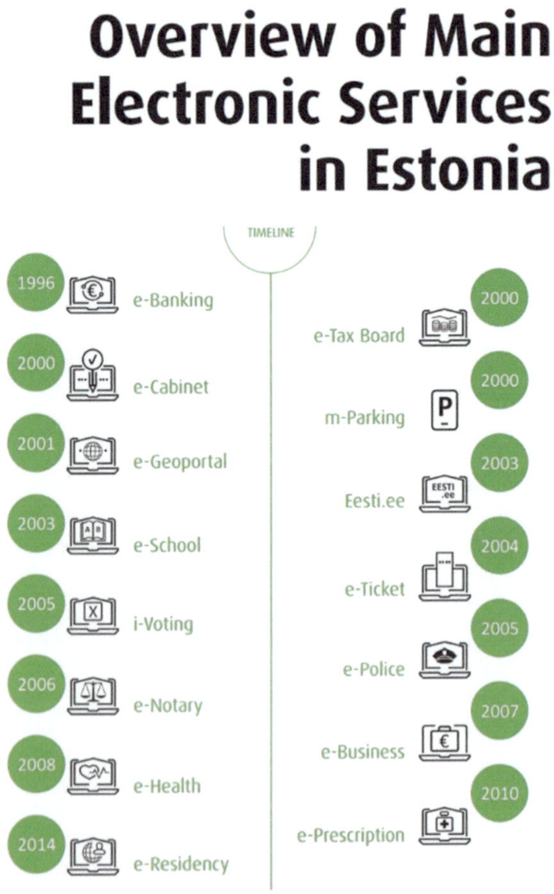

Fig. 60 E-governance services of Estonia by chronology

government, a proactive ICT sector and a switched-on, tech-savvy population. Estonia boasts the world's leading IT infrastructure and e-services.

The Estonian government has enacted many e-government legislations to support these services.

They are

Public Information Act
Digital Signatures Act
Archives Act
Population Register Act
Identity Documents Act
Personal Data Protection Act
Information Society Services Act
Electronic Communications Act
Public Procurement Act
State Secrets and Foreign Classified Information Act

All services are available online 24 h a day and seven days a week from any country in the universe and use them (Table 4).

E-Governance can be global. As shown in the figure, Estonia tried to link with other European countries, such as Finland, share resources, and integrate global government activities.

The main principles adopted by Estonia for electronic services are as follows.

- Centralised policy development
- Decentralised implementation
- Transparent and efficient public sector
- Neutrality of technological platforms
- Citizen/customer orientation
- Functioning model for the protection of personal data
- Measures against digital divide.

All these legislations paved the way for the achievement indicators given below. The figure below shows some evaluations as indicators regarding Estonia's e-government (Fig. 61).

2.9.2 What Are the Proposed and Existing EU Acts that Estonians May Accept as Necessary for E-Digital Economic Community Gateways?

Estonia is part of the European Union (EU), a political and economic union of 27 member states that share common values and goals, such as democracy, human rights, and the rule of law. The following four acts and other approaches are essential for E-Gateway global governance: the Digital Market Act, the Digital Services Act, the Artificial Intelligence Act, and the Data Protection Act.

Table 4 E-governance functions and benefits

E-Governance services [110–169]	Functions	Benefits
E-Banking	99.8% of Banking is online in Estonia	No need to visit the Bank in person Transactions are cheaper if made online Save time and money for customers and service providers Accessible 24/7 Secured access from any place on earth with an internet connection
E-Cabinet	The information system of Government Sessions in Cabinet enables Ministers to prepare and conduct whole sets of procedures for Ministers without papers	Ministers are better organised It is possible to attend meetings remotely There was no need to carry a large volume of papers Average session times were reduced to 4 to 5 h There was a lower environmental impact Decision-making was transparent and understandable
E-Geoportal	Link to electronic cadastral registers containing information on the value, natural status, and use of land	Easy to get all property information Convenient access to land registry documents saves time and administrative costs and creates paper-free administration Transparent and curtail real estate fraud and corruption
E-School	To make study-related material easily accessible. Facilitate the work of teachers. Cultivate strong engagement with parents	Kids get an overview of grades and homework assignments Parents get more involved with their child's education Improved communication with parents and teachers More efficient record-keeping and organisation for teachers

(continued)

Table 4 (continued)

E-Governance services [110–169]	Functions	Benefits
I-Voting	Internet voting (I-voting or online voting) is a system that allows voters to participate in national or local elections by casting ballots online via an Internet-connected computer from anywhere in the world. It is used as an additional voting method to improve accessibility to elections. It should not be confused with other electronic Voting systems used elsewhere that rely on special voting devices set up at the polling station	The Estonian solution is simple, convenient, and secure, allowing voters to cast their ballots from a location of their choosing (home, office, abroad) without going to a polling station. There is a designated pre-voting period during which the voters can log onto the system using their ID cards or mobile IDs, establishing the voter's identity. However, upon casting a ballot, the voter's identity is removed from the ballot before it reaches the final counting stage performed by the National Electoral Commission, and thus, each vote remains anonymous
E-Notary	The e-notary system is online, the platform explicitly created for notaries, and helps them in their everyday work, enabling electronic communication with government agencies (including performing Registry queries)	Keeping a schedule of notarising activities, preparing agreements, allowing digital signatures, making reliable queries to the state registries, forwarding records to state registries, minimising paperwork, printing, and repeated data entries, reducing red tape between notaries and clients, and a faster way of doing business with notaries
E-Health	The Electronic Health Record is a nationwide system that gathers data from different healthcare providers to generate a single electronic file, providing a comprehensive record for each patient. The system contains information on diagnoses, doctor visits, tests, hospital treatments, prescribed medications, etc. Moreover, in emergencies, doctors can use a patient's ID card to review time-critical information, such as blood type, allergies, recent treatments, ongoing medication, or pregnancy	Significantly reduce administrative bureaucracy for doctors Access to time-critical information during emergencies Patient-friendly health care is provided Increasing the efficiency of the health care system. Healthcare trend data is automated

(continued)

Table 4 (continued)

E-Governance services [110–169]	Functions	Benefits
E-Residency	Estonia made history by launching the first supranational e-residency scheme in the world—a state-issued electronic ID for non-residents that allows secure authentication and digital signature of documents—thus moving towards the idea of a country without borders	Digitally signed documents and contracts Verify the authenticity of signed documents Encrypt and Transmit documents securely Establish a company in Estonia online and administer from anywhere in the world Conduct E-Banking and Remote Money Transfer
E-Tax Board	Electronic Tax filing system for Tax and Customs for corporate tax returns, value-added tax returns, excise duty return declaration, customs declaration	Easy and Convenient to use, Lower administrative costs, tax returns only require 3–5 min to fill in, taxpayers get a refund in five days, and transparent and efficient tax collection
m. Parking	A convenient way to pay for parking using a mobile	Convenient parking form There is no need for parking infrastructure, and it saves municipalities money
estee.ee	Offers access to 815 e-services to about 200 countries with 7 million views, which is five times the population of Estonia in 2014	Single Point Contact. Platform to promote new services, and more accessible access to e-services
E-Ticket	Personalised ticketing using mobile phones. Tourists only buy paper tickets	No need for paper tickets. It is more straightforward to check the validity of tickets upon each entry
E-Police	The e-police system comprises two main tools: a mobile workstation installed in each Patrol car and a positioning system used in the headquarters, showing each officer's location and status. As a result, each police vehicle is equipped with a computer in the luggage compartment, a monitor, a positioning device, and access to a digital map. In addition, the mobile workstations installed in patrol cars give officers in the field nearly instantaneous access to vital information, enabling aggregated queries from police databases, the Citizen and Migration Board Motor Vehicle Registry, Traffic, and Insurance Fund, etc	The control centre knows the location of each patrol car at any given time. This has improved efficiency, e.g., a 70% increase in offence reports handled daily, road fatalities decreased by over 400%, and a 1000% increase in vehicle queries performed monthly

(continued)

Table 4 (continued)

E-Governance services [110–169]	Functions	Benefits
E-Business	The e-Business Register is an online platform linked to the official database comprising the real-time data on all legal entities registered, in Estonia	Review the company's general data and tax arrears data Access to annual reports, statutes, and personal and commercial pledges. Data, etc., and perform real-time monitoring of processing data and record amendments to verify business and entrepreneurship prohibitions of Estonian persons, to visualise relations between various companies and persons
E-Prescription	Digital prescription is one of the key innovations in Estonia's cutting-edge e-healthcare system. A 2015 survey revealed that electronic prescription was the most popular e-service among citizens	Medical appointments are not required for routine refills. Saving time Patients need not keep track of paper prescriptions It reduced paperwork in hospitals and pharmacies

Source Compilation by Author

1. **The Digital Market Act (DMA)** [170–172]

The Digital Markets Act (DMA) is a proposed regulation by the European Union to promote fair and open digital markets. It targets explicitly large online platforms, known as 'gatekeepers,' that wield significant control over market access. The DMA outlines a set of obligations and prohibitions for gatekeepers, ensuring they operate fairly and transparently. Notably, it prohibits gatekeepers from unfairly ranking their products or services, mandates data portability, and requires them to allow third-party software to interoperate with their services. The DMA has garnered praise for its potential to level the playing field for smaller businesses and foster innovation. While it has faced criticism for being overly prescriptive, its positive impact on the tech sector's growth potential cannot be overlooked. The DMA is currently under review by the European Parliament and Council, and its final form is yet to be determined.

2. **The Digital Services Act (DSA)**

The Digital Services Act (DSA) is a proposed regulation by the European Union to modernise the legal framework for digital services. It introduces new rules and responsibilities for online platforms, including social media, marketplaces, and other intermediaries. The DSA's primary objective is to enhance transparency, accountability, and, most importantly, user safety in the digital realm. Key provisions include the requirement for platforms to remove illegal content promptly, the push for greater transparency in online advertising, and the empowerment of users to exercise more

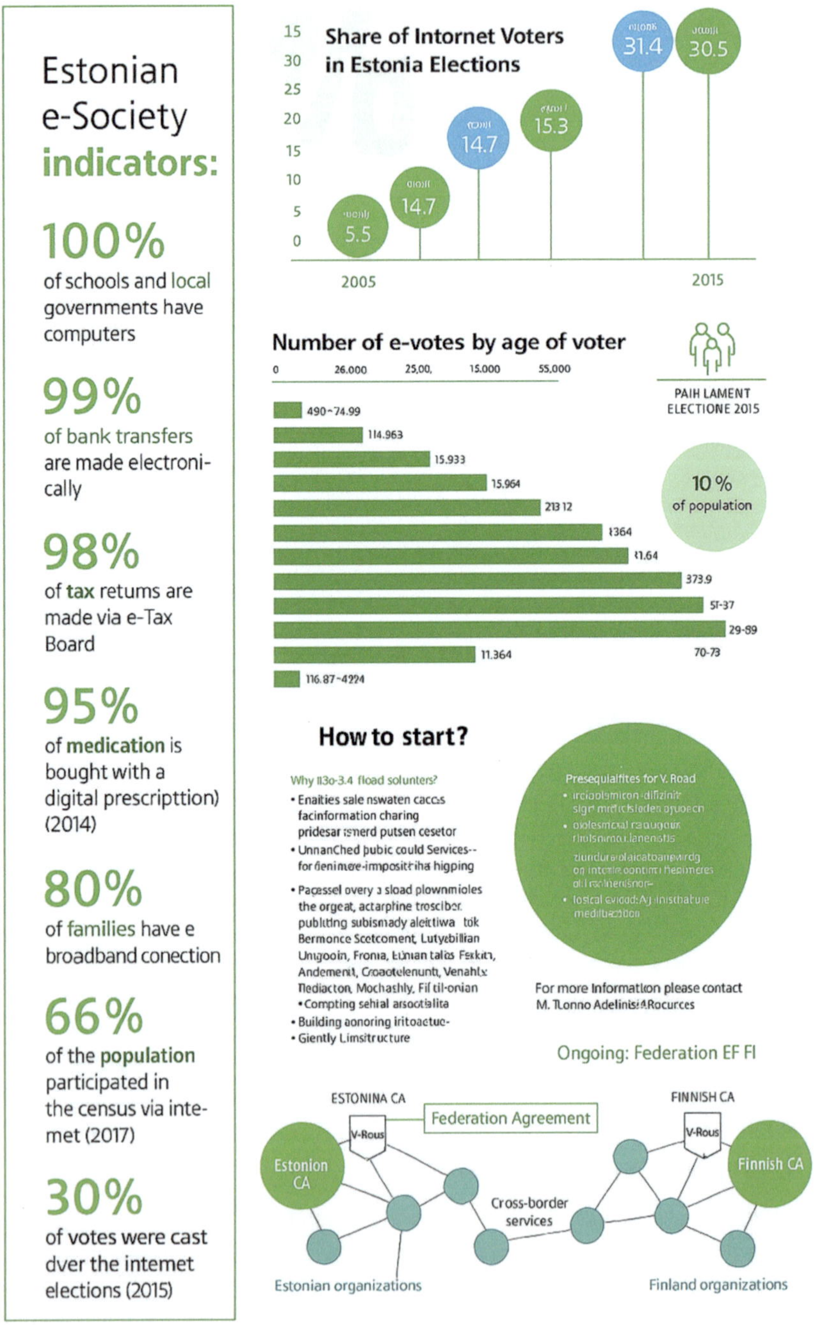

Fig. 61 Estonian E-Society Indicators. *Source* Government of Estonia

control over their data. The DSA is currently in the European Parliament and Council review stage, and its final form has yet to be determined.

3. **The Artificial Intelligence Act (AIA)**

The Artificial Intelligence Act is a proposed regulation by the European Union to establish a legal framework for artificial intelligence (AI). The proposal, released in April 2021, is under review by the European Parliament and Council. The AI Act lays down rules and requirements for AI systems, focusing on transparency, accountability, and user safety. It also introduces a risk-based approach to AI regulation, with more stringent rules for high-risk AI systems. The AI Act has been commended for its potential to foster innovation and trust in AI while safeguarding fundamental rights. However, it has also faced criticism for its perceived inflexibility and potential to stifle innovation in the AI sector.

4. **The Data Protection Act (DPA)** [173–176]

One of the EU's key objectives of the Data Protection Act is to ensure high protection of individuals' data within its territory and beyond. Personal data is any information that relates to an identified or identifiable natural person, such as name, address, email, phone number, health records, or online identifiers. The EU has adopted a comprehensive and coherent legal framework for data protection, which consists of the following main instruments:

- The General Data Protection Regulation (GDPR), which entered into force in May 2018, applies to processing personal data by controllers and processors in the EU and transferring personal data to third countries or international organisations
- The ePrivacy Directive complements the GDPR and regulates the processing of personal data in the context of electronic communications, such as cookies, direct marketing, or location data.
- The Data Protection Law Enforcement Directive sets out the rules for processing personal data by competent authorities to prevent, investigate, detect, or prosecute criminal offences or execute criminal penalties (Data Protection Law Enforcement Directive, 2016).
- The Data Protection Regulation for EU Institutions applies to processing personal data by EU institutions, bodies, offices, and agencies.

The EU data protection framework also grants several rights to data subjects, who can exercise them against the controllers and processors of their data. These rights include:

- **The right to be informed**: Data subjects have the right to receive clear and concise information about the processing of their data, such as the identity and contact details of the controller, the purposes and legal basis of the processing, the recipients and transfers of the data, the retention period, and the rights of the data subjects
- **The right of access**: Data subjects have the right to obtain confirmation from the controller as to whether their data are being processed and, if so, to access their data and receive a copy of it

- **The right to rectification**: Data subjects can request the controller to rectify inaccurate or incomplete personal data concerning them without undue delay
- **The right to erasure**: Data subjects have the right to request the controller to erase their data without undue delay under certain conditions, such as when the data are no longer necessary, when the consent is withdrawn, when the data are unlawfully processed, or when there is a legal obligation to erase them.
- **The right to restriction of processing**: Data subjects have the right to request the controller to restrict the processing of their data under certain conditions, such as when the accuracy, lawfulness, or necessity of the processing is contested or when the data are no longer needed but required for the establishment, exercise, or defence of legal claims
- **The right to data portability**: Data subjects have the right to receive their data in a structured, commonly used, and machine-readable format and to transmit it to another controller under certain conditions, such as when the processing is based on consent or contract, and when the processing is carried out by automated means
- **The right to object**: Data subjects have the right to object to the processing of their data under certain conditions, such as when the processing is based on legitimate interests or public interest or when the processing is for direct marketing purposes
- **The right not to be subject to automated decision-making**: Data subjects have the right not to be subject to a decision based solely on automated processing, including profiling, which produces legal effects or similarly significant effects on them, unless the decision is necessary for a contract, authorised by law, or based on explicit consent

5. **Digital Sovereignty and Human Centric Approach** [177–184]

The European Union (EU) has adopted a strategic vision for the digital transformation of its economy and society based on two fundamental principles: digital sovereignty and human-centricity. Digital sovereignty refers to the ability of the EU and its member states to shape their digital policies and regulations in line with their values and interests and to ensure their autonomy and resilience in the face of global challenges and threats. Human-centricity means that the digital transition should be driven by the needs and rights of the people and that the benefits of digital innovation should be widely shared and accessible to all.

One of the main instruments to achieve this vision is the development of E-Gateways, which are digital platforms and infrastructures that facilitate the cross-border exchange of data, services and transactions within the EU and with its partners. E-Gateways aims to create a digital market where citizens and businesses can access and offer online products and services across the EU without barriers or discrimination. E-Gateways also support the EU's external action and cooperation by promoting its values and standards and enhancing its digital connectivity with third countries and regions.

Some examples of E-Gateways are:

- The European Data Spaces are secure and interoperable data-sharing environments that enable the flow and use of data across sectors and borders while ensuring data protection and privacy.

The Digital Green Certificate, digital proof that a person has been vaccinated against COVID-19, tested negative, or recovered from an infection, can facilitate safe and free movement within the EU during the pandemic.

- The eIDAS Regulation establishes a legal framework for electronic identification and trust services, such as electronic signatures, seals, and timestamps, that can access online public and private services across the EU. The European platform is a digital library, museum, and archive that provides access to over 50 million items of European cultural heritage, such as books, paintings, films, and music.
- The European Blockchain Services Infrastructure is a network of distributed ledger technology nodes that provides cross-border public services based on blockchain, such as self-sovereign identity, diploma verification, and audit trails.

These E-Gateways demonstrate the EU's commitment to building a sovereign, human-centric, and inclusive digital economic community that contributes to global digital governance and cooperation. All these EU Acts strengthen the E-Gateways.

It is vital that other Indo-Pacific countries, both core and peripheral, adopt all legal measures to strengthen the E-Gateway Legal Architecture further and make it human-centric and easy to use for business transactions.

6. **2030 Digital Compass: A Vision for Europe's Digital Transformation: An analysis of the European Commission's strategy for achieving digital sovereignty and leadership by 2030**

The 2030 Digital Compass is a strategic document that outlines the European Commission's vision and goals for the digital transformation of the European Union by 2030. It builds on the previous initiatives of the Digital Single Market and European Digital Strategy. It aims to ensure that Europe becomes a global digital economy and society leader while respecting its values and principles.

The 2030 Digital Compass is based on four cardinal points: digital skills, digital infrastructure, digital transformation of businesses and public services, and digital rights and values. The Commission proposes concrete targets and indicators for each area to measure progress and performance. The Commission also proposes a governance framework to monitor and coordinate the implementation of the strategy involving the Member States, the European Parliament, the private sector, civil society, and international partners.

The 2030 Digital Compass is part of the broader European recovery plan, Next Generation EU, which allocates 20% of its budget to digital investments. The Commission also proposes launching a Digital Decade Policy Program, a multi-stakeholder platform to foster dialogue and cooperation on the digital agenda. The Commission invites the European Council to endorse the 2030 Digital Compass and adopt a Declaration on Digital Principles, a set of shared values and rights that should guide Europe's digital transformation.

The Four Cardinal Points of the 2030 Digital Compass: This section summarises the four cardinal points of the 2030 Digital Compass, highlighting each area's main objectives and challenges.

3 E- Gateway Development Initiatives of the EU

Digital skills: The Commission aims to ensure that by 2030, at least 80% of adults and 20 million employed ICT specialists in the EU have basic digital skills. The Commission also wants to reduce the gender gap in digital skills and professions and to promote digital education and lifelong learning. The Commission recognises the importance of digital skills for European citizens' employability, competitiveness, and social inclusion and the need to address the skills gap and labour market mismatch. The Commission proposes to support the development of digital skills through various actions, such as the Digital Education Action Plan, the European Skills Agenda, the updated Europa's platform, the European Qualifications Framework, and the European Social Fund Plus.

Digital infrastructure: The Commission aims to ensure that by 2030, all European households have gigabit connectivity and that all populated areas are covered by 5G. The Commission also wants to develop and deploy the next generation of digital technologies, such as quantum computing, artificial intelligence, cloud computing, blockchain, and cybersecurity. The Commission acknowledges the importance of digital infrastructure for the competitiveness, innovation, and resilience of the European economy and society and the need to address the existing digital divide and dependency on foreign providers. The Commission proposes to support the development of digital infrastructure through various actions, such as the Connecting Europe Facility, the European Electronic Communications Code, the European Cloud Initiative, the European High-Performance Computing Joint Undertaking, and the Digital Europe Programme.

Digital transformation of businesses and public services: The Commission aims to ensure that by 2030, at least 75% of companies use cloud computing, big data, and artificial intelligence and that more than 90% of SMEs reach a basic level of digital intensity. The Commission also wants to increase the EU's share of the global data economy and foster the development of digital ecosystems and platforms. Moreover, the Commission aims to ensure that by 2030, all vital public services are available online, all citizens have access to their e-medical records, and 80% of citizens use e-government services. The Commission emphasises the importance of digital transformation for the productivity, growth, and sustainability of European businesses and public services and the need to address the existing barriers and challenges in the digital single market. The Commission proposes to support the digital transformation of companies and public services through various actions, such as the European

Industrial Strategy, the European Data Strategy, the Digital Services Act, the Digital Markets Act, the European Green Deal, and the Single Market Programme.

Digital rights and value-centric approach [185–187] respect and promote the fundamental rights and values of the EU, such as democracy, the rule of law, privacy, data protection, non-discrimination, and fairness. The Commission also wants to enhance the digital sovereignty of the EU by ensuring its strategic autonomy and resilience in the digital domain and fostering its global leadership and influence in digital governance. The Commission stresses the importance of digital rights and values for the European digital society's trust, legitimacy, and social cohesion and the need to address the existing threats and challenges in the digital environment. The Commission proposes to support digital rights and values through various actions, such as the European Democracy Action Plan, the Digital Education Action Plan, the General Data Protection Regulation, the ePrivacy Regulation, the Artificial Intelligence Act, and the European Cybersecurity Strategy.

3.1 ASEAN Trade Facilitation Framework (ATFF) [188]

The ASEAN Trade Facilitation Framework (ATFF) is a tool to increase trade facilitation in the ASEAN region. It focuses on implementing existing ASEAN commitments, obligations, and instruments related to trade facilitation and aims to reduce trade transaction costs.

The ATFF includes initiatives such as:

Harmonising standards and technical requirements

Mutual Recognition Arrangements (MRAs)

Developing a single regulatory regime, directive, or agreement

The ATFF focuses on reforms such as improving time efficiency through technology, promoting electronic transactions, enhancing transparency and certainty in intra-regional trade, aligning with international best practices or regional certifications, and improving coordination among government agencies.

ASEAN facilitates trade through Free Trade Agreements (FTAs) and Closer Economic Partnership (CEP) agreements. The ASEAN Free Trade Area (AFTA) was established in 1992 to eliminate tariffs on trade in goods, investments, and services within the ASEAN region.

AFTA consists of three agreements: the ASEAN Trade in Goods Agreement (ATIGA), the ASEAN Framework Agreement on Services (AFAS), and the ASEAN Comprehensive Investment Agreement (ACIA).

ASEAN's FTAs offer many opportunities for foreign investors, including:

Access to Singapore's financial services hub

Manufacturing operations in low-cost regions like Vietnam, Indonesia, and Malaysia

Reduced importer costs

Improved custom clearances

Increased access to products eligible for preferential treatment

Tax holidays and deductions

ASEAN has also signed six free trade deals with other regional economies.

The Regional Comprehensive Economic Partnership (RCEP) is a significant trade agreement for ASEAN that offers the potential for an integrated region to leverage technology, communication, and natural resources to enhance trade. ASEAN reduces trade barriers through the Agreement on the Common Effective Preferential Tariff (CEPT) Scheme for the AFTA. This agreement requires that tariffs on most products traded within the region be reduced to no more than 5%. It also involves the elimination of quantitative restrictions and other non-tariff barriers. In 2010, ASEAN's first six member countries eliminated tariffs; in 2015, the newer members partially eliminated tariffs. In January 2018, the remaining tariffs on specific items were eliminated.

The ASEAN Free Trade Area (AFTA) has several benefits, including:

Trade facilitation: AFTA's streamlined trade rules and simplified customs processes improve efficiency, which benefits businesses.

Lower prices: Lower tariffs on raw materials used by the region's producers should result in lower consumer product prices.

Greater market access: Companies can reach a broader consumer base and take advantage of the expanding middle class in the area.

Increased competitiveness: Small- and medium-sized businesses can leverage the various FTAs ASEAN offers to make their exports more competitive.

Greater access to high-quality goods: ASEAN consumers have a wider choice of better-quality consumer products.

Foreign investment: AFTA aims to attract foreign direct investments.

3.2 Cross-border E-Commerce Framework (CBEFM) by Asia Pacific Economic Community APEC [189]

The Cross-Border E-Commerce Framework (CBEFM) by APEC is a set of standards for customs administrations, government agencies, and e-commerce stakeholders to use when implementing cross-border e-commerce.

The APEC Principles on Trade Facilitation, endorsed in 2001, guide the framework and the APEC Blueprint for Action on e-commerce. The framework's objectives include promoting cross-border e-commerce across APEC.

The framework defines cross-border e-commerce as online ordering, sale, communication, payment, cross-border transactions and shipments, and physical goods destined for a consumer or buyer.

The APEC Cross-border Privacy Enforcement Arrangement (CPEA) is a framework for regional cooperation in enforcing privacy laws for cross-border data flows. The CPEA was endorsed in 2009, began in 2010, and was updated in 2019. Eleven APEC economies participate in the CPEA, with 27 Privacy Enforcement Authorities from Australia, Canada, Chinese Taipei, Hong Kong, Japan, Korea, New Zealand, the US, Mexico, Singapore, and the Philippines.

China's cross-border e-commerce market was valued at 15.7 trillion yuan in 2022 and is expected to reach 16.3 trillion yuan in 2023. In 2023, China's cross-border e-commerce imports and exports reached 2.38 trillion yuan ($330.9 billion), a 15.6% increase year-on-year. Exports were 1.83 trillion yuan, up 19.6% from the previous year, while imports were 548.3 billion yuan, up 3.9%.

China's cross-border e-commerce market has grown at a compound annual rate of about 18% over the past five years. In 2021, China's cross-border e-commerce imports and exports increased 15% year-on-year to $273 billion. In 2020, China's cross-border e-commerce imports were valued at $88.2 billion, a 16.5% increase from 2019.

China's e-commerce market is the world's largest, with total sales reaching $1.8 trillion in 2019. About a third of this volume is cross-border e-commerce, products imported to China from other countries.

In 2021, the global cross-border B2C e-commerce market was valued at $764.73 billion, with a projected revenue of $5154.16 billion by 2030.

3.3 *Trade Lens (Developed by Maersk and IBM) [190]*

Trade Lens was a blockchain-enabled global trade platform developed by IBM and Maersk in 2018 to digitise and simplify global supply chains. The platform aimed to connect supply chains on a blockchain to make inter-corporate trade more efficient and save money for member shipping companies.

Trade Lens used an electronic shipping ledger to record details of cargo shipments as they left their origin, arrived in ports, were shipped overseas, and were eventually received. The platform facilitated information exchange for more than 65% of containerised trade and built a network of over 300 members, including ocean carriers Alianca and CMA CGM.

In November 2022, Maersk and IBM announced that they would discontinue Trade Lens by early 2023, stating that the need for full global industry collaboration had yet to be achieved.

The tradelines do not have wider acceptance and are likely to be closed.

3.4 Building Trust and Collaboration Among Stakeholders [191]

Building trust and cooperation between stakeholders is essential for successful leadership and collaboration.

Here are some ways to build trust and collaboration:

Identify stakeholders

Identifying stakeholders is a crucial step in stakeholder management and is essential for the success of any project or organisation.

Communicate openly and frequently

Open and continuous communication is essential for building trust within a project team.

Deliver on your promises

Consistency is vital to building trust with stakeholders because it enables them to plan and work more efficiently.

Involve and empower your stakeholders

Proactive stakeholder engagement is pivotal in establishing trust among stakeholders and society.

Recognise and appreciate your stakeholders

Build rapport and relationships with your stakeholders

Manage conflicts and expectations with your stakeholders

Conflict resolution brings people together and promotes teamwork and trust among the members.

Enable collaborative relationships

Collaborative decision-making can build trust among stakeholders by allowing them to collaborate and consider each other's perspectives.

4 Emerging Trends in E-Gateway Development: Blockchain [192–206]. Machine Learning and Artificial Intelligence [207–210]

Blockchain is a database that links information in a chain and is shared and managed by multiple parties. It can help e-commerce businesses in some ways, including:

Product Authenticity

Blockchain can help create transparent supply chains, enabling customers to verify the authenticity of products like luxury goods, pharmaceuticals, and high-end fashion.

Inventory control

Retailers can decrease items from inventory and order replacement stock when pre-defined thresholds are reached.

Digital Ownership

Retailers can own all digital assets, including digital storefronts, product photos/videos, and reviews.

Shipping costs

Blockchain can help lower shipping costs, a significant factor in bulk and unit pricing.

Customised products

Blockchain can help sellers provide customised products that appeal to unique tastes or functional requirements.

Peer-to-peer transactions

Using blockchain technology for peer-to-peer transactions, companies can eliminate intermediaries and save time and money.

Supply chain management

All supply chain participants can use the distributed ledger that blockchain technology provides to track and verify transactions.

Guaranteed delivery

Blockchain guarantees that items are delivered in the anticipated timeframe since less time is required for data validation.

1. **Some companies that use blockchain in e-commerce include**:

Starbucks

It uses a blockchain-based system to track the journey of coffee beans from farms to stores.

De Beers

It uses blockchain in its supply chain to track the provenance and authenticity of diamonds.

2. **Artificial intelligence is increasingly becoming a part of everyday life**

    - Generative AI is building content immediately.
    - Predictive AI is peaking in the future.
    - Deep learning is producing human-like personalities.
    - Chatgpt 3.5, 4 and O, etc., are household names, as are Microsoft Copilot Pro, Google Gemini and Notebook LLM. Free versions are also available for limited use.
    - Big data is a "thing" now.

This also expands to e-commerce, with AI producing new opportunities to understand customers, interact with them, and automate essential processes.

- Global spending on AI in e-commerce will reach over $8 billion by 2024.
- Let's dive deeper into AI's impact on e-commerce and how it can improve online stores.
- The main types of AI that e-commerce companies can employ.
- "Artificial intelligence" is a broad concept that includes multiple use cases. These types show business functions that can be impacted or improved.

3. **Data mining**

Most e-commerce companies have abundant data to draw from. AI makes collecting, analysing, and leveraging data easier to produce actionable information. For example, consider producing recommended products based on past sales behaviours.

4. **Machine learning** [211, 212]

Self-learning algorithms can enhance inventory management, forecast sales trends and interactions, and automate logistics to provide a superior customer experience.

5. **Natural language processing (NLP)**

NLP is a type of AI that focuses on understanding and generating human language. E-commerce companies use NLP to improve chatbots, allowing them to answer frequently asked questions and product inquiries better.

This means customers can get fast and accurate answers to their questions, even outside business hours.

6. **Computer vision**

Computer vision is AI technology that enables computers to recognise and classify visual images. E-commerce companies are using this technology to improve product search and recommendations. For example, a customer can take a picture of a product they are interested in and use an app that incorporates image recognition to find the product quickly.

4.1 *Information Dissemination and Networking*

Digital directory: Create an online platform listing community members, their businesses, contact details, and skills.

News and updates: We share relevant news, events, and opportunities through websites, social media, and mobile app notifications.

Virtual forums and discussions: Facilitate online talks, knowledge sharing, and collaboration among members.

4.2 *Market Linkage and Promotion. E-commerce Platform*

Enable online sales of community-made products and services, reaching a wider audience.

Digital marketplace: Facilitate connections between community members and potential buyers/clients within and beyond the city.

Online marketing tools: Offer training and resources for digital marketing through social media and search engine optimisation.

4.3 *Resource Mobilization and Financial Services*

Crowdfunding platform: Facilitate collective funding for community-driven projects and initiatives. Digital payment systems: Implement cashless transactions for more effortless and secure financial interactions.

Access to financial services: Partner with banks and financial institutions to offer microloans, credit facilities, etc.

4.4 *Capacity Building and Skill Development*

Online training programs: Offer skill development courses relevant to community needs and market demands.

Virtual knowledge hubs: Provide access to educational resources, mentorship programs, and expert consultations.

Collaboration with educational institutions: Facilitate skill-based training and certification programs.

4.5 Advocacy and Policy Engagement

Digital petition platforms: Mobilize community support for advocacy campaigns and policy changes.

Online surveys and feedback tools: Gather community voice and data for evidence-based advocacy. Interactive platforms with government agencies: Facilitate communication and collaboration with relevant authorities.

4.6 Additional Considerations

Language accessibility: Ensure information and platforms are accessible in local languages.

Digital literacy training: Offer basic digital skills training to bridge the digital divide.

Data privacy and security: Implement robust data protection measures to build trust and ensure responsible data use.

Partnerships and collaborations: Partner with local NGOs, government agencies, and private sector actors for a wider reach and to impact trends in E-Gateway development:

5 Potential Challenges for E-Gateways in the Indo-Pacific Region [213–215]

5.1 Language Barriers and Intercultural Communication

The Indo-Pacific region is linguistically and culturally diverse, with numerous languages spoken across countries. E-Gateways must address language barriers to facilitate seamless communication and collaboration.

5.2 Time Zone Differences

The Indo-Pacific's vast geographical expanse means considerable time zone variations, and coordinating real-time interactions, such as video conferences or customer support, is challenging.

5.3 Cybersecurity Risks

E-Gateways involve the exchange of sensitive data and transactions. Robust cybersecurity measures are crucial to protect against cyber threats, data breaches, and unauthorised access.

5.4 Infrastructure Development and Connectivity

Some regions within the Indo-Pacific may need more digital infrastructure, including reliable internet connectivity. E-Gateways require robust networks and a reliable power supply for uninterrupted operations. Inadequate digital infrastructure (e.g., internet connectivity, power supply) affects project scalability and reliability.

5.5 Economic Dependence on China

China plays a significant role in the Indo-Pacific economy. Balancing economic ties with China while diversifying partnerships with other countries can be complex.

5.6 Resource Constraints, Cost Overruns, and Budget Constraints

Limited financial resources and technical expertise may hinder the establishment and maintenance of efficient E-Gateways. E-government projects may exceed their initial budgets due to unforeseen expenses or scope changes. Proper financial planning and risk management are essential.

5.7 Regulatory Frameworks

Harmonising regulations across diverse countries is challenging. E-Gateways must navigate varying legal requirements for data privacy, taxation, and cross-border transactions.

5.8 Political and Geopolitical Factors

Political instability, territorial disputes, and geopolitical tensions can impact E-Gateway operations. Navigating these complexities requires strategic planning.

5.9 Diversion of Funds

Prioritising investments in E-Gateways amid other pressing national priorities (such as defense or healthcare) can be challenging.

5.10 Local Acceptance and Adoption

Effective communication and awareness campaigns are required to convince local businesses, governments, and communities to embrace E-Gateways.

5.11 Lack of Trust. Building Trust Between End Users and the Government is Crucial

Skepticism about data security, privacy, and transparency can hinder project adoption.

5.12 Resistance to Change. People Often Need Help Transitioning from Traditional Processes to Digital Platforms

Overcoming this resistance requires effective change management strategies.

5.13 Political Will and Leadership

Strong political commitment is necessary for successful implementation. Lack of support from top management can lead to project delays or abandonment.

5.14 Stakeholder Engagement

Involving relevant stakeholders (government agencies, citizens, businesses) ensures the project meets their needs and expectations.

5.15 Legal and Regulatory Barriers

Legal complexities related to data privacy, intellectual property, and cross-border transactions can hinder progress.

5.16 Interoperability Challenges

Integrating diverse systems and databases across government departments requires standardised protocols and collaboration.

5.17 User Adoption and Training

Training programs ensure that end users (citizens and officials) understand and effectively use the system.

5.18 Changing Requirements and Scope

Evolving needs and shifting priorities can lead to scope changes during project execution.

5.19 Vendor Management

Managing relationships with technology vendors, contractors, and consultants is vital for project success.

5.20 Special Issues of Island Communities of the Indo-Pacific

Island communities must be carefully grouped to develop joint shared E-Gateways creation and management.

6 Digital Economic Global Enterprise Ecosystem Design

A digital ecosystem is a network of interconnected information technology resources that can function as a unit. It comprises stakeholders, systems, and an enabling environment that allows people and communities to use digital technology to access services, engage with each other, and pursue economic opportunities. A digital ecosystem comprises various elements, such as software, hardware, data, and people, facilitating digital transactions, communication, and collaboration. It can benefit business processes and efficiency, and some benefits include faster adoption of technology, Reliable communication among customers and trading partners, allowing enterprises to leverage new and legacy technologies, and building automated processes around technologies. The components of a digital economy include e-business and e-business infrastructure. A digital ecosystem can be organised as a "traditional" ecosystem where individual entrepreneurs, organisations, and other agents operate autonomously to optimise the idea-to-venture process and maximise the market potential of digital solutions.

What is the ecosystem of the digital economy?

6.1 What is a Digital Economic Ecosystem?

A digital economic ecosystem is a group of interconnected information technology resources that can function as a unit. It comprises suppliers, customers, trading partners, applications, third-party data service providers, and various technologies.

6.2 What is an Enterprise Digital Economic Ecosystem?

The digital economic ecosystem is a dynamic, interconnected network that necessitates reliable communication among customers and trading partners. When integrated, enterprises can leverage new and legacy technologies and build automated processes around them to grow their businesses continually.

6.3 What Are the Three Main Components of the Digital Economy?

- E-business.
- E-business infrastructure.
- E-commerce.

6.4 What Are the Three Pillars of the Digital Economy?

Digital enabling infrastructures. Infrastructure is one of the most critical factors that is accelerating the economy.

E-commerce. Individuals familiar with e-commerce may need to understand that the digital economy and e-commerce are the same.

Digital media

6.5 What is an Example of a Digital Enterprise? [216]

One example of a digital business is Amazon.com. Amazon.com started as an online bookstore and has since expanded to become a powerful e-commerce platform. Amazon.com uses digital technology to connect buyers and sellers and to facilitate transactions.

6.6 How Big is the Global Digital Economy?

The global eco-digital economy will double in the next five years.

Capgemini foresees a significant growth spurt for the global eco-digital economy. Last year, it was valued at $16.6 trillion; by 2028, it is expected to reach approximately $32.9 trillion (Figs. 62 and 63).

6.7 What Innovations Shape Digital Economic Enterprise Ecosystem

The digital economy is shaped by technological innovation, hyper-connectivity, and changing customer expectations. Some key trends and technologies that are expected to shape the digital economy include:

Artificial intelligence (AI) and machine learning

AI, including machine learning, deep learning, and neural networks, is expected to have a growing role in the digital economy.

Digital connectivity

Robust broadband connectivity and infrastructure are crucial to the success of the digital economy.

Internet of Things (IoT)

The IoT is a technological trend in the industrial sector's digital transformation.

Big Data Analytics (BDA)

BDA is a technological trend in the industrial sector's digital transformation.

Robotics

Robotics is a technological trend in the industrial sector's digital transformation

Cloud computing

Cloud computing is a technological trend in the industrial sector's digital transformation.

Data-driven decision-making

Data-driven decision-making is a crucial driver of digital transformation.

Competitive edge

A competitive edge for a business over others is a crucial driver of digital transformation

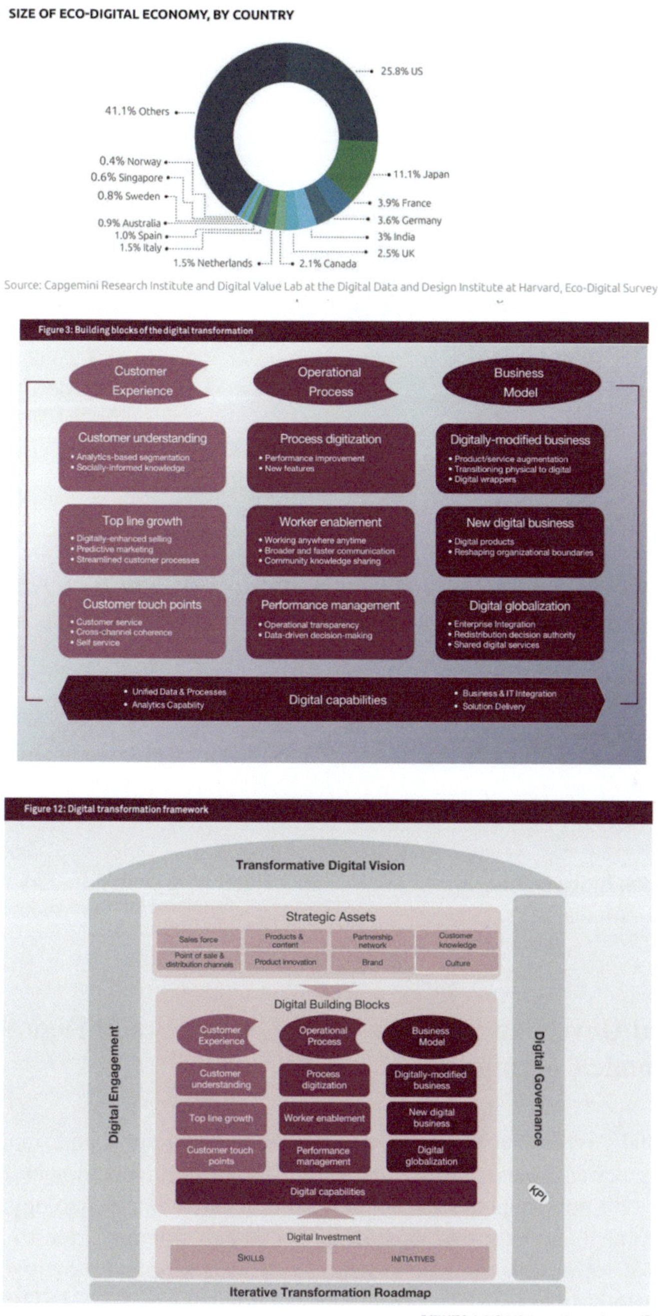

Fig. 62 The size of the digital economy by country and digital transformation

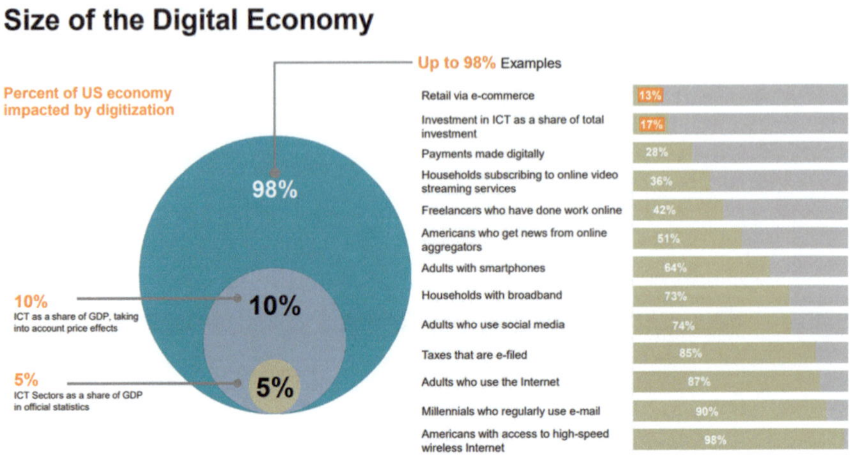

Fig. 63 Size of digital economy in the US

6.8 Key Players in a Digital Economic Ecosystem Enterprises

Platform owners: Responsible for defining the platform's purpose, creating the technological infrastructure, establishing rules, and more.

Large digital conglomerates shape competition in the digital sector.

Potential entrepreneurs: Important players in creating and keeping the ecosystem healthy.

Deal-makers: Individuals with valuable social capital with deep fiduciary ties within regional economies.

7 Global Governance Design for E-Gateways and Digital Economic Community [217, 218]

Global digital governance encompasses the norms, institutions, and standards that shape the regulation around the development and use of these technologies. Electronic governance or e-governance is the application of information technology for delivering government services, exchange of information, communication transactions, integration of various stand-alone systems between government-to-citizen (G2C), government-to-business (G2B), government-to-government (G2G), The National E-Governance Plan (NeGP), of India initiated in 2006, aimed to make government services accessible, efficient, transparent, and affordable for all citizens. Several initiatives, including Digital India, Digi-Locker, Mobile Seva, myGov.in, and more, have been launched under NeGP to promote e-governance across India.

E-Governance in India has steadily evolved from the computerisation of Government Departments to initiatives that encapsulate the finer points of Governance, such as citizen centricity, service orientation and transparency.

7.1 Global Economic Governance

The Global Economic Governance Programme was established to foster research and debate into how global markets and institutions can better serve the needs of people in developing countries.

Definition of e-Governance. Electronic governance or e-governance implies that the government functions with the application of ICT (Information and Communications Technology). Hence, e-governance moves towards SMART governance, suggesting simple, moral, accountable, responsive, and transparent governance. Global governance manifests itself in various forms: International agreements (such as the trade rules of the GATT) comprise one aspect of global governance, and international organisations such as the World Trade Organisation (WTO), the World Bank, and the International Monetary Fund (IMF) comprise another. Global governance is a process of cooperative leadership that brings together national governments, multilateral public agencies, and civil society to achieve commonly accepted goals. It provides strategic direction and marshals collective energies to address global challenges.

7.2 Challenges of Global Governance

Climate change

States, NGOs, and IGOs work together to address climate change, and international agreements set goals and responsibilities to reduce its effects. However, climate governance is a complex global issue that no government can tackle alone.

Triple Planetary Crisis

The complex and intertwined effects of climate change, biodiversity loss, and pollution are a significant challenge for global governance reform.

Sustainable development

Sustainable development is a long-term process affecting the environment, economy, and society locally and globally. It is a major global challenge in the twenty-first century.

Corruption

Corruption is a global concern that affects both rich and poor countries. It contributes to instability and poverty and can drive fragile countries towards state failure.

Other challenges of global governance include:

- Migration
- Health
- Food security
- Frequent leadership changes
- Lack of ideology
- Policy reversals
- Weak institutional setups
- Limited leadership experience.

Effective global governance requires practical international cooperation, which can promote shared values and interests and reduce vulnerabilities.

Other challenges that the digital economy faces include:

Lack of awareness

People may not be aware of the scope of e-governance and may rely on intermediaries to use it.

Cost

In developing countries, cost can play a significant role in regulating the use of e-governance.

Digital illiteracy

Digital illiteracy can be a problem in the successful implementation of digital programs.

Poor infrastructure

Poor infrastructure can be a problem in the successful implementation of digital programs.

Low internet speed

Low internet speed can be a problem in successfully implementing digital programs.

Lack of coordination

Lack of coordination among various departments can be a problem in successfully implementing digital programs.

Taxation issues

Taxation issues can be a problem in the successful implementation of digital programs.

7.3 *The Need for Global Governance of Digital Economies*

Global digital governance requires international institutions to design regulations for digital services. It has been growing in response to globalisation and digitalisation and requires global cooperation to address challenges and achieve economic benefits. Digital governance can offer many advantages for development, including improving public service delivery, increasing transparency and accountability, reducing corruption and fraud, enhancing civic engagement and social cohesion, and promoting innovation and learning.

Global digital governance covers issues such as:

The Internet, Digital tax, Cross-border data flow, Global trade, Telecommunication systems, financial inclusion, Automated trading, E-governance, E-commerce, and Cybersecurity. The United States, the European Union, India, and China are influential in setting the global agenda for digital governance.

7.4 *Limitations of Traditional Models*

Existing governance structures often need help to address the complexities of the digital economy. Traditional governance models can struggle to address the complexities of the digital economy because they need to account for the critical role of interdependence and the resulting complexity. The digital economy often operates in non-traditional market structures, and digital platforms may have network effects and economies of scale that existing frameworks can't efficiently address.

7.5 *Dispute Resolution Process in Global Governance*

Mediation

A voluntary method of dispute resolution involves a neutral third party called a mediator.

Arbitration

An alternative to the judicial method of dispute resolution.

Negotiation

A style of conflict resolution that involves competing, compromising, collaborating, avoiding, or accommodating.

Conciliation

Sometimes, it is considered a subset of mediation.

Litigation

A method of dispute resolution that can be costly and time-consuming.

Adjudication

A quick and cost-effective alternative to court that helps resolve payment claims and money owed.

7.6 Eight Steps to Effective Conflict Management

- Be aware.
- Conflict can arise at any time.
- Be proactive.
- Prevention and early resolution are the most effective.
- Seek to understand all sides of the issue.
- Initiate dialogue.
- Know when to ask for help.
- Assess your options.
- Act.
- Reflect on the situation.

7.7 Feasibility Assessment and Actions by the Community at the Local Level

Selected program initiatives of QUAD, I2U2 and other Bilateral Trades in the Indo-Pacific that can be implemented locally.

7.8 Digitally Embrace the Fourth Pillar of Tax and Anti-Corruption in the Indo-Pacific Economic Framework

"We are committed to promoting fair competition by enacting and enforcing effective and robust. Taxes, anti-money laundering, and anti-bribery regimes align with existing multilateral obligations, standards, and agreements to curb tax evasion and corruption in the Indo-Pacific region. This involves sharing expertise and seeking ways to support the capacity-building necessary to advance accountable and transparent systems" QUAD Summit.

8 The Future of Work and Continuing Education in a Digital Economic Community [219–221]

The future of work and education in a digital economic community is expected to see the rise of remote work and online learning. Technology is expected to drive the growth of higher-wage roles like software developers and financial risk specialists. The educational system may also evolve to incorporate the gig economy, focusing more on practical, job-specific skills, flexible learning, and technology.

Here are some trends that may shape the future of work and education in a digital economic community:

Remote work

Advances in technology have made it possible for many workers to work from home or anywhere in the world.

Online learning

Mobile platforms have increased access to online learning, and virtual reality (VR) may become more common.

Adaptive learning systems

Digital tutors may understand a learner's strengths, learning pace, and areas for improvement and provide personalised learning paths.

Digital media

Students can create digital portfolios to showcase their growth and development, which can be helpful for academic and professional purposes.

8.1 The Changing Nature of Work Requires Workers to Develop New Skills to Remain Competitive in the Job Market

The changing nature of work requires workers to develop new skills to remain competitive in the job market. This includes technical skills, soft skills, the ability to learn continuously, and the flexibility to adapt to new work arrangements.

8.2 Explore Innovative Approaches to Teaching and Training to Equip Individuals with the Skills Needed

Digital literacy skills are about navigating, understanding, and effectively using digital technologies. Suppose you're looking to increase your digital literacy skills.

In that case, you must develop your knowledge and skills in various areas, such as online communication, information literacy, and digital content creation.

Knowing how to learn online is essential to improving your digital literacy skills. This process involves using digital tools and platforms to access educational resources, connect with others, and participate in online communities. As more and more education and training occur online, it's becoming increasingly important to feel comfortable using digital technologies to learn.

Another vital component of increasing your digital literacy skills is learning from others. Try finding mentors and role models who can provide guidance and support as you develop your digital skills. It could also involve engaging with peers and colleagues to learn from their experiences and perspectives. Collaborative learning and knowledge-sharing can broaden and deepen your understanding of digital technologies and how they can help achieve different goals.

8.3 How to Improve Digital Literacy in the Workplace

By improving your digital literacy, you can increase productivity, efficiency, and competitiveness and open new opportunities for growth and innovation. Here are some ways that business owners and employees can improve their digital literacy skills in the workplace:

Business owners can:

Encourage employees to participate in digital literacy training programs and provide opportunities for ongoing professional development.

Foster a culture of experimentation and innovation, encouraging employees to try out new tools and technologies that can help streamline processes and improve workflows.

Provide resources like online tutorials, guides, and tools to support employees in developing their digital literacy skills.

Employees can:

Seek digital training programs and courses to learn new skills and keep up with emerging technologies.

To stay updated with industry news and trends, follow relevant publications and social media accounts, and attend conferences and events.

Practice using digital tools and platforms in their daily work to improve efficiency and productivity.

By implementing these strategies and utilising available resources, business owners and employees can improve their digital literacy skills in the workplace, leading to tremendous success and growth for the business.

8.4 How to Adapt Social Safety Nets to Address Potential Disruptions Caused by Automation and Changing Job Markets

About half of the activities (not jobs) carried out by workers could be automated.

Analysis of more than 2000 work activities across over 800 occupations shows that specific categories of activities are more easily automatable than others. These include physical activities in highly predictable and structured environments and data collection and processing, which account for roughly half of people's activities across all sectors. The least susceptible categories include managing others, providing expertise, and interfacing with stakeholders.

Nearly all occupations will be affected by automation, but only about 5% of occupations could be fully automated by currently demonstrated technologies. Many more occupations have automated portions of their constituent activities: about 30% of the activities in 60% of all occupations could be automated. Most workers—from welders to mortgage brokers to CEOs—will work alongside rapidly evolving machines. The nature of these occupations will likely change as a result (Fig. 64).

Automation can impact the job market by displacing low-skilled jobs and creating new ones that require new skills. For example, machines can automate repetitive tasks in manufacturing, data entry, and customer service, which can lead to job displacement. However, automation can create jobs in other sectors, such as data analysis, digital marketing, and robotics maintenance. The World Economic Forum estimates that by 2022, 75 million jobs may be displaced by AI automation, while 133 million new jobs may emerge. As automation takes over routine tasks, employers increasingly seek workers with strong technical skills, such as programming, data analysis, and digital literacy. Soft skills such as critical thinking, problem-solving, and communication are also becoming more important. By automating, we can eliminate the need for human labour, and by augmenting, we can use AI to enhance the intelligence and performance of human beings. People are finding innovative ways to utilise artificial intelligence in the workplace.

9 Conclusion

9.1 A Roadmap for Designing and Implementing Your Digital Economic Global Enterprise Ecosystem

Creating Digital Vision and Goal

Vague goals lead to ambiguous results. You must create a digital vision for your organisation to create a successful enterprise digital transformation program. What are you going to achieve by going digital?

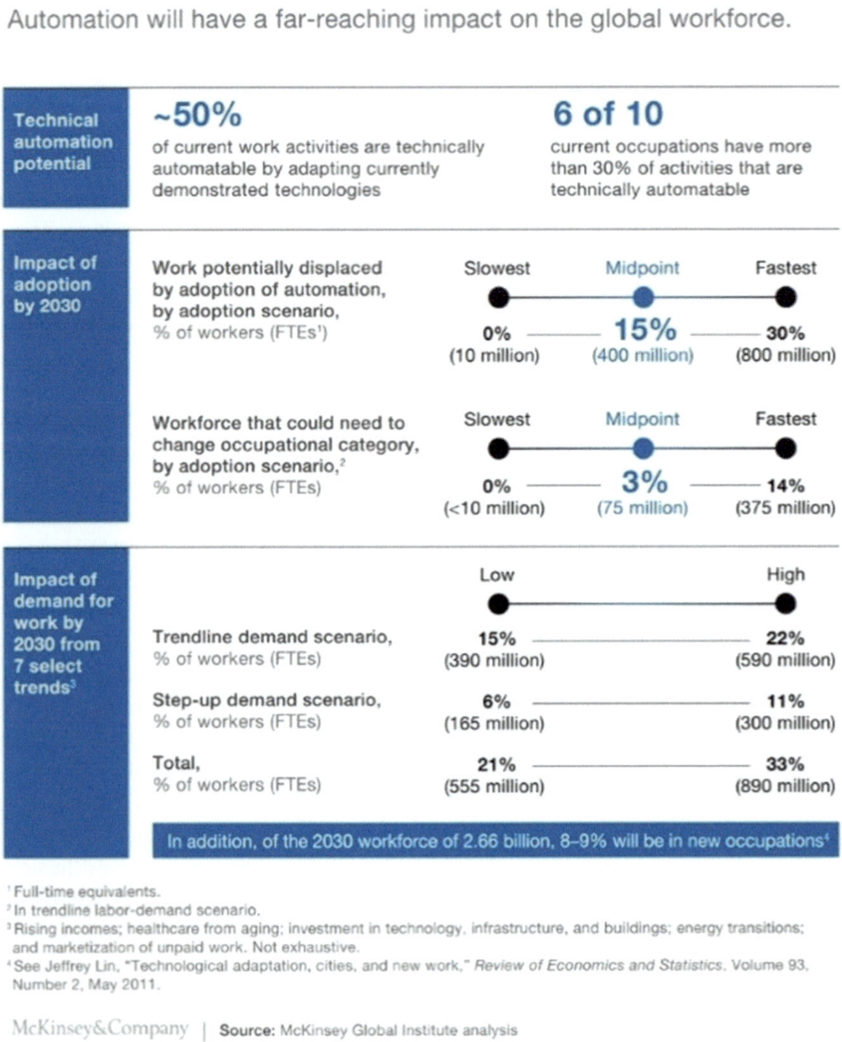

Fig. 64 Impact of automation

What are your goals? Please clarify them before proceeding to the next steps of your digital journey. Digital economic communities aspire to enhance their living conditions, which they will fix in concrete terms. To achieve this, the digital economic community requires high-value-adding work.

Stakeholder Identification

An enterprise digital transformation is about re-imagining your business and needs serious commitment from management. You need to identify a digital champion in your organisation who will steer the entire program over the years. This can be the

CEO themselves or a Chief Digital Officer who will report directly to the CEO, so that you always see the bigger picture. You need to identify the stakeholders from both business and IT who will be responsible for the various facets of your digital program. Combine this with a governance program that monitors all the nuts and bolts of the project through the steering committee, which ensures the project is on track and always on the prime agenda of management.

Design Thinking

An organisation is its people, and design thinking is one of the best ways to foster innovation. Design Thinking is a systematic process of thinking that empowers even the most traditional thinker to develop new, innovative solutions to the problem at hand. It empathises with your customers' issues and the people in your organisation who will be responsible for solving those problems. A design thinking-led innovation technique can create synergies between business and IT to look at business challenges in a newer light.

Ecosystem approach to business

If you look at the born-digital companies and the factors leading to their success, you will find a common theme. They have all created an ecosystem to connect the suppliers and providers. This is a significant shift from the traditional pipe-based models, where the approach was linear. At the superficial level, the ecosystem enables a network of value seekers and providers, systems that allow value exchange between them and underlying data that powers value matching across participants. So, for your enterprise to be digital, it must accommodate a larger ecosystem that makes it scalable.

Platform Strategy

This may be the most crucial step in your digital transformation roadmap. The enterprise needs to reimagine itself as a platform business to complement the ecosystem approach. Amazon, Uber, and Airbnb are all examples of platform businesses that serve customers through the value they bring with their models. Thus, your platform is a connected ecosystem of customers, companies, vendors, and partners, with intelligent decisions driven by data generated by the ecosystem and open, flexible, scalable systems that enable innovation and execution. Platforms leverage data generated from the ecosystem to drive intelligence that enhances value.

Choosing the Right Digital Transformation Partner

Your enterprise may need in-house technical skill sets for digital transformation. Hence, it would be best to rely on a trusted digital transformation partner to walk alongside you on your digital transformation journey. While evaluating a digital transformation partner, you must assess its technology capability, industry expertise, and ability to successfully execute a platform-led digital strategy. For example, Sonata Software specialises in executing such platform-based digital transformation programs. The Platform formation approach helps client re-imagine their business as open, connected, scalable and intelligent.

Organisational Change Management

A platform-based digital transformation approach needs awareness, education, and training. Communication plays a key role here. The first step can be a workshop that brings all the stakeholders together to open minds to design thinking, persona mapping, and platform mindset. This workshop allows you to align and uncover potential themes for your digital transformation journey.

Digital transformation roadmaps are not accessible. There are ups and downs as you progress towards your own. But they are essential if you want to stay relevant and continue to provide value to your customers.

9.2 Building a Thriving Digital Economic Community

Here are some ways to build a thriving digital economic community:

Create an inclusive digital society.

Ensure that everyone has access to the digital economy, regardless of age, gender, disability, or location.

Foster innovation and entrepreneurship

Support startups and small businesses to help them reach a wider audience, access funding, and scale their operations.

Build trust

Create a regulatory environment that protects consumer rights, ensures data privacy and security, and promotes fair competition.

Invest in infrastructure

Develop the infrastructure to support a digital economy, such as broadband internet access, payment solutions, and e-commerce strategies.

Priorities in social care

Use technology to connect employees, manage overwork, and monitor well-being.

Create a region-wide framework

Collaborate to create a region-wide digital economy framework that will engender higher levels of trust.

Enhance trade

Use digital technologies and e-commerce to enhance trade across sectors and industries.

Build digital literacy

Ensure that individuals and businesses have digital literacy, technological expertise, and adaptability to thrive in the digital era.

Support digital transformation

Governments, businesses, and organisations must assess their current digital capabilities, set clear objectives, and develop a roadmap for implementation.

Build digital trust

Building digital trust in the world's digital economy could unleash trillions of opportunities. A thriving digital economy can create a vibrant ecosystem that drives economic growth, innovation, and entrepreneurship. By creating an inclusive digital society, fostering innovation and entrepreneurship, building trust, and investing in infrastructure, we can make a digital economy that benefits everyone.

9.3 Potential Challenges and Opportunities for the Future of Global Governance and the Digital Economy

"Challenges such as data privacy concerns, cybersecurity threats, and the need for digital literacy need to be addressed for the digital economy to thrive," says Saurabh Srivastava, co-founder of Fintech India.

Simply put, it's an interconnected world where data is the currency, and innovation is the driving force. Opportunities in the Digital Economy Enable Global Reach. One of the most compelling advantages of the digital economy is the ability to connect with a global audience.

Five factors—individual empowerment, increasing awareness of human security, institutional complexity, an international power shift, and a liberal world political paradigm—will mainly shape the future of global governance.

Rising geoeconomic fragmentation and the surge in trade-restrictive and industrial policy measures also harm medium-term growth prospects. Trade linkages are already changing as a result, with potential efficiency losses. The net effect is to make the global economy less, not more, resilient.

The following key trends and technologies are expected to shape the future of the digital economy: AI and machine learning. AI, including machine learning, deep learning, and neural networks, is advancing and will play a growing role in the digital economy.

Mobile technology, business analytics, cloud computing, and social media influence the growth of the digital economy. Other factors include labour, land, capital, scientific and technological progress, and information and knowledge. Global governance can provide a 'rule of law' (whether formally codified or not) at the international level, which can protect the weak from the strong and provide a stable basis for the development of global institutions.

9.4 Challenges of Geopolitics

The most critical challenge is the ever-changing geopolitical scenarios that emerge for various reasons that affect international trade and the global economy, like the Trump-Tariff war. Today, friendly countries can be tomorrow's enemies or countries under competition and, therefore, under sanctions. A country that was not sanctioned yesterday may be sanctioned tomorrow. These sanctions change the supply chain. It is challenging to change supply chains overnight, and many countries' Global economies and lives are affected. Sanctions throw away the conventional wisdom of trading with those who give the lowest price, trading with countries with lower transportation costs, and trading with friends and not with economies under sanction. The term friend-shoring originated in the days of sanction competition and derisking. This means there is a demand for rapid changes in existing supply chains. The digital economy, community, and nations can change supply chains faster than those not embraced by digital. The map measures the relative geopolitical risk of stability of countries by rank in the Indo-Pacific (Fig. 65).

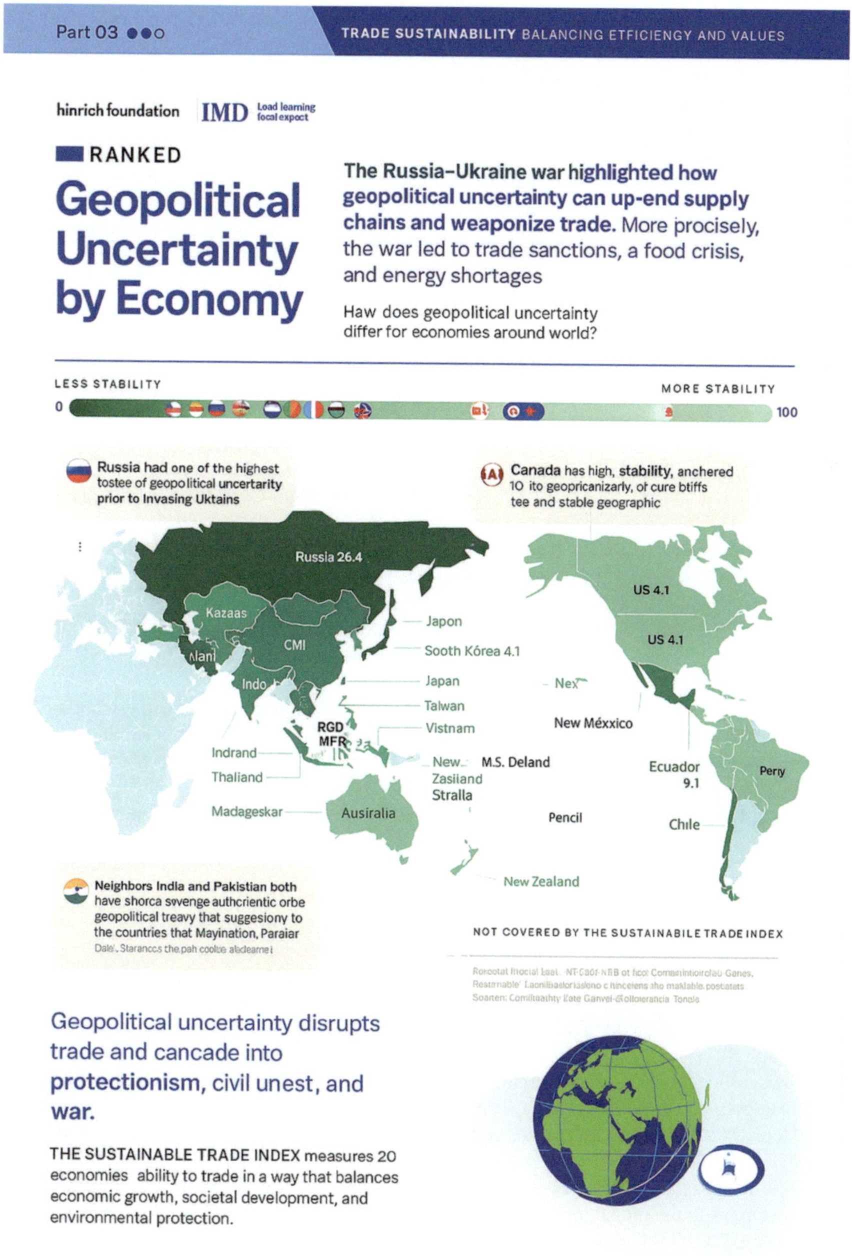

Fig. 65 Geopolitical Uncertainty by Economy and Challenges to International Trade

References

1. Doyle T, Rumley D (2020) The rise and return of the Indo-Pacific. OUP, Oxford
2. Austin MR (ed) (2020) Asia's new geopolitics: essays on reshaping the Indo-Pacific. Hoover Institution Press, Stanford
3. Cannon BJ, Rossiter A (eds) (2018) Rising powers quarterly 3(2). Special issue on "The "Indo-Pacific"—Regional Dynamics in the 21st Century's New Geopolitical Center of Gravity"
4. Rory M (2020) Indo-Pacific Empire: China, America, and the contest for the world's pivotal region (Excerpt)
5. https://Mapsofworld.com
6. https://lki.lk/wp-content/uploads/2020/01/imagel.png
7. Jaishankar D (2019) Acting east: India in the Indo pacific, Brookings India
8. Vinod Kumar TM (ed) (2023) Indo-Pacific system of smart megacity: emerging architecture and megacity case studies. Springer, Singapore
9. www.populationpyramid.net
10. Government of India (2014) "Smart cities concept note" New Delhi
11. https://ceoworld.biz/indo-pacific CEO World magazine Indo pacific Region
12. https://www.state.gov/wp-content/uploads/2019/11/Free-and-Open-Indo-Pacific-4Nov2019.pdf
13. The White House (U.S.) National security strategy of the United States of America, December 2017, p 2. https://www.whitehouse.gov/wp-content/uploads/2017/12/NSS-Final-12-18-2017-0905.pdf
14. The U.S. Department of Defense, Indo-Pacific Strategy Report, June 1, 2019, https://media.defense.gov/2019/Jul/01/2002152311/-1/-1/1/DEPARTMENT-OF-DEFENSE-INDO-PACIFIC-STRATEGY-REPORT-2019.PDF
15. PwC [PricewaterhouseCoopers Private Limited] (2014) Future of India: the winning leap. http://www.pwc.in/en_in/in/assets/pdfs/future-of-india/future-of-india-thewinning-leap.pdf
16. http://www.pwc.in/en_in/in/assets/pdfs/future-of-india/future-of-india-the-winning-leap.pdf. [PricewaterhouseCoopers Private Limited] (2014) Future of India: The Winning Leap
17. Vinod Kumar TM (ed) (2018) Smart metropolitan regional development: economic and spatial design strategies. Springer, Singapore
18. Vinod Kumar TM (ed) (2021) Smart global megacities: Tokyo, Mumbai, New York, Hong Kong-Shenzhen, Kolkata New York. Springer, Singapore
19. Vinod Kumar TM (ed) (2021) Smart global megacities: Chennai, Kochi-Kannur. Springer, Singapore
20. Beaverstock JV, Smith RG, Taylor PJ (2004) World-city network: a new meta geography? Ann Assoc Am Geogr 90(1):124–125; Beaverstock JV, Smith RG, Taylor PJ, Walker DRF, Lorimer H (2000) Globalization and world cities: some measurement methodologies. Appl Geogr 20:47
21. van der Ploeg F, Poelhekke S (2008) Globalization and megacities rise in the developing world trade policy. CESifo working paper no. 2208 category 7
22. Alankar (2015) India's megacities and climate change: explorations from Delhi and Mumbai. STEPS working paper 79
23. Bourdeau-Lepage L, Huriot J-M (2005) The metropolis in retrospect. From the loosely trading metropolis to the global metropolis. Rech Econ Louvain—Louvain Econ Rev 71(3)
24. Davis DE (2005) Cities in a global context: a brief intellectual history. Int J Urban Reg Res 29(1):99
25. Hill RC, Kim JW (2000) Global cities and the developmental states: New York, Tokyo, and Seoul. Urban Stud 37(12):2168
26. Castells M (2000) The rise of the network society, the information age: economy, society, and culture, vol 1, 2nd edn. Oxford, Cambridge
27. Savitch HV (2001) What is new about globalisation, and what does it portend for cities? Int Soc Sci J 54(172):180–181

28. Vinod Kumar TM (ed) (2016) Smart economy in smart cities, International Collaborative Research: Ottawa, St Louis, Stuttgart, Bologna, Cape Town, Nairobi, Dakar, Lagos, New Delhi, Varanasi, Vijayawada, Kozhikode, Hong Kong. Springer, Singapore
29. Anukoonwattaka W, Romao P, Bhogal P, Bentze T, Lobo RS, Vaishnav A (2021) Digital economy integration in Asia and the Pacific: insights from DIGISTRII 1.0. Asia-Pac Sustain Dev J 28(2):113–148
30. ASEAN (2021a) ASEAN digital Masterplan 2025. Association of Southeast Asian Nations, Jakarta
31. Bukht R, Heeks R (2017) 'Defining, conceptualising and measuring the digital economy', working paper series, No. 68. University of Manchester, Centre for Development Informatics, Manchester
32. Loh D (2021) 'ASEAN's digital economy projected to hit $1tn by 2030', Nikkei Asia, 10 November. https://asia.nikkei.com/Business/Business-trends/ASEAN-s-digital-economy-projected-to-hit-1tn-by-2030
33. ASEAN (2021b) 'Bandar Seri Begawan Roadmap: an ASEAN digital transformation agenda to accelerate ASEAN's economic recovery and digital economy integration'. Association of Southeast Asian Nations, Jakarta
34. ASEAN (2021c) Economic community: overview. https://asean.org/our-communities/economic-community/
35. ASEAN and USAID (2021) ASEAN digital integration index: measuring digital integration to inform economic policies. Association of Southeast Asian Nations and United States Agency for International Development, Jakarta
36. ASEAN (2021d) Work plan on implementing the ASEAN agreement on electronic commerce. Association of Southeast Asian Nations, Jakarta
37. Bajt D, Duval Y (2020) 'Paperless trade facilitation in 2019: global and Asia-Pacific State of Play'. UNCTAD Transp Trade Facil Newsl 85 (Q1)
38. Deloitte (2020) Deloitte Indonesia business and industry updates. https://www2.deloitte.com/id/en/pages/about-Deloitte/articles/business-and-industry-updates.html
39. Elms D (2022) 'Using trade agreements to foster trade in digital services', presentation for the multi-year expert meeting on trade, services and development, hybrid (online and Geneva), 4–6 July
40. Google, Temasek, and Bain & Company (2021) e-Conomy SEA 2021. https://www.bain.com/globalassets/noindex/2020/e_conomy_sea_2020_report.pdf
41. Ha SH, Lim SW (2014) The progress of paperless trade in Asia and the Pacific: enabling international supply chain integration', ADB working paper series on regional economic integration, No. 137. Asian Development Bank, Manila
42. Knickrehm M, Berthon B, Daugherty P (2016) Digital disruption: the growth multiplier. Accenture Research and Oxford Economics. https://www.anupartha.com/wp-content/uploads/2016/01/Accenture-Strategy-Digital-Disruption-Growth-Multiplier.pdf
43. Sawatari T (2022) Intra-ASEAN cooperation in the field of digital trade (in Japanese). Fukuoka Univ-Commer Rev 67(1):57–77
44. Tomassian CC (2009) Physical infrastructure and regional integration. Bulletin FAL, vol 280, no 12. Economic Commission for Latin America, Santiago
45. UNECE and UN/CEFACT (2018) White paper: paperless trade, version 1. United Nations Economic Commission for Europe and United Nations Centre for Trade Facilitation and Electronic Business. https://unece.org/fileadmin/DAM/cefact/Guidance%20Materials/Whitepapers/WP-PaperlessTrade_Eng.pdf
46. Yean TS (2021) Digital commitments in ASEAN's free trade agreements. ISEAS Perspect 2021(163). ISEAS—Yusof Ishak Institute, Singapore
47. Bhuvan (2011) Bhuvan. Retrieved 2019, from Bhuvan
48. Census of India (2001) Retrieved from primary census abstract, census of India
49. Census of India 2011 (2011) District census handbook
50. Department of Town and Country Planning, Govt. of Kerala (2011) District Spatial Plan Thrissur. Thrissur

51. Google Earth (2019) Google Earth. Retrieved October 12, 2019, from Google Earth
52. Govt. of India (2018) Brief Industrial Profile of Thrissur District 2017–18
53. Govt. of Kerala (2019) Economic Review 2018. Thiruvananthapuram
54. Govt. of Kerala (2019) Thrissur District. Retrieved October 13, 2019, from Thrissur District: https://thrissur.nic.in/
55. UNDP (2005) Human development report
56. UNDP (2018) Human development report
57. Fact Sheet: Quad Leaders' Summit—Press Information Bureau
58. https://pib.gov.in/PressReleaseIframePage.aspx?PRID=1757944
59. The Four Pillars of Successful Project Delivery—QUAD. https://slcontrols.com/four-pillars-successful-project-delivery-quad/
60. Quad Charts: A Visual Approach to Strategic Planning. https://balancedscorecard.ltslean.com/Quad-chart
61. What You Need to Know About the I2U2—United States Institute of Peace. https://www.usip.org/publications/2022/07/what-you-need-know-about-i2u2
62. I2U2: The Aspects and the Prospects—Indian Council of World Affairs … https://icwa.in/show_content.php?lang=1&level=1&ls_id=7760&lid=5166
63. India's engagement with West Asia through I2U2: Challenges and prospects. https://www.sociologyjournal.in/assets/archives/2023/vol5issue2/5027-1688465363824.pdf
64. ADB (2019) Asia-Pacific forum on financial inclusion. Retrieved from [URL]
65. ADB (2019) Technical assistance and capacity building. Retrieved from [URL]
66. APEC (2019) APEC taxation administration forum. Retrieved from [URL]
67. APG (2019) Asia/Pacific group on money laundering. Retrieved from [URL]/
68. DFAT (2019) Indo-Pacific Centre for health security. Retrieved from [URL]
69. EITI (2019) The extractive industries transparency initiative. Retrieved from [URL]/
70. FATF (2018) International standards on combating money laundering and the financing of terrorism & proliferation. Retrieved from [URL]
71. FATF (2019) FATF report to the G20 leaders' summit. Retrieved from [URL]
72. GFAR (2019) Global forum on asset recovery. Retrieved from [URL]
73. IMF (2019) Regional technical assistance centers. Retrieved from [URL]
74. JICA (2019) JICA's activities in the Indo-Pacific Region. Retrieved from [URL]
75. OECD (2017) OECD/G20 base erosion and profit shifting project. Retrieved from [URL]/
76. OECD (2019) Digitalisation and tax administration. Retrieved from [URL]
77. OECD (2019) OECD/Korea Policy Centre. Retrieved from [URL]/
78. StAR (2019) Stolen asset recovery initiative. Retrieved from [URL]/
79. U.S. Department of State (2019) Indo-Pacific transparency initiative. Retrieved from [URL]/
80. U.S. Department of State (2018) A free and open Indo-Pacific: advancing a shared vision. Retrieved from [URL]
81. UNDP (2019) Financing for development: tax and illicit financial flows. Retrieved from [URL]
82. UNODC (2018) United Nations convention against corruption. Retrieved from [URL]
83. World Bank (2019) Digital government. Retrieved from [URL]
84. World Bank (2019) Taxation. Retrieved from [URL]
85. ADB (2021) Southeast Asia prepares shift to digital payments, ADB Southeast Asia Development Solutions (SEADS), 14 June. https://seads.adb.org/solutions/southeast-asia-prepares-shift-digital-payments
86. TEMPO.CO (2022) QRIS payment service to be available in 4 ASEAN countries, 7 July. https://en.tempo.co/read/1609774/qris-payment-service-to-be-available-in-4-asean-countries
87. Mahesh A, Bhat G (2021) Digital payment service in India—a case study of unified payment interface. Int J Case Stud Bus IT Educ (IJCSBE) 5(1). ISSN: 2581-6942
88. Banking Regulations Act 1949
89. Payment and Settlement Systems Act, 2007 (PSS Act)
90. Section 43A of IT Act, 2000 and the IT Rules, 2011
91. Section 25 of the Payment and Settlement Systems Act, 2007 (PSS Act)

92. Ministry of Communications and Information Technology (Department of Electronics and Information Technology) NOTIFICATION, Information Technology (Preservation and Retention of Information by Intermediaries Providing Digital Locker Facilities) Rules, 2016. New Delhi, the 21st July 2016 Digit Locker
93. ONDC (2021) Strategy paper-open network for digital commerce, democratising digital commerce in India
94. ONDC (2022) Building trust in the ONDC network consultation paper 20 Sep 2022
95. Mahesh KM, Aithal PS, Sharma KRS (2022) Open network for digital commerce-ONDC (E-commerce) infrastructure: to promote SME/MSME sector for inclusive and sustainable digital economic growth. Int J Manag Technol Soc Sci (IJMTS) 7(2). ISSN: 2581-6012
96. http://www.zdnet.com/blog/estonia/
97. Ernsdorff M, Berbec A (2007) "Estonia: the short road to e-government and e-democracy. In: E-Government in Europe. In: Nixon PG, Koutrakou VN (eds) Re-booting the state. Routledge, London, pp 171–183
98. E-User (2005) E-Government Country Brief: Estonia. E-User: Public Online Services and User Orientation, 8/2005
99. Kilvits K (1999) Innovation and restructuring in Estonian industry. In: Ennuste U, Wilder L (eds) Harmonisation with the Western Economics. Estonian economic developments and related conceptual and methodological frameworks. Estonian Institute of Economics at Tallinn Technical University, Tallinn
100. Landler M (December 13, 2005) Hot technology for chilly streets in Estonia. New York Times
101. Lumiste R, Pefferly R, Purju A (2007) Estonia's economic development: trends, practices, and sources. A Case Study, Washington
102. Riismaa K (2007) Implementing paper-free document exchange in the Estonian public sector, DEXA E-Government conference in Regensburg, Germany SESS
103. Landler M (2005) Hot technology for chilly streets in Estonia. New York Times, 13 Dec 2005
104. Riismaa K (2007) Implementation of paper-free document exchange in the Estonian public sector, DEXA E-Government Conference in Regensburg, Germany SESS
105. http://www.abc.net.au/news/2014-11-25/estonia-to-offer-e-residency-online-to-non-citizens/5917240
106. https://www.youtube.com/watch?v=XOg0o3nmCYQ
107. http://archive.org/stream/TallinnManual/TallinnManual_djvu.txt
108. http://journal.georgetown.edu/?s=estonian+e-residency
109. https://dl.acm.org/cms/attachment/842d95ee-9c34-49d1-8581-b77b99d8c08a/uf1.html
110. Alvarez RM, Hall TE, Trechsel AH (2009) Internet voting in comparative perspective: the case of Estonia. PS Polit Sci Polit 42(3):497–505
111. Bochsler D (2010) Can internet increase political participation? Remote electronic voting and turnout in the Estonian 2007 parliamentary elections. Paper presented at the conference 'Internet and Voting', Fiesole, Italy
112. Bochsler D (2010) Can internet increase political participation? Remote electronic voting and turnout in the Estonian 2007 parliamentary elections. Paper presented at Conference "Internet and Voting", Fiesole, Italy
113. Antiroikko A-V, Malkia M, Savolainen RE (2004) Transformation in governance: new directions in government and politics. Idea Group Publishing, London (Autoregistrikeskus. Paberivaba ARK. 2007). http://www.ark.ee (Cited January 2008)
114. Vinod Kumar TM (ed) (2015) E-Governance for smart cities. Springer, Singapore
115. Clift S (2003) E-democracy, E-Governance and public net-work. Publicus.Net. http://www.publicus.net/articles/edempublicnetwork.html
116. Denzau A, North D (1994) Shared mental model: ideologies and institutions. Kyklos 47(1):3–31
117. Dimitrova DV, Chen YC (2006) Profiling the adopters of e-government information and services: the influence of psychological characteristics, civic mindedness and information channels. Soc Sci Comput Rev 24:172–188
118. Drechsler W (2006) Dispatch from the future. Washington Post, 5 Nov 2006, sec B01

119. Drüke H (ed) (2004) Local electronic government: a comparative study. Routledge, London
120. Dyker D (1996) The computer and software industries in the East European economies. A bridgehead to the global economy? Eur Asia Stud 48(6):130–915
121. Economist Intelligence Unit (2004) E-Government in Central Europe. Rethinking Public Administration. Economist Intelligence Unit, London
122. Economist Intelligence Unit (2005) E-readiness index. Economist Intelligence Unit and IBM Institute for Business Value, London
123. .Kindly note that citation for Ref. [123] has been provided but source for the respective reference is missing. Please check and provide.
124. Commission E (2005) Online availability of public services: how is Europe progressing? European Commission Directorate General for Information Society and Media, Luxembourg
125. Gil-Carcia R (2005) Enacting state websites: a mixed method study exploring E-government success in multiorganizational setting. In: Haertsch P, Schubert P, Selz D (eds) Ph.D. dissertation, State University of New York at Albany. 2002. Digital Elforgreich. Springer, Berlin
126. IDABC (2005) E-government in Estonia. European Communities, Brussels
127. IDABC (2005) E-government in Latvia. European Communities, Brussels
128. IDABC (2005) E-government in Slovakia. European Communities, Brussels
129. IDABC (2005) E-government in Slovenia. European Communities, Brussels
130. Kallas R (2009) Tööturuamet paneb töökuulutused internetti. Äripäev, 2 Mar 2009
131. Kitsing M, Howard PN (2009) Turning dirt roads into information highways: the conceptual misformation of the internet diffusion. University of Washington: World Information Access Project Working Paper, 2009.2, Seattle
132. Krull A (2003) ICT infrastructure and E-readiness assessment report. Praxis Center for Policy Studies, Tallinn
133. Lumiste R, Pefferly R, Purju A (2007) Estonia's economic development: trends, practices, and sources. A case study. World Bank Commission on Growth and Development, Washington
134. Lustsik O (2003) E-pangandus eestis: Kiire kasvu põhjused ja tegurid. Kroon Ja Majandus 3:23–34
135. Merill S, Grofman B (1999) A unified theory of voting: directional and proximity spatial models. Cambridge University Press, Cambridge
136. Moon D (1990) What you use depends on what you have: information effects on the determinants of electoral choice. Am Polit Q 18:3–24
137. Moon D (1992) What you use still depends on what you have: information effects in presidential elections, 1972–1988. Am Polit Q 20:241–427
138. Nixon PG, Koutrakou VN (eds) (2007) E-government in Europe. Rebooting the State. Routledge, London
139. North DC (1990) Institutions, institutional change and economic performance. Cambridge University Press, Cambridge
140. Osalusveeb. Data and Information. In Riigikantselei [database online]. Tallinn (2009). http://www.osale.ee (cited July 1, 2009)
141. Ott A, Siil I (2003) Eesti e-riik tulgu. Eesti Päevaleht, 7 Feb 2003, sec Comment
142. Raun A (2009) E-riik komberdab edulugudest hoolimata. Postimees, November 22, 2009. http://www.postimees.ee/?id=191347
143. Roth D (2004) Catch us if you can. Fortune, 16 Feb 2004
144. Rudi H (2009) E-valimised tõid juurde 11 222 ID-kaardi kasutajat. Postimees, 23 June 2009
145. Salem F (2007) Benchmarking the e-Government Bulldozer: beyond measuring the tread marks. Meas Bus Excell 11(4):9–22
146. Antiroikko A-V, Malkia M, Savolainen RE (2004) Transformation in governance: new directions in government and politics. Idea Group Publishing, London
147. Autoregistrikeskus. Paberivaba ARK (2007) [cited January 2008]. Available from http://www.ark.ee
148. Dimitrova DV, Chen YC (2006) Profiling the adopters of e-government information and services: the influence of psychological characteristics, civic mindendness and information channels. Soc Sci Comput Rev 24:172–188

149. Dyker D (1996) The computer and software industries in the east European economies. A bridgehead to the global economy? Eur-Asia Stud 48(6):915–30
150. Unit and IBM Institute for Business Value (2004) E-government in central Europe. Rethinking public administration. Economist Intelligence Unit, London
151. Ernsdorff M, Berbec A (2007) Estonia: the short road to e-government and e-democracy. In: Nixon PG, Koutrakou VN (EDS) E-government in Europe. Re-booting the state. Routledge, London, pp 171–183
152. Gil-Carcia R (2005) Enacting state websites: a mixed method study exploring e-Government success in multiorganizational setting. PhD dissertation. State University of New York at Albany
153. Haertsch P, Schubert P, Selz D (eds) Digital elforgreich. Springer, Berlin
154. Kilvits K (1999) Innovation and restructuring in Estonian industry. Harmonisation with the Western Economics. Estonian economic developments and related conceptual and methodological frameworks, Ennuste U, Wilder L (eds). Estonian Institute of Economics at Tallinn Technical University, Tallinn
155. Kilvits K (1999) Innovation and restructuring in Estonian industry. Harmonisation with the Western Economics. Estonian economic developments and related conceptual and methodological frameworks, Ennuste U, Wilder L (eds) Estonian Institute of Economics at Tallinn Technical University, Tallinn
156. Kitsing M, Howard PN (2009) Turning dirt roads into information highways: the conceptual misformation of the internet diffusion. Seattle
157. University of Washington: world information access project working paper, 2009.2
158. Nixon PG, Koutrakou VN (eds) (2007) E-government in Europe. Re-booting the state. Routledge, London
159. Osalusveeb. Data and information. In Riigikantselei [database online]. Tallinn, 2009 [cited July 1, 2009]. Available from www.osale.ee; Ott A, Siil I (2003) Eesti e-riik tulgu Eesti Päevaleht, 7 Feb 2003, sec Comment
160. Siil I (2001) Estonia: preparing for the information age. International Council for the Information Technology in Government Administration, ICA Information No. 74
161. Täna Otsustan Mina (2007) Information and data about the portal [cited October 202007]. Available from: https://www.eesti.ee/tom/ideas.py/avaleht
162. Tiigrihüppe Sihtasutus (1999) Tiigrihüppe programm 1997–1999. TiigrihüppeSihtasutus, Tallinn
163. United Nations (2008) UN E-government survey 2008. From E-government to connected governance. United Nations, New York
164. United Nations (2005) Global E-government readiness report 2005: from E-government to inclusion. United Nations, New York, UNPAN/2005/14
165. United Nations (2004) Global E-government readiness report 2004: towards access for the opportunity. United Nations, New York, UNPAN/2004/11
166. https://asean.org/our-communities/economic-community/trade-facilitation/
167. https://www.apec.org/meeting-papers/annual-ministerial-meetings/2017/2017_amm/annex-a
168. https://www.maersk.com/news/articles/2022/11/29/maersk-and-ibm-to-discontinue%20tradelens#:~:text=The%20TradeLens%20platform%20was%20announced,efficient%20and%20secure%20global%20trade
169. Agranoff R, McGuire M (2003) Collaborative public management: new strategies for local governments. Georgetown University Press
170. European Commission (2020) Proposal for a regulation of the European Parliament and the Council on contestable and fair markets in the digital sector (Digital Markets Act)
171. European Parliament (2021) Digital markets act: European Parliament kicks off its work: https://eur-lex.europa.eu/legal-content/EN/TXT/?uri=COM:2020:842:FIN, https://eur-lex.europa.eu/legal-content/EN/TXT/?uri=COM:2020:842:FIN
172. https://www.europarl.europa.eu/news/en/press-room/20210603IPR05528/digital-markets-act-european-parliament-kicks-off-its-work, https://www.europarl.europa.eu/news/en/press-room/20210603IPR05528/digital-markets-act-european-parliament-kicks-off-its-work

173. GDPR (2016) Regulation (EU) 2016/679 of the European Parliament and of the Council of 27 April 2016 on protecting natural persons concerning the processing of personal data and on the free movement of such data, and repealing Directive 95/46/EC (General Data Protection Regulation). Off J Eur Union L 119:4–88
174. ePrivacy Directive (2002) Directive 2002/58/EC of the European Parliament and of the Council of 12 July 2002 concerning the processing of personal data and protecting privacy in the electronic communications sector (Directive on privacy and electronic communications). Off J Eur Communities L 201:37–47
175. Data Protection Law Enforcement Directive (2016) Directive (EU) 2016/680 of the European Parliament and of the Council of 27 April 2016 on the protection of natural persons with regard to the processing of personal data by competent authorities for the purposes of the prevention, investigation, detection or prosecution of criminal offences or the execution of criminal penalties, and on the free movement of such data, and repealing Council Framework Decision 2008/977/JHA. Off J Eur Communities L 119:89–131
176. Data Protection Regulation for EU Institutions (2018) Regulation (EU) 2018/1725 of the European Parliament and of the Council of 23 October 2018 on the protection of natural persons with regard to the processing of personal data by the Union institutions, bodies, offices and agencies and on the free movement of such data, and repealing Regulation (EC) No 45/2001 and Decision No 1247/2002/EC. Off J Eur Communities L 295:39–98
177. Bauer M, Erixon F, Lamprecht P, Lee-Makiyama H (2020) Europe's digital sovereignty: from rulemaker to superpower in the age of US-China rivalry. European Centre for International Political Economy
178. European Commission (2015) A digital single market strategy for Europe. COM(2015) 192 final
179. European Commission (2018) Communication on a European strategy for data. COM(2020) 66 final
180. European Commission (2019) European blockchain partnership: Building the EU blockchain services infrastructure
181. European Commission (2020) Shaping Europe's digital future. COM(2020) 67 final
182. European Commission (2021) A trusted and secure European e-ID for the EU. COM(2021) 118 final. Europeana. (2021). About us. [URL]
183. European Union (2014) Regulation (EU) No 910/2014 of the European Parliament and of the Council of 23 July 2014 on electronic identification and trust services for electronic transactions in the internal market and repealing Directive 1999/93/EC. Off J Eur Union L:257/73
184. Vesnic-Alujevic L, Breitegger M, Nascimento S (2020) Human-centric digital age: how to foster citizens' engagement and participation in the EU. EUR 30347 EN, Publications Office of the European Union
185. European Commission (2021) 2030 Digital compass: the European way for the Digital Decade. COM(2021) 118 final. Retrieved from [URL]
186. European Commission (2020) Shaping Europe's digital future. COM(2020) 67 final. Retrieved from [URL]
187. European Commission (2015) A digital single market strategy for Europe. COM(2015) 192 final. Retrieved from [URL]
188. Vinod Kumar TM (ed) (2017) E-democracy for smart cities. Springer, Singapore
189. Qi R, Feng C, Liu Z, Mrad N (2017) Blockchain-powered Internet of Things, E-Governance and E-Democracy. In: Vinod Kumar TM (ed) 2017 E-Democracy for smart cities. Springer, Singapore
190. Nuñez Mencias A, Dillenberger D, Novotny P et al (2018) An optimized blockchain solution for the IBM z14. IBM J Res Dev 62(2/3)
191. https://www.ibm.com/topics/blockchain-ai
192. Gaur N, Desrosiers L, Novotny P et al (2018) Hands-on blockchain with hyperledger: building decentralised applications with hyperledger fabric and composer. Packt Publishing

193. Wolfson R (2018) Understanding how IBM and others use blockchain technology to track global food supply chain. Forbes, 11 July 2018
194. Roberts JJ (2017) IBM and Stellar are launching blockchain banking across multiple countries. Fortune, 16 Oct 2017
195. Wood EE (2018) Secure key announces blockchain-based identification service. itbusienss.ca, 8 Feb 2018
196. Gemological Institute of America (2018) Chow Tai Fook and GIA Bring Diamond Grading Reports to Consumers via Blockchain, May 23, 2018. Available: https://www.gia.edu/gia-news-press/chow-tai-fook-gia-blockchain; 8. del Castillo M (2018) IBM-Maersk blockchain platform adds 92 clients as part of global launch. Forbes, 9 Aug 2018
197. Dinh TT, Wang J, Chen G, Liu R et al (2017) BLOCKBENCH: a framework for analysing private blockchains. In SIGMOD 2017
198. Ethereum Blockchain App Platform. https://www.ethereum.org
199. Croman K, Decker C, Eyal I, Gencer AE, Juels A, Kosba A, Miller A, Saxena P, Shi E, Gün Sirer E, Song D (2016) On scaling decentralized blockchains—(a position paper), Financial Cryptography Workshops
200. Neureal, Open-source, peer-to-peer, AI supercomputing, live data stream prediction powered by blockchain, infinitely scalable
201. IBM Corporation, IBM Blockchain Platform: https://www.ibm.com/blockchain/platform
202. Astigarraga T, Chen X, Chen Y, Gu J, Hull R, Jiao L, Li Y, Novotny P (2018) Business-level blockchain users with a rules framework for smart contracts. In: The 16th international conference on service-oriented computing (ICSOC)
203. Vinod Kumar TM (ed) (2022) Smarter master planning: case studies of digital innovations. Springer, Singapore
204. Tiwari V, Chatterjee A (2022) Application of artificial neural network in master planning—a case of simulating the land use and land cover changes in Bhopal. In: Vinod Kumar TM (ed) Smarter master planning: case studies of digital innovations. Springer, Singapore
205. TraneAI: Decentralized Network for Artificial Intelligence at Scale, Whitepaper: https://github.com/TraneAI/Whitepaper/blob/master/TraneAi-TPAIWhitepaper.pdf
206. Jagielski M, Oprea A, Biggio B et al (2018) Manipulating machine learning: poisoning and countermeasures for regression learning. Computer Science Cryptography and Security Archive (cs.CR)," Cornell University, April 2018
207. Kurtulmus AB, Daniel K (2018) Trustless machine learning contracts; evaluating and exchanging machine learning models on the ethereum blockchain. CoRR abs/1802.10185. arXiv:1802.10185 http://arxiv.org/abs/1802.10185
208. Bonawitz K, Ivanov V, Kreuter B, Marcedone A, McMahan HB, Patel S, Ramage D, Segal A, Seth K (2017) Practical secure aggregation for privacy-preserving machine learning. ACM CCS
209. https://eias.org/policy-briefs/enter-eu-the-challenges-and-cooperation-potential-of-the-indo-pacific-strategy/
210. https://www.csis.org/analysis/challenges-and-opportunities-indo-pacific-water-sector
211. https://www.nbr.org/publication/one-region-multiple-strategies-how-countries-are-approaching-the-indo-pacific/
212. https://www.imd.org/reflections/digital-ecosystems/#:~:text=A%20digital%20ecosystem%20is%20complex,different%20from%20traditional%20business%20ecosystems.
213. https://www.newamerica.org/planetary-politics/reports/governing-the-digital-future/the-global-digital-governance-map/
214. https://international-partnerships.ec.europa.eu/policies/global-gateway/global-gateway-overview_en
215. https://t20japan.org/task-forces/the-future-of-work-and-education-for-the-digital-age/#:~:text=They%20will%20bring%20immense%20economic%20opportunities%2C%20such,jobs%2C%20higher%20GDP%20growth%2C%20and%20better%20living
216. https://www.linkedin.com/pulse/tomorrows-classrooms-digital-education-trends-future-talentsprint-hakrc/

217. https://www.capgemini.com/wp-content/uploads/2017/07/Digital_Transformation__A_Road-Map_for_Billion-Dollar_Organizations.pdf
218. Evans D, The Internet of Things how the next evolution of the internet is changing everything, CISCO, San Jose, CA, USA, White Paper
219. Taylor, The next generation of the internet revolutionizing the way we work, live, play, and learn, CISCO, San Francisco, CA, USA, CISCO Point of View
220. https://www.glueup.com/blog/digital-transformation-roadmap
221. https://www.sonata-software.com/blog/digital/7-steps-start-your-digital-transformation-roadmap

Prof. T. M. Vinod Kumar has 55 years of experience in Urban Planning as an author, editor, teacher, researcher, and adviser/consultant. He has worked in many Indo-Pacific countries such as India, China, Bhutan, Nepal, Malaysia, Indonesia, and Hawaii, USA. He was Dean of Studies, Head of the Department of Urban Planning, Head of the Centre for Systems Studies and Analysis, Head of the Centre for GIS and Remote Sensing, and Centre for Urban Studies of the School of Planning and Architecture, New Delhi; Visiting Professor National Institute of Technology, Calicut, and Institute of Technology Bandung, Indonesia and Professional Associates, East–West Resources Systems Institute Honolulu, Hawaii, Fellow Centre for the Study of Developing Societies, Delhi, Project Manager in Council for Social Development, New Delhi, Regional Program Coordinator at the International Centre for Integrated Mountain Development (ICIMOD) and Planner-Engineer at the Ford Foundation. He is the author of many books and journal articles. Innovation" (ed and edited "Geographic Information System for Smart Cities" (Copal:2014), "E-Governance for Smart Cities" (Springer:2015) "Smart Economy in Smart Cities" (Springer:2016) and "E-Democracy for Smart Cities" (Springer:2017), "Smart Metropolitan Regional Development: Economic and Spatial Design Strategies" (Springer:2018).," Smart Environment for Smart Cities" (Springer:2019)., "Smart Living for Smart Cities Case studies" (Springer:2020)" Smart Living for Smart Cities Community study and Ways and Means" (Springer:2020), "Smart Global Megacities Collaborative Research-Tokyo, Mumbai, New York, Hong Kong-Shenzhen, Kolkata"(Springer 2021) and Smart Global Megacities Collaborative Research-Chennai and Kochi-Kannur", (Springer:2021); Smart Master Planning: Case Studies of Digital Innovation"(Springer: 2022), Smart Master Planning: Case Studies on Domain Innovations" (Springer:2022). COVID-19, Containment, Life, Work, and Restart, Urban Studies (Springer: 2022). COVID-19, Containment, Life, Work, Restart, Regional Studies (Springer: 2022). The Smart Megacity System of Indo-Pacific: Emerging Architecture and Megacities Studies (Springer: 2023), Indo-Pacific Core and Peripheral Digital Economic Communities (Springer: 2025).

Okun's Law and Digitalization in ASEAN-10 Economies

Martin Boďa, Mariana Považanová, and Katarína Vitálišová

Abstract Whilst the economic success of ASEAN countries is frequently attributed to the transformation of their economies towards digitalization and technical innovation, little is known about the unemployment-output behaviour of ASEAN economies in the course of the business cycle. One aspect of this conjunctural behaviour is studied in this chapter in the hope of providing insights into how the unemployment-output relationship–enacted in Okun's law—unfolds in conjunction with the structural changes of their economies and other factors. A two-stage procedure is applied to that end. First, for a period of three decades, 1991–2022, a time-varying version of Okun's law is estimated for each ASEAN economy. Second, the estimated Okun coefficients are matched against a set of explanatory variables including various metrics of digitalization, sectoral and labour market characteristics. The results indicate that in the majority of ASEAN economies unemployment was mostly insensitive to output, or that the unemployment-output relationship was constant and otherwise of a negligible magnitude, which corroborates the resilience of economic growth in the Southeast Asian region. Nonetheless, this sensitivity is heightened by the reorientation towards medium and high-tech industries whilst the adoption of digital technologies by the population has actually no effect per se.

Keywords Okun's law · ASEAN · Digitalization · Difference version · Gap version

---

M. Boďa · M. Považanová (✉) · K. Vitálišová
Faculty of Economics, Matej Bel University in Banská Bystrica, Banská Bystrica, Slovak Republic
e-mail: mariana.povazanova@umb.sk

M. Boďa
e-mail: martin.boda@outlook.com

K. Vitálišová
e-mail: katarina.vitalisova@umb.sk

M. Boďa
Faculty of Social and Economic Sciences, Jan Evangelista Purkyně University in Ústí nad Labem, Ústí nad Labem, Czech Republic

T. M. Vinod Kumar (ed.), *Indo-Pacific E-Gateways*, Advances in 21st Century Human Settlements, https://doi.org/10.1007/978-981-95-4824-8_2

1 Introduction

Integration of digital technologies is a characteristic feature of current economic and social life, fundamentally altering how businesses operate, societies function, and economies develop. It contributes significantly to structural changes in economic systems, shifts in labour markets, evolution of business models, and reconfiguration of societal interactions. In the contemporary global context, digitalization has emerged as a driving force for economic growth, productivity enhancement, and competitive advantage across nations and regions.

Certainly, there are many countries that benefit from the rapid development of digitalization, and a good example of such countries is a regional grouping known as the Association of Southeast Asian Nations (ASEAN). ASEAN is formed on economic principles, and seeks to promote not only economic cooperation, but also to provide security to its ten members, viz. Brunei Darussalam, Cambodia, Indonesia, Laos, Malaysia, Myanmar, the Philippines, Singapore, Thailand, and Viet Nam. The economic success of this regional block is testified by the fact that ASEAN has achieved remarkable economic significance by accounting for 3.6% of global GDP in 2023, and ranking fifth worldwide [1, 2, p. 32]. In the field of digital economy, the annual report [3] reveals that the ASEAN region reached US$ 100 billion in revenue across all digital economy sectors growing at 27% annually since 2021. To compare it globally, it is expected that the world digital economy will reach an annual growth rate of 6.9% in the period of six years from 2023 to 2028 [4]. This statistic confirms that the ASEAN region is a region with the fastest-growing internet economy. Its growth is even faster as predicted, and it is expected to reach US$ 1 trillion by 2030 [5]. In comparison to the expected volume of the global digital economy in 2028 that should be US$ 16.5 trillion [4], the share of ASEAN digital economy is expected to be around 6%. This immense progress is a result of the strategical approach adopted by the ASEAN economies in 2014 that recognizes the role of the internet economy for economic development and that promotes digital development and facilitates technological advances [6].

The digitalization of ASEAN economies manifests itself in two critical dimensions. First, ASEAN economies are undergoing substantial structural shifts especially in advanced technology sectors—such as fintech, healthtech, and edtech—claiming increasingly significant portions of value-added and employment [7]. Second, societal life is more accustomed to digital technologies, including information and communication technologies, e-commerce and fintech. According to [8], online/mobile banking and e-wallets belong to the most used apps after social media in ASEAN countries. In 2017, nearly 90% of ASEAN's population was literate, and over 50% was under 30 years of age, with most having internet access [2]. This young demographic is positioned to drive digital transformation in a form of digital innovation, digital spending and accelerating digital adoption in workplace environments. The working population in ASEAN has sufficient digital skills for basic computer use, coding, and digital reading, these skills are often self-taught and acquired through practical experience rather than formal training. Indeed, there are some signs of digital convergence in the reliance of the ASEAN population upon digital technologies.

Moreover, not only digitalisation, but also other global changes and regional trends (e.g. climate challenges, demographic development, changes in consumer behaviour) have contributed to the significant structural transformation of ASEAN during last decades. The ASEAN economies shift their core axes from low productivity and labour-intensive sectors to higher productivity and skills-intensive sectors. This process is characterized by the gradual movement of the workforce from labour-intensive sectors to skill-intensive ones, with a reallocation of economic activity and labour from agriculture to industry and services [9].

Digitalisation undoubtedly contributes to economic growth, alters mechanisms with which economies work and translates into recognizable structural changes: both in terms of the shares of individual sectors in total value added, and in terms of the sectoral shares in total employment (e.g. [10, 11]). The impact of digitalisation on the labour market is reflected also in the (dis)usage of labour force in the economy. Unemployment, characterising conditions in which there are people in the economy willing and able to work, yet without an employment, is a persistent global problem despite the fact that much effort has been devoted to eliminating it. A widely accepted measure characterising the extent of unemployment is the unemployment rate (despite some of its shortcomings, which are discussed later). Two principal mechanisms with which digitalization contributes to unemployment are lively debated (e.g. [12]). On the one hand, digitalisation may stimulate creation of new jobs by which it may act against unemployment. On the other hand, some positions are lost in consequence of digitalisation, which in turn may lead to a rise in unemployment.

Nonetheless, digitalization is not the only factor influencing the development of the labour market and it also affects total production of the economy. In order to understand better the impact of digitalization on the economy, it is necessary to examine its impact on the development of the relationship between output and unemployment, and match it against digitalization and its trends. This is the empirical content of this chapter. The labour market is here condensed into one of the most representative metrics of its performance, viz. the unemployment rate, and economic production is represented by the most complex workable measure, viz. real gross domestic product. The relationship between these two metrics is in this chapter considered through temporal changes or cyclical fluctuations as embodied by Okun's law.

It spite of the fact that Okun's law is merely an empirical relationship between output changes and unemployment changes, it is in fact a standard and globally recognized analytical framework for examining the stimuli going between economic performance and unemployment. In the face of the existence of several studies exploring the restructuring of ASEAN economies [13–15], research on the relationship between output and unemployment in the region and how it is affected by digitalisation remains limited.

This study addresses the said research gap by focusing on Okun's law across the ten ASEAN countries from 1991 to 2022 and explaining it, *inter alia,* by digitalization tendencies. The research employs a two-stage approach that is enacted first by estimating time-varying Okun coefficients using the VC method devised by [16, 17], then by explaining variations in these coefficients through regressions incorporating digitalization metrics, labour market characteristics, and sectoral indicators. Digitalization is represented by measures of its social dimension, mainly connectivity to networks facilitating the use of digital services, and by measures of its economic size, as the share of medium and high-tech manufacturing upon value added creation. By this dual inclusion of digitalization into the analysis, this study extends the traditional scope of explanation adopted in research on Okun's law [18, 19] and provides deeper insights into how ASEAN's digital transformation is reshaping the fundamental economic relationships.

The remaining text consists of further six sections. Section 2 describes concepts of digitalization and operationalizes its measurement with respect to available data. Section 3 explains the logic and econometrics behind Okun's law and elucidates the framework adopted in its explanation. Section 4 identifies links through which digitalization shapes Okun's law. Section 5 briefly describes the data set sourced from the World Bank online database and presents the results. Section 6 comments trends in digitalization followed by the ASEAN economies and compares them with trends in the seven Group of Seven (G7) countries to obtain a better interpretational perspective. Finally, Sect. 7 is concluding, it discusses and offers insights.

2 Digitalization and Its Measurement

The purpose of this section is not to engage in efforts to develop suitable definitions of digitalization or the digital economy since authoritative sources themselves concede that there is hardly a consensus on the matter (e.g. [20], pp. 18–19, [21]). To some degree, the lack of definition is associated with the evolving nature, scope and construal of what actually these two terms subsume (e.g. [22], p. 6). Rather, considerations of digitalization are limited here to the extent of its conceivable bearing on the Okunian unemployment-output relationship and to the extent of the availability of data for ASEAN economies.

With no aspiration for completeness or exhaustiveness, digitalization is recognized here in two different, if interconnected layers: digitalization of society and digitalization of the economy.

Digitalization of society is understood here as the widespread proliferation and penetration of digital technologies or tools within a society. It means that the population is becoming more accustomed to information and communication technologies (ICT), digital platforms or digital solutions. People are starting to be more dependent on them in everyday life and require them. In consequence, they act as consumers of digital technologies on the demand side of the economy. A digitalized society pushes towards a digitalized economy with its digital population likely to abound with skills requisite to service and operate digital solutions. Aspects of digitalization of society are best captured by indicators that measure use of, or access to, the digital infrastructure, and such indicators are represented in the Digital User's Index utilized by the International Monetary Fund (IMF) [23, p. 29] to measure adoption of digital technologies, or in the connectivity dimension of the Digital Economy and Society Index (DESI) developed for European Union member states [24]. This kind of digitalization was reported in ASEAN societies to be on the rise as early as 15 years ago in spite of the existing vast cross-country differences with the availability and use of the internet and digital technologies [25].

Digitalization of the economy means structural changes in the sectoral make-up and reorientation from resource-based low-technology activities to activities highly reliant upon digital technologies generating presumably higher value added. Effects of this economic digital development may have a destructive or a constructive flavour: some activities are abandoned or suppressed in favour of others with a promise of higher productivity and value added. This kind of digitalization leads directly to the digital economy as a part of the economy (not necessarily a single sector) whose activities are driven by advanced technological innovations reliant or enhanced notably by digital technologies.

The digital economy is intrinsically tied to the rapid emergence and proliferation of information and communication technologies across all domains of human activity, necessitating new frameworks for understanding the factors that influence economic development and success. It acts as a catalyst for economic advancement and signifies a paradigmatic shift from traditional models, highlighting the centrality of data, connectivity, and digital infrastructure. As a rapidly expanding sector, it holds a vast potential for future growth. A pivotal role in the rise and progress of the digital economy must be ascribed to digital technologies that helped to shape old economic functions and develop new ones [26], and to the internet that redefined markets and organisational structures [27]. According to [28], digitalisation significantly contributes to GDP growth by enhancing productivity and unlocking new market opportunities. The digital economy is now increasingly underpinned by intuitive technologies that simplify communication and interaction.

The digital economy can be characterized through its three main components. The first one is a supporting infrastructure that relates to telecommunications, networks, PC devices, hardware, software, and to other technical elements. The reason being, the accessibility and quality of digital infrastructure significantly influence a country's ability to participate in the digital economy [27]. The second component is e-business, presented by various processes and economic activities implemented through computer networks. The last one is e-commerce, i.e. the sale of goods and services online [29]. Some studies highlight how e-commerce platforms facilitate market expansion, reduce transaction costs, and enhance consumer choice [30]. Integration of digital payment systems further streamlines digital transactions, fostering economic growth [31].

All these components have to be compatible and create one unique system that forms the digital ecosystem. It requires an institutional, technical, technological, business, governance, commerce framework, as well as corresponding human skills to operate it [32]. An institutional framework and e-government are critical for a resilient, secure, and sustainable digital ecosystem. They must establish clear rules, processes, policies and strategies (including financial and innovation support) in order to guide digital transformation. Whilst effective strategic management is essential, it depends heavily upon having skilled professionals with solid education. Its implementation must be fostered by a well-developed technological and technical infrastructure including internet access and reliable connectivity. This is a precondition for implementing the digital transformation of all economic sectors and for the long-anticipated shift to society 5.0, which integrates digital and physical environments allowing people to make better use of network-collected information through human analysis.

The digital economy uses as a key tool electronic commerce (e-commerce) [33], which is a term that refers to buying or selling goods and services through electronic transactions made over the internet or other computer networks. While the ordering process happens online, payment and final delivery can occur online or offline. These transactions may take place between businesses, consumers, individuals, governments, or other public and private organizations [34, 35]. It initiates the business models (such as business-to-business, business-to-consumer, consumer-to-consumer, business-to-government, mobile commerce) based on the multi-sided markets and platforms [33, 36]. Specialized intermediaries, electronic gateways (e-gateways), facilitating all the electronic process, are needed for the operability of e-commerce. E-commerce platforms hinge upon e-gateways to manage complex technical and security tasks behind every sale, and these e-gateways succeed by providing a scalable, secure bridge that supports the continuous growth of online transactions.

The digital economy brings many benefits that have reshaped how businesses and individuals operate. Benefits of the digital economy have been discussed and analysed in the previous literature [37–42] and they may be summarized as follows. Increased efficiency and improved productivity belong amongst the most referred advantages. By automating routine tasks, digital tools reduce the need for manual labour and minimize human error. Cloud computing and data analytics allow organizations to optimize resource allocation and make more informed decisions, whilst digital platforms streamline workflows and facilitate faster communication and collaboration. As a result, organizations can operate more swiftly and accurately than ever before. Another positive aspect of the digital economy is broader market access and the encouragement of innovation. Small businesses and entrepreneurs can reach global markets through e-commerce platforms without making large upfront investments, which sometimes acted as an insurmountable barrier to entry some decades ago. Digital marketplaces reduce geographic limitations, allowing niche products to find buyers around the world. From a consumer's point of view, the digital economy creates space for tailored offerings as well as better for their decision-making based on online reviews and comparisons. Digital platforms as part of e-commerce often provide personalized recommendations and tailored experience, and the outcome flexible offers enriched by varieties from traditional business models. Another positive aspect is access to information owing to advanced technologies that should be reflected in the creation and implementation of business strategies. Real-time data sharing accelerates collaboration and fosters sustainable development by introducing innovative solutions that increase quality of life and by employing advanced methodologies to safeguard the business environment, while ensuring the financial viability of the organization. Moreover, digitalisation of some services (e. g. mobile banking and digital wallets) allows including those who were previously excluded for various reasons. Thanks to new technologies enables more precise quality control, more thorough division of labour, and a wider span of supervisory control. It results in cost reductions for both businesses and consumers. Furthermore, the digital economy has also given rise to new job opportunities in emerging sectors. Demand for tech-focused roles—such as software development, data science, cybersecurity, and digital marketing—has surged. Gig-economy platforms and freelance marketplaces provide flexible earning opportunities, while fields like artificial intelligence, blockchain, and the Internet of Things create demand for specialized skills.

As opposed to these advantages, the same authors also identify challenges or threats of the digital economy [37–42]. One of the most pressing issues is the digital divide, which exacerbates inequality. Unequal access to high-speed internet leaves rural and low-income regions, lacking in digital literacy, excluded from full participation in the digital economy. Another issue is the skills gap on the part of users requisite for the adequate use of new ITC infrastructure. Automation and artificial intelligence can replace routine positions in manufacturing or retail as well as basic office functions. More sophisticated work positions require advanced skills and require further education and training. Albeit educational programs strive to reflect changing workforce demands, technological progress and adoption is even more rapid. An important challenge for the digital economy is privacy and data security. The extensive accumulation of personal and commercial data on different platforms by both private organizations and governmental bodies significantly increases the risk of unauthorized access, data breach, and data misuse. Advanced surveillance technologies—employed by corporate entities and governmental agencies—pose substantial risks to individual privacy and civil liberties. It is influenced by the fragmented nature of international data protection legislation, resulting in considerable variation in regulatory standards across jurisdictions. This challenge also entails greater exposure to cybersecurity threats resulting from an inadequate technical infrastructure and poorly implemented safety measures. Heightened exposure to cyberattacks appertains to actors in all spheres: individuals, enterprises and governments. In the case of critical infrastructure sectors, cyberattacks can lead to large-scale service disruptions. For small and medium-sized enterprises, cyberattacks can have irreversible and destructive consequences. Furthermore, hard and soft infrastructure has its own vulnerabilities, which necessitates adoption and implementation of offline backup and proper safety rules; to say nothing about the fact that rapid technological change can render existing investments obsolete, forcing thus organizations to continually upgrade their systems. From the market perspective, digitalization can hinder competition and innovation due to the domination of oligopolies (or strong leaders) on digital marketplaces with network effects that reinforce their market positions. Smaller vendors may struggle to negotiate fair terms or gain visibility on these platforms, which inevitable limits consumer choice and restricts entrepreneurial growth. Furthermore, the rapid expansion of the digital economy strongly supports globalization in consumer consumption at the expense of specific local products with consequences seen in the erosion of local economies and in the loss of cultural homogeneity.

To conceptualize a digital economy for statistical purposes, it is possible to count as digital such industries or activities (i.) that participate directly in producing digital outputs (narrowly), or (ii.) that use digital inputs as part of their production process (broadly), or (iii.) that rely on digital inputs (narrowly), or (iv.) that are significantly enhanced by digital inputs (broadly), whilst understanding that there may not be clear-cut borders between fully digital and non-digital [21, pp. 36–37]. In fact, the OECD framework of measuring the digital economy converges to a tiered definition: economic activity from producers of digital content, ICT goods and services yields a core measure, extended by economic activity derived from producers reliant on digital inputs supplies a narrow measure, and extended by economic activity from producers significantly enhanced by digital inputs generates by a broad measure [21, p. 40]. An instructive overview of older approaches is provided by [43, Table 1], where three tiers of the digital economy are nicely distinguished ranging from the core through a narrowly-scoped to a broadly-scoped digital economy [43, Fig. 3]. The core is made up by the ICT sector. The core plus digital services, the platform, sharing and gig economy form a narrowly-scoped digital economy, which extended by e-business, e-commerce, industry 4.0, precision agriculture and the algorithmic economy goes into a broadly-scoped digital economy. This somewhat resembles the approach adopted by the United States Bureau of Economic Analysis that includes in the digital economy three components: the digital-enabling infrastructure (the ICT sector), e-commerce as the digital transactions that take place by using this infrastructure, and digital media as the content created by this infrastructure [22, p. 7]. Having developed an actionable accounting definition of the digital economy, the size and importance of digital activities may be measured by nominal gross value added, employee compensation or employment [22, p. 11], out of which the first is amongst the high priority indicators for measuring digital activity within the OECD framework [21, p. 65]. In contrast, the United Kingdom digital economy is typically described via gross value assed or employment [44, 45].

3 Okun's Law: Its Framework and Explanation

Okun's law came to existence into the 1960s during the golden era of the US prosperity when macroeconomic policy was dominated by Keynesian thoughts. The US economy was on a path of secular growth, and some politicians and economists even fantasized that business cycles had ceased to exist or could be avoided [46, pp. 31–33]. As the US economic policy makers guided by their Keynesian advisers moved to the concept of potential output to assure that that the economy operates at its full-employment capabilities, the policy concern zoomed in to the issue of unemployment that was still perceived to be high and its costs were elaborated in detail [47, 48, pp. 24–25]. Professional economists represented at the US Council of Economic Advisers assisting the administrations of Presidents John F. Kennedy and Lyndon B. Johnson hoped to decrease unemployment through a suitable mix of fiscal and monetary policy measures. Following the Keynesian tradition, the US economic policy was formulated to follow the foal of full employment, whose level was set then by confronting historical records at an unemployment rate of 4%, which was actually the level of unemployment in the USA in the mid-1950s [48, p. 24]. Okun [49] experimented with the idea that there might be a perhaps stable relationship between unemployment and output and that this relationship might be used to predict potential output. The reason being, if economic policy is capable through its concentrated effort to suppress unemployment to the 4% full-employment level, it should further boost economic production. Okun [49] sought to develop a workable framework for short-term forecasting of output that would be sustainable under full employment, hence potential output. This is how Okun's law as a formula, be it statistical or economic, originated.

It no longer serves as a statistical device of potential output prediction, albeit there have been attempts to update it in this regard [50]. Currently, Okun's law features as a crucial ingredient of contemporary macroeconomic models [51, pp. 414–415] since it connects the labour market and economic production. When it is extended by the new classical version of the Phillips curve augmented by expectations, one immediately has links between the labour market and inflation, or between economic production and inflation (in the latter case, an aggregate supply function of the Lucas-Rapping type). The former stems, for instance, from the formulation in [52, p. 99], the latter is then visible, for instance, from the derivation in [53, p. 38]. For this reason, the Okun equation or suchlike representation frequently appears in structural macroeconomic models alongside the Phillips curve [54–56]. Whereas the underlying structural mechanism between unemployment and output that it describes is intuitive, many aspects associated with its formulation, interpretation and applicability are less so. In fact, there are two basal and standard versions of Okun's law, but these are multiplied by a variety of econometric approaches that endeavour to incorporate observed deviations from one universal static Okun equation. As the negative correlation between unemployment and output predicted by Okun's law tallies with economic wisdom and intuition, it is viewed as an empirical regularity (e.g. [57], p. 176), even though its strengths and other features vary over time (e.g. [50, 58, 59], [60], pp. 252–253).

There are two outlooks on the causation behind Okun's law. On the one hand, to an economist adhering to Keynesian principles, Okun's law, in whatever form the Okun equation is represented, embodies a causal relationship going from output to unemployment, as the Keynesian doctrine claims that it is output that determines employment and in turn affects unemployment. On the other hand, from a statistical viewpoint, the technique with which Okun [49] approached the unemployment-output relationship was rather creative, swapping the "dependent" and "independent" variables as he considered it fit. By intending to develop a tool for potential output prediction, he began by claiming that labour (under)utilization was a factor of the potential output level, then he reversed the position and estimated regression relationships where unemployment was dependent on output. This causes some confusion in the interpretation, in spite of the originally declared philosophy and intention, and leads many to assert that Okun's law is a correlational relationship (e.g. [61], p. 4, [62], p. 2).

Confined to the statistical perspective and accepting that both variables appearing in the Okun equation are endogenous, it may be confusing that Okun's law cannot be interpreted as a causal link between unemployment and output, but merely as a correlational relationship. Although Okun [49] adopted a Keynesian and policy oriented viewpoint and studied how unemployment reacts to output, the arrangement can be reversed and the Okun equation may be easily reformulated into a format where output is related to unemployment. This arrangement of the Okun equation is more aligned with a neo-classical position, in which output is generated by mixing sundry factors of production.

Given the simultaneity of both unemployment and output noted in the preceding paragraphs and discussed at length in [63], in essence it does not matter whether it is output or unemployment that is specified as the regressand. This is usually dictated by the anticipated purpose with which Okun's law is estimated. At no loss of generality, the Okun equation is formulated here in accordance with the Keynesian view that states unemployment as the regressand on the left-hand side and output positions as the regressor on the right-hand side. The nabla symbol is applied in statistical expressions to denote variations or cyclical fluctuations. In this vein, ∇y_t represents variations (fluctuations) in output for time t and ∇u_t stands for variations (fluctuations) in unemployment for time t. Whilst the difference version of Okun's law relates temporal variations of both variables and measures them as period-on-period changes, the gap version actually confronts cyclical fluctuations of these variables and measures them as deviations from long-term trends. All these variation and fluctuation metrics are represented typically as percentages on an annual basis, less frequently over a quarterly period (e.g. 3% annually, or –2% points over a quarter). Hence, with the difference version, ∇y_t is the rate of output growth defined for time t by the expression $\nabla y_t = \log(y_t/y_{t-1})$, and ∇u_t is the absolute change in the unemployment rate measured for time t as $\nabla u_t = u_t - u_{t-1}$. In contrast, to operationalize the gap version, suitable estimates of long-term trends of output and unemployment are needed. These are denoted here as ${}^{lt}y_t$ and ${}^{lt}u_t$, respectively, both being stated for time t. A suitable approach to handle the estimation procedure is required, although it is likely to be fully exposed to an easy critique (e.g. [64]). In scholarly studies on Okun's law, but also massively elsewhere, atheoretical filters and time series decomposition techniques are most commonly applied [65, p. 1469], out of which the Hodrick-Prescott filter outstands as the usual and conventional choice [66, p. 27]. Under the current interpretation of the gap version, ∇y_t is the output gap defined for time t as $\nabla y_t = \log(y_t/{}^{lt}y_t)$ and ∇u_t is the absolute change in the unemployment gap measured for time t as $\nabla u_t = u_t - {}^{lt}u_t$. The gap version has an advantage of being consistent with the sustainability concerns of economic policy making [67, p. 40, 68, pp. 1–2]. Simulaneously, the gap version answers more readily to the Keynesian interpretation of economic policy that should seek the realization of potential in place of the mere realization of expansion [46, pp. 40–41].

The accepted pro-Keynesian arrangement of the Okun equation with ∇u placed as the regressand and ∇y as the regressor dominates research on Okun's law as documented by [66, p. 26]. For some time instance t, an equation representing Okun's law in a linear format encompassing both the difference and gap version takes the generic form

$$\nabla u_t = \alpha_t + \beta_t \cdot \nabla y_t + e_t, \tag{1}$$

where the intercept and slope, α_t and β_t, are possibly time-varying, and where e_t is a random term that behaves like a white noise sequence uncorrelated with all values of the regressor ∇y. The time index t in Eq. (1) goes consecutively over some discrete set T. Of interest to economic policy is the Okun coefficient that in Eq. (1) is the slope coefficient β_t. At the risk of some simplification, it can be thought of as the coefficient of co-elasticity of unemployment and output over time (for the difference version) or over the business cycle (for the gap version). Under stabilized normal economic conditions the Okun coefficient is expected to be significant and negative. Conventional patterns of the business cycle imply either an increase in output associated with a drop in unemployment ($\nabla y > 0$, $\nabla u < 0$), or a fall in output accompanied with a rise in unemployment ($\nabla y < 0$, $\nabla u > 0$). The Okun coefficient β_t in Eq. (1) either captures the average benefits of the positive former situation or measures the average losses of the undesirable latter situation.

It must be emphasized that the Okun equation in (1) is dynamic as it allows both key parameters α_t and β_t to vary in time. If the time indices were removed from the regression parameters, and one would have $\alpha = \alpha_t$ and $\beta = \beta_t$ for each time t, the Okun equation in (1) would become static and less realistic. Although this is an approach that is fairly common even in canonical studies on Okun's law (e.g. [69–71]), it collides with the credible reality of changing and non-constant economic processes (e.g. [72], pp. 285–286). Instead, economic processes rather evolve gradually and adjust to shocks and absorb structural changes [73]. Furthermore, the dynamic representation with time-varying α_t and β_t given in (1) suffices to handle typical departures of a simple static Okun equation from the otherwise observed reality. Over the years, it has been established that the Okun coefficient β_t is instable in time (e.g. [74]), alternates with stages of the business cycle (e.g. [50]), or a typical Okun equation suffers from some general form of non-linearity (e.g. [75]).

Unfortunately, it is not common in scholarly studies to consider a dynamic format of Okun's law, and the adoption of a non-constant-coefficient framework is a relatively recent disposition in Okunian research (e.g. [74, 76, 77]). It seems that the VC (varying coefficient) method of Schlicht [16, 17] has won favour with late studies (e.g. [74, 78]). An indisputable advantage of the VC method is that it combines in an analytically tractable manner the principles of least squares estimation with the notion of slow temporal adjustments of the regression equation. The method considers Eq. (1) to be an observation equation and furnishes it with state equations that represent the intercept and slope as random walks. Instead of employing the traditional Kalman filter that requires more expertise and more inputs for the initial steps of estimation, the VC method is works with an extended least squares criterion that accounts for the minimization of both the observation errors and the state innovations. There are four principal advantages for adopting the VC approach. A first advantage is that the VC method produces estimates of Okun coefficients that are directly comparable with other studies. The reason being, when the trajectory of the Okun coefficient in Eq. (1) is averaged for the entire time period T in question, the result is a quantity that is comparable and mostly identical with the Okun coefficient constant for the entire period. A second advantage is that it comes with a full inference framework. It supplies not only estimates of regression coefficients, but also their standard errors. The reason being, the VC method yields point-wise estimates and standard errors $([b_t, SE(b_t)])_{t \in T}$ that answer to the trajectory $(\beta_t)_{t \in T}$. One may construct an approximate confidence band in a usual form and check a to whether it includes zero. For instance, for a 95 level of confidence, the confidence band is approximately $(b_t - 2 \cdot SE(b_t), b_t + 2 \cdot SE(b_t))_{t \in T}$. A third advantage is that there is full software support that allows easy replication of the results. The VC method is available from the author of the method, E. Schlicht, in a free package TVC to program gretl. A fourth advantage is that it received praise in a comparative study [79] who assessed the performance of four rival algorithms for estimation of parameters in a time-varying regression set-up.

A conventional finding of studies on Okun's law is that it varies in magnitude. For advanced countries with stable economic environments, the Okun coefficient is usually statistically significant and negative (e.g. [70, 71]), which implies that unemployment decreases alongside rising output during an upturn, and vice versa, that unemployment increases alongside falling output in an downturn stage of the business cycle. Whether these movements are thought of in terms of period-on period changes or are deemed as fluctuations around a long-term trend depends upon whether the difference or gap version is adopted, respectively. For less developed countries, the Okun coefficient is frequently found insignificant or sometimes significant and positive (e.g. [71]), which runs at odds with economic intuition. One might thus surmise that a stable economic environment and some specific configuration of factors is needed to assure that *co*movements between the labour market and aggregate output are lined up. A question inevitably arises what explains the size of Okun coefficients across economies that are measures of the magnitude with which Okun's law presents itself. Candidate factors are recruited amongst sundry labour market, sectoral, institutional, sociodemographic and broader societal indicators that may explain the observed differentiated *co*rresponsiveness of unemployment and output across the business cycle.

A two-stage procedure is typically applied in studies on Okun's law (e.g. [18, 19, 71, 74]). In a first stage Okun coefficients are estimated by a suitable econometric method, in a second stage the estimated Okun coefficients are regressed upon a set of well-reasoned potential explanatory factors This generally accepted and statistically valid two-stage procedure is applied also in this chapter. Stage 1 is operationalized here by the VC method applied to Eq. (1). The result is a panel of pooled time-varying estimates and their standard errors, say $\Pi = \{[b_{it}, SE(b_{it})]\}_{i \in I, t \in T}$, where the index i denotes one of economies in the index set I and the index t designates one time instance in the index set T. In stage 2, the panel of interest Π is confronted with another panel, in this case of explanatory variables, say $Z = \{[z^1{}_{it},\ldots, z^K{}_{it}]\}_{i \in I, t \in T}$, where $z^k{}_{it}$ is one of K explanatory variables measured for economy i at time t. They may be formed in a manner that amends for possible heteroskedasticity (e.g. by using the logarithmic transformation), possible endogeneity issues (e.g. by backshifting some variables) or addresses some forms of non-linearities (e.g. by using squares or other functions of original variables). Stage 2 does not suffer if some observations are absent and the combined panel [Π, Z] is unbalanced. The regression equation of stage 2 for economy i and time t writes as

$$b_{it} = d_i + \delta_1 z_{it}^1 + \cdots + \delta_K z_{it}^K + v_{it}, \tag{2}$$

where d_i is a fixed effect, $\delta_1,\ldots, \delta_K$ are regression parameters of interest that embody the average influence of the factors $z^1,\ldots, z^K$ upon Okun coefficients in the block of economies, and where v_{it} is a random term. Fixed effects d_i represent economy-specific characteristics that are omitted from Eq. (2) and may be arbitrarily correlated with the regressors $z^1,\ldots, z^K$. Random terms v_{it} are assumed to be generated from a white noise process uncorrelated with the regressors $z^1,\ldots, z^K$. Equation (2) presumes a linear format of the relationship and that $\delta_1,\ldots, \delta_K$ are constant. Admittedly, there is some inconsistency between Eqs. (1) and (2), but the experience of the authors is such that they can be handled comfortably, at least as a first approximation, as constant.

However, using estimated Okun coefficients as regressands injects heteroskedasticity into Eq. (2) as each estimated coefficient comes with its own uncertainty or sampling variance measured by standard error (e.g. [80, 81]). The study [81] summarized and assessed four approaches to handle such regressions based on ordinary least squares (OLS) or weighted least squares (WLS) that are employed or are useful in stage 2, with pure OLS being least appealing. In the empirical analysis this chapter adopts the WLS procedure advocated by Saxonhouse [80], as an improvement over the simplistic OLS approach. Albeit the explanation second-stage regression dates back as far as 1991 [82], sometimes the issue of heteroskedasticity is overlooked even in recent studies and OLS is used with simple standard errors (e.g. [71]). The estimation procedure adopted for Eq. (2) in the present case applies a dual estimation approach in a comparative manner with fixed effects: simple WLS with endogeneity of some regressors $z^1,\ldots, z^K$ neglected, and weighted two-stage least squares (W2SLS) with endogeneity fully catered for.

4 Link of Digitalization to Aggregate Activity and Okun's Law

Both concepts of digitalization considered affect (un)employment and economic output, where obviously digitalization of society has intermediated impact, whereas digitalization of the economy exerts its impact directly. Depending on the specificity of the economic environment and its initial conditions, sectoral make-up, possible skills mismatch, there is no reason to believe that digitalization may affect the labour force and economic performance equally, and that it must have a neutral effect upon the Okun's law over time or across the business cycle. On the contrary, it may even stir up cyclical fluctuations that may be likened to Juglar fixed-investment waves or Kuznets infrastructural-investment swings (e.g. [83], pp. 189–190, [84], pp. 9, 14). The reason being, digitalization of economic activities and their reallocation across sector will have differentiated effects upon employment and labour productivity in diverse sectors. Computers and digital solutions are anticipated to fully substitute routine tasks in the workplace, be they basal analytic and interactive, or manual (e.g. [85], pp. 118–121). In this regard, Fossen and Sorgner [12] speak of destructive and transformative effects of digitalization. Destructive digital transition means displacement of human labour in consequence of automation and robotization of specific tasks or activities. This leads to dismissals of employees and economic effects are then seen in increased productivity and increased operation costs that are, however, offset by savings on personnel costs. There are pressures upon unemployment or flows outside the labour force. Transformative digital transition happens when human labour is aided, but not displaced, by digital technologies and the improved interaction between human labour and digital elements facilitates increased labour productivity. Transformative digitalization may even stimulate new job positions and—unless there are coordination failures in the matching process—their filling may alleviate unemployment. The balance of destructive (substitutive) and transformative (complementary) effects cannot be established a priori for it is obviously specific to each economy and may vary in time or with the level of economic development. It transpires that a noteworthy factor shaping how digitalization affects the covariation between unemployment and output is the employment structure of the economy as low-skill and middle-skill employments are easy to substitute by computers, robots or digital solutions. Some signs of the upcoming patterns are detectible even now. For instance, the study [86] provided evidence for the United States labour market that jobless recoveries in recent recessionary periods in the era of digitalization were associated with the disappearance of routine occupations and middle-skill employments. Another study [87] reported that whilst these jobless recoveries did not obstruct Okun's law and the relationship survived, it adapted with changed coefficients. Even demographic conditions may matter in the process. For instance, for economies with ageing societies digitalization may be welcome and may help address shortages of skilled labour (e.g. [85], pp. 132–133), encouraging sterility in the labour market in the course of the business cycle.

For one thing, whereas there is a widespread contention amongst political elites of the world that digitalization catalyses or boosts economic growth (e.g. [88], pp. 7, 9), often supported by scholarly research (e.g. [10, 11]), the picture may be somewhat varied. The nexus may be positive for some countries and negative for others (e.g. [89, 90]), or the evidence is inconclusive (e.g. [91]) or suggests only marginal benefits [92]. The relationship between digitalization and economic growth may be more intricate with feedbacks looping from one to another [93, p. 126]. If there really exists a transmission channel between digitalization and output, any reconfiguration of its links or any disturbance in the form of exogenous shocks or changes in the institutional environment could modify Okun's law. This modification could lead either to its deterioration or intensifying.

Amidst intensive discussions how digitalization should be conceived or measured and amidst scattered academic attempts to devise a suitable metric of digitalization (e.g. [94], Table 4), the limiting factor remains availability of data. The data for this analysis were sourced from the World Development Indicators subset of the World Bank's database (https://databank.worldbank.org), which were not always ideally populated for ASEAN countries. The societal dimension of digitalization was proxied by its connectivity to networks facilitating use of digital services. Five metrics were actually available: fixed broadband take-up, fixed telephone take-up, mobile cellular take-up, proportion of the population using the internet, and secure internet servers. The first three uptake metrics were available as subscriptions per 100 people and the last metric was expressed per 1 million people. In spite of tendencies by official authorities or academics to compact multi-faceted phenomena into a single comprehensive metric, such as through principal component analysis, multi-criteria decision methods or data envelopment analysis, no such synthesization was attempted here for no less than two reasons. First, it is fairly obvious that these metrics are rather non-homogenous and do not cover all conceivable aspects of a digitalized society. For instance, a fairly comprehensive metric in this respect is the Digital Economy and Society Index (DESI) described by [24], but its partial indicators are for ASEAN countries currently nonreachable. Second, any such synthesization induces a loss of partial information in spite of the level of sophistication. Hence, these five metrics were considered each separately as explanators of Okun coefficients in second-stage regressions as described around Eq. (2). The economic dimension of digitalization was proxied by the share of medium and high-tech manufacturing upon value added creation. The reason being, higher-tech manufacturing sectors are usually reliant on the latest technologies and participate in the production of digital goods (computers, televisions, communication devices, digitally-enhanced consumer goods). This measure overlooks many segments of the economy characteristic of high digital intensity (e.g. [21], Appendix A), but is vindicated by the belief that digital transformation, to all intents and purposes, diffuse evenly across a wide array of economic sectors. This metric was also included as a separate full-blown explanatory variable for Okun coefficients.

Once again, it is fully conceded that these metrics are partial and can be criticized on being not sufficiently descriptive of all aspects of digitalization or being unbalanced. Yet, their selection is not unconventional as confronted with other studies (e.g.

[94], Table 4), and is rendered with the faith that trends signalized by these metrics are *pari passu* with the impact of digitalization across other areas of the society and economy.

5 Data and Results

The entire analysis was conducted with the use of data sourced from the the World Bank (https://databank.worldbank.org) in February 2025. The time frame covers 32 years from 1991 to 2022 and could not be extended on either side for unavailability of key variables. Albeit the set of variables considered at the onset was richer—as explained below—, the set of actually used variables was leaner and is presented in Table 10 in the Appendix. The effective set of observations varied for each ASEAN economy and differed across analytical steps, but is always reported as part of the tabular output.

In stage 1, Okun coefficients were estimated for each ASEAN economy separately by using the VC method of Schlicht [16, 17]. In step with [74, 78], the specification presumed a constant intercept and a time-varying Okun coefficient and was applied with the difference version and two variants of the gap version of Okun's law. The difference version confronted unemployment changes (g_y) with real output growth rates (Δu), whereas the gap version juxtaposed unemployment gaps (gap_u) with output gaps (gap_y). The Hodrick-Prescott filter employed in the estimation of gap variables was blended with an extended procedure described in the notes to Table 10 in the Appendix in order to mitigate possible end-point distortions, for which it is heavily criticised (e.g. [95]). Furthermore, in response to the discussions concerning the choice of the smoothing parameter λ (e.g. [96]), the gap version was implemented with two different smoothing weights: $\lambda = 6.25$ as opposed to the conventional choice for annual data $\lambda = 100$ [97, 98].

Organized for each ASEAN economy alone, Table 11 in the Appendix displays the unemployment and output variables needed to estimate the Okun equation specified in (1), but with a time-invariant intercept. It is natural and expected that annual variation in unemployment and output (g_y, Δu) will differ from cyclical fluctuations in unemployment and output (gap_y, gap_u), as the former captures temporal development and the latter measures overheating or downturns of the economy. What is somewhat unpleasant is that the choice of the smoothing parameter in the application of the Hodrick-Prescott filter matters and leads to quantitatively different gap estimates. In most cases, the differences cannot be disregarded as negligible, especially for the estimated output gaps. They are perhaps more striking for the estimated output gaps of Cambodia, Indonesia, Myanmar or Thailand.

Table 11 evinces that the ASEAN economies are not on a par with the dynamics of their aggregate activity, but for the period of the three decades 1991–2022 showed sometimes notable differences. For instance, whilst Bruneian real GDP rose only at an annual average 0.44%, the other nine ASEAN economies managed to achieve more encouraging growth paths between 3.56 and 6.97% annually. As seen in both the standard deviation and the range of annual growth rates, in Cambodia and Myanmar the growth trajectory was tempestuous, marked with year-on-year turbulences. That said, the ASEAN labour markets were all much calmer and more stable as the average variation in unemployment ranged between an annual –0.04% (for Laos) and 0.07% (for Myanmar). All in all, for the examined three decades, the ASEAN economies did not display unusual variation and yearly changes in unemployment, but their real aggregate output followed appealing growth trajectories (except Brunei where on average real GDP rather stagnated).

Given the varied mosaic of the estimated gap variables and contradictions stemming from choosing the smoothing parameter λ either 6.25 or 100, it is difficult to identify convincing patterns. Nonetheless, focusing on situations where the gap estimates roughly agree, it appears that the economies of Laos, Malaysia, the Philippines, Singapore and Viet Nam operated during the years 1991–2022 on average below potential, albeit in a exceptional few years showing considerable overheating. In like manner, in Laos and Malaysia unemployment was on average above its normal long-term level, whilst in Indonesia and Singapore unemployment tended on average to be below its long-term level, suggesting thus some kind of overemployment. Furthermore, the dispersion of estimated output gaps is for each economy safely higher than that of estimated unemployment gaps, suggestive of more pronounced fluctuations in economic output than in labour (under)utilization.

Hence, in spite of the noted differences between the economies, Table 11 reveals that in the investigated period ASEAN economies attained an upward secular growth path with somewhat volatile real GDP levels, but relatively stable unemployment levels.

The estimated trajectories of Okun coefficients are presented visually in Fig. 1 and numerically in Table 12 in the Appendix. Not only does Fig. 1 display the time-varying patterns, but particularly identifies their statistically significant portions in thick red. The procedure to determine statistical significance was footed on 95% approximate confidence intervals constructed around point-wise estimates of time-varying Okun coefficients in the manner depicted in Sect. 2. To ensure comparability across the ASEAN economies, vertical axes have all equal ranges. It is somewhat surprising that for most economies the estimated trajectories show similar patterns irrespective of the variation of Okun's law, and differences are only numerical or in areas of (in)significance.

- For *Brunei Darussalam*, for most of the period unemployment and output were not corresponsive, and Okun's law was only activated for a few years in the late 2010s, depending on the variation applied. For instance, for the gap version with $\lambda = 6.25$, the Okun coefficient was significant for six consecutive years between 2014 and 2019 with an average value of –0.158, ranging then from –0.191 to –0.117. For the gap version with $\lambda = 100$, the significance span was somewhat different with the average value of the significant Okun coefficient more than twofold, –0.376, and a somewhat volatile trajectory between 2012 and 2022.

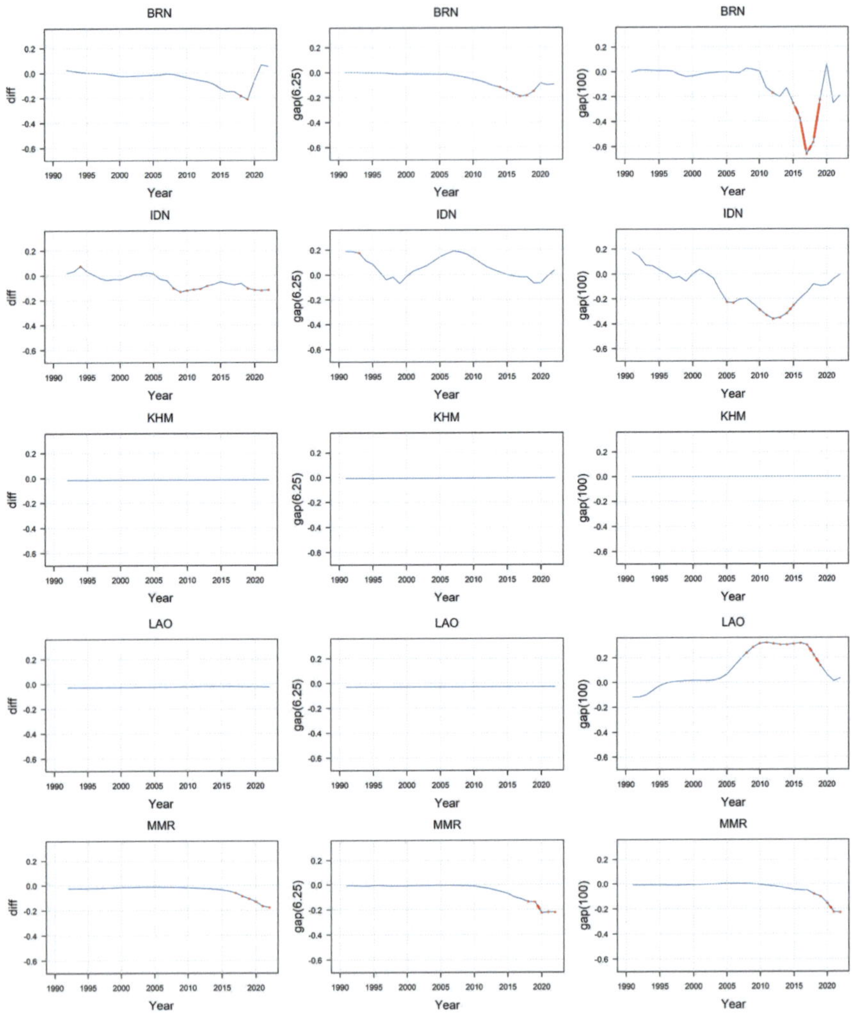

Fig. 1 Time-varying trajectories of Okun coefficients

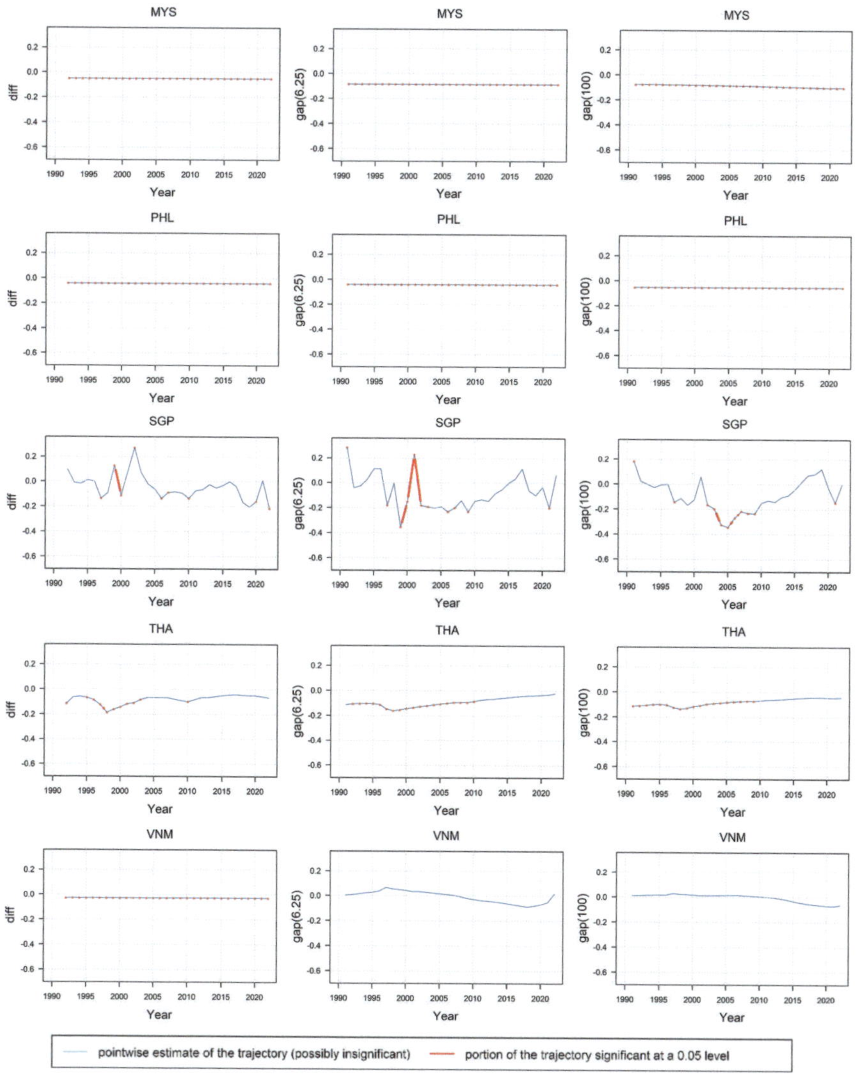

Fig. 1 (continued)

- Okun's law was found not to hold in *Cambodia* as the Okun coefficient was tested insignificant for the entire period with each variant of Okun's law applied, being thus zero.
- In contrast, validity of Okun's law and its time invariance was established for the entire period for *Malaysia* and *the Philippines*, where constant trajectories, but now significant, were estimated with each variant of Okun's law. For Malaysia, the Okun coefficient is estimated as constant –0.050 for the difference version

and about –0.085 for the gap versions. Likewise, for the Philippines, this value is about –0.040 or –0.050, depending on the variant applied.

- For *Indonesia*, non-zero and significant unemployment-output *co*sensitivity manifested itself actually for the difference version for 11 years (1994, 2008–2013, 2019–2022) with an average value of the Okun coefficient at –0.093 (inclusive of one occasional positive value estimated for 1994) and for the gap version with $\lambda = 100$ for 8 years (2005–2006, 2010–2015) with an average Okun coefficient of –0.295. Otherwise, albeit insignificant, the estimated trajectories point to turbulent conditions affecting Okun's law during the period.
- There are doubts about the validity of Okun's law in *Laos* as the difference version and the gap version with $\lambda = 6.25$ were estimated with an entirely insignificant (constant) trajectory of the Okun coefficient. For the gap version with $\lambda = 100$ the Okun coefficient was tested significant only for a dozen years from 2008 until 2019, but its value was always positive, contrary to economic expectations, oscillating about its average 0.280.
- In *Myanmar*, unemployment *co*varied with output mere 6 years for the difference version and 5 years for the two gap variants at the end of the decades. Yet, if significant, the Okun coefficient was always found negative, and for the difference version averaged to –0.121, whilst for the gap version either to –0.128 or –0.158. A similar pattern was detected for *Thailand* except that the Okun coefficient was found significant at the beginning of the period 11 years for the difference version with an average significant negative Okun coefficient of –0.117, and 19 years for the gap version with an average significant negative Okun coefficient of either –0.116 or –0.100. Hence, whilst in Myanmar Okun's law began to hold after about 2018, in Thailand, it held until about 2009.
- Discordant and fairly tumultuous patterns were observed for *Singapore*, where the Okun coefficient was found significant for 9 years for the difference version with an average value of –0.065, and for 11 years for either implementation of the gap version with an average value of –0.128 or –0.190. Nonetheless, there were two cases of a positive Okun coefficient for the difference version, and one case for each gap variant of Okun's law. The erraticism of the estimated Okunian trajectories is discernible also in the fact that years of significant Okun coefficients were not consecutive
- Finally, in *Viet Nam* Okun's law was confirmed only for the difference version and the Okun coefficient was found constant for the entire period at –0.028. In contrast, for the gap version the Okun coefficient was found insignificant over the entire period regardless of the smoothing factor used.

To sum up, at the risk of some simplification, the patterns communicated by Fig. 1 may be translated into a categorization of ASEAN economies into six unevenly populated categories. They are condensed into Table 1. Only Cambodia, Malaysia and the Philippines represent pure categories where Okun's law either failed to hold for the entire period (with unemployment and output being unrelated), or it held as predicted by the simplification framework behind Okun's law (with unemployment and output being negatively correlated).

Table 1 Categories of ASEAN economies suggested by the pattern of the estimated Okunian trajectory

Category	Countries
Okun's law insignificant over the entire period (and the Okun coefficient zero)	Cambodia
Okun's law significant over the entire period and the Okun coefficient negative (as expected)	Malaysia, the Philippines
Okun's law significant only in some (typically consecutive) years and the Okun coefficient negative (as expected)	Brunei Darussalam, Indonesia, Myanmar, Thailand
Okun's law significant only in some scattered years and the Okun coefficient mostly negative (as expected)	Singapore
Okun's law significant only for one variant in some consecutive years with the Okun coefficient positive (contrary to expectations)	Laos
Okun's law significant only for one variant over the entire period with the Okun coefficient negative (as expected)	Viet Nam

Furthermore, the interpretation of an estimated Okun coefficient may be illustrated for the case of the Malaysian economy where Okun's law was found short of irregularities and constant for both the difference version and gap version. The value of the Okun coefficient for the difference version –0.050 suggests that between 1992 and 2022 with an annual increase of Malaysian real GDP by one percentage point unemployment would tend to decrease annually by a mere 0.05% points. Likewise, the value –0.085 for the gap version indicates that between 1991 and 2022 an acceleration of Malaysian real GDP by one percentage point above potential would be accompanied by a decrease of unemployment by a mere 0.085% points below its long-term level. Naturally, this pattern works also vice versa in the decrease/increase direction. At any rate, the values –0.050 and –0.085 are not much, and the detected unemployment-output *co*rresponsiveness is feeble. For the Philippines this mutual responsiveness is even smaller.

Nonetheless, to all intents and purposes, the Okunian trajectories presented in Fig. 1 and summarized by their mean levels in Table 12 in the Appendix are all low, suggesting either a marginal and negligible level of the *co*rresponsiveness between labour markets and economic activity, or that Okun's law simply does not hold (in the sense that the Okun coefficient is zero). This is a finding that characterizes many less developed countries, although the actual reasons for the non-existence of Okun's law will likely differ from case to case. Still, in that event, unemployment changes are driven mostly by other reasons than variations in economic activity, and themselves do not contribute significantly to such variations. Indeed, Ball et al. [71, 74] in their global comparative studies established considerable heterogeneity in Okun's law across countries, with unemployment being more responsive in advanced economies. [74], p. 667] attributed this to a feebler employment response over the business cycle as a large share of the labour force is either unaccounted for or works for the informal sector.

By the classification of world economies adopted by the IMF [99, pp. 121–123], only Singapore is recognized as an advanced economy, whereas the remaining nine ASEAN economies fall amongst "Emerging and Developing Asia". In comparison to Singapore, the estimated Okun coefficients for the other ASEAN economies are near zero, to say nothing of their frequent insignificance. The exception to this rule is Brunei Darussalam and Indonesia, albeit in short periods. That said, Singapore itself showed for most years of the analysed period insignificant Okun coefficients. These results agree with the stylized fact that in developing countries Okun's law is weaker than in advanced countries. An explanation for this pattern is the unusual proliferation of non-standard employment and precarious work in ASEAN countries [100] as well as informal employment, which in 2024 was estimated to reach 69.3% of total employment in the ASEAN region [101, p. 6]. Out of these, the former form is underrepresented statistically in reported employment and the latter form is not properly identified in official statistics. Hence, labour market changes over time or across the business cycle may not materialize as unemployment changes.

It must also be emphasized that the period of three decades, 1991–2022, scrutinized in the analysis is marked by structural transformation that transcends not only individual sectors, but also labour markets all over the ASEAN region. With this phenomenon, the labour force is gradually diverted from labour-intensive sectoral activities to skill-intensive sectoral activities, which is discernible frequently by withdrawing labour from agriculture and deploying it to industry and services [102, pp. 4–6]. Economic development marching in unison with digitalization in the ASEAN region [103, p. 16] paves way to higher agricultural productivity and skill-based and knowledge utilization of labour elsewhere, and the structural transformation of employment is necessarily reflected in the strength or validity of Okun's law.

Furthermore, one may make cases studies of ASEAN economies and their labour markets in order to indicate which factors might have a say in the discovered patterns (e.g. [9], pp. 13–25). For instance, *Viet Nam* succeeded in sustaining a fast growth rate for the recent decades with swift adjustments of the labour market that precluded labour mismatches. A specific asset in Viet Nam's economic success is a persistent high female labour force participation rate, much higher than in other ASEAN economies, for which it owes to its Socialist past. In contrast, *Malaysia*'s growth was slower, and is attributed to the rise in sectoral labour productivity without any apparent contribution of labour reallocation between sectors, hence structural transformation [104]. Malaysia also relies heavily on foreign labour with undocumented workers estimated as numerous as regular migrant workers [105, pp. 17–18]. Whether official or shadow, the Malaysian labour market enjoys a higher level of flexibility that bypasses unemployment statistics. As another case study, economic conditions in *Indonesia* differ from other ASEAN countries in two notable respects. Albeit the country is one of the countries that "officially" industrialized in the 1970s, the transition to agriculture to services had happened before the industry sector matured [106, p. 1] and led to an extremely polarized society with a thin middle-class labour force and employment concentrated in low-productivity or high-productivity sectors

[9, p. 19]. A somewhat different example is *Thailand*, where structural transformation took on a gentle, constrained pace. Agricultural employment remained relatively stable for years or only slowly reduced when confronted with the needs of the industry and services sectors, whilst Thai agriculture's relative contribution to GDP diminished notably. This contributed to a strong urban/rural divide where "the long-standing and politically toxic disparities between rich and poor regions" persisted [121, p. 113]. A final example is the *Philippines*, which outstands from other ASEAN countries by falling behind in industrialization efforts. The core issues for the stagnant pace of industrialization are high labour underutilization, slow poverty reduction, and stagnant investment [9, p. 24]. The third issue is in fact a traditional narrative explaining the slow and unbalanced economic growth of the Philippine economy. For instance, Lanzona [107, pp. 2–3] blames the failure of the industrialization program on the absence of a conceptual and systemic framework of technical investments to that would satisfy even basic areas of structural transformation. Necessarily, this also indicates that now the Philippines are likely to be lagging also in digitalization efforts for the gap may be difficult to close. An alternative explanation has lately been offered by Brines [108] where the Philippines is seen entrenched in "an agricultural trap", characterized by low levels of transformation in agriculture with extremely poor productivity and a high utilization of the labour force that is then missing in areas generating higher value added.

For the sake completeness, Table 2 juxtaposes Okun coefficients estimated for ASEAN economies by two relevant reference studies, [71, 74], framed as global comparative studies. It was not possible to make a meaningful comparison with other studies of relevant quality that would have at least a partial coverage of ASEAN economies. For instance, when Okun's law was estimated for a panel of countries in [109], sundry economic environments were inadequately mixed into one artificial one-size-fits-all Okun coefficient. A modified Okun equation was applied in [110] in an attempt to account for asymmetric responses across the business cycle. An employment-based formulation of Okun's law was utilized in [111]. Likewise, the concept of asymmetric cointegration was employed in [112] so that the estimated Okun coefficients were rendered incomparable with conventional counterparts. Finally, a reverse arrangement of the Okun equation was posited in [113] where the output variable is utilized as a response. In addition, no other study published under acceptable scientific conditions have reported the full coverage of the ten ASEAN economies.

Table 2 evinces that the current findings are not altogether in dissonance with the findings of the two relevant studies. In Indonesia and Singapore, Okun's law is observed non-operational, in Malaysia or Thailand the estimated Okun coefficients are roughly of the same magnitude. Differences arise for Viet Nam or the Philippines, and they are related to nominal sizes. The current estimates are much lower. To some extent, this is a consequence of the fact that [71, 74] analysed a period from the 1980s to the mid-2010s based on data from the IMF, whereas now the temporal focus is on the period from 1991 to 2022 by using data from the World Bank.

In stage 2, the estimated Okun coefficients regardless of their significance were regressed upon a set of conceivable explanatory factors available in the World Bank

Table 2 Comparison of the estimated Okun coefficients with relevant studies

Country	Ball et al. [71]		Furceri et al. [74]	The present study (Table 12 in the Appendix)		
	Diff	Gap (100)	Gap (100)	Diff	Gap (6.25)	Gap (100)
BRN			0.087^{ns}	-0.038^{ns}	-0.054^{ns}	-0.098^{ns}
KHM				-0.016^{***}	-0.007^{***}	0.000
IDN	-0.017^{ns}	-0.041^{ns}	-0.005^{ns}	-0.045^{ns}	0.061^{ns}	-0.106^{ns}
LAO				-0.025^{***}	-0.029^{***}	0.111^{ns}
MYS	-0.118^{***}	-0.105^{***}	-0.096^{***}	-0.050^{***}	-0.084^{***}	-0.087^{***}
MMR				-0.041^{ns}	-0.047^{ns}	-0.034^{ns}
PHL	-0.230^{***}	-0.175^{***}	-0.149^{***}	-0.042^{***}	-0.041^{***}	-0.053^{***}
SGP	-0.015^{ns}	-0.012^{ns}	0.002^{ns}	-0.044^{ns}	-0.072^{ns}	-0.089^{ns}
THA			-0.107^{***}	-0.082^{*}	-0.090^{*}	-0.080^{**}
VNM	-0.297^{**}	-0.169^{ns}	-0.251^{***}	-0.028^{***}	-0.007^{ns}	-0.005^{ns}

Notes Both reference studies applied an Okun equation where the unemployment variable was positioned on the response side of the equation and where cyclical variables for the gap version were estimated by the Hodrick-Prescott filter. The choice of the smoothing parameter is indicated in a conventional format. Whereas the estimates from the reference studies were reported directly either with significances [71, Tables 1 and 4], or they were derived from their reported standard errors [74, Table 1], the significance levels of the Okun coefficients estimated by this study were calculated as approximate Gaussian by confronting average trajectories with their standard deviations as they are reported in Table 12 in the Appendix

Legend Significance labels displayed at Okun coefficients convey the following meaning: *** for p-values ≤ 0.001, ** for p-values ≤ 0.01, * for p-values ≤ 0.05, $^{\bullet}$ for p-values ≤ 0.10, and ^{ns} otherwise

database. This source was also utilized in this subsequent stage to ensure comparability and integrity of the analysis. This explanatory stage could reckon with 310 Okun coefficients for the difference version, and a total of 320 for either variant of the gap version. As expounded in Sect. 3, the procedure of Saxonhouse [80] was applied in order to take into account the degree of (in)significance in Okun coefficients by applying weights inversely proportional to (squared) standard errors produced by the VC method. The regression analysis exploited the panel format of data, relied upon a suitable informative selection of regressors, accounted for the likely endogeneity associated with some regressors, and addressed concerns about the differentiated stationarity features of variables. A few notes are needed in respect of the adopted procedure.

- The initial set of regressors included six metrics of digitalization as presented in Sect. 3, that is, fixed broadband take-up, fixed telephone take-up, mobile cellular take-up, proportion of the population using the internet, secure internet servers, medium and high-tech manufacturing value added to GDP. They were seconded by diverse metrics capturing broad environmental, societal, labour market and economic conditions, that is, share of agricultural land, agriculture/manufacturing/trade relative to GDP, employment share of agriculture/industry/services, total/

youth unemployment, per-capita GDP in 2015, urban population share, population share of megacities, access to electricity of the total/urban/rural population, patent applications. Some other variables were considered as potential instruments, and their list included population density, life expectancy, rule of law, political stability, corruption control, regulatory quality, carbon dioxide emissions, particulate matter emissions, precipitation, humidity, temperature, foreign exchange rate and its change, foreign direct investment to GDP. The admission of variables in each group into the analysis was guided by the requirement of having no less than 200 observations, which assures nominally at least $200/320 = 62.5\%$ of the potential sample size.

- The explanation was performed exclusively in a one-way fixed-effect framework since random effects do not conceptually apply to the present set-up of 10 ASEAN economies.
- The initial set of 22 candidate regressors was reduced to the following 5 regressors: medium and high-tech manufacturing value added to GDP (medhighmanuf), share of agricultural land (agriland_perc), total unemployment rate (unemp_t), employment share of services (emp_serv_t), mobile cellular take-up (mobcell). Their definition is available in Table 10 in the Appendix, and their statistical summary is reported in Table 13 in the Appendix. A general-to-specific search strategy was conducted for each variant of Okun's law separately, where an adequately informative model was identified by minimizing the Bayesian/Schwarz information criterion (BIC). Only linear specifications were sought, and the final specification was determined by taking a union of the regressors resulting from the three searches. This final specification was applied identically for each variant of Okun's law in a non-instrumented fashion by using a weighted version of the within estimator (weighted least squares, WLS), and the estimation was then repeated in an instrumented fashion by using an instrumented analogue of the within estimator (weighted two-stage least squares, W2SLS). Rule of law (ruleoflaw) and political stability (polstab) were used as instruments for total unemployment, whereas foreign direct investment to GDP (fdi) and population density (popdens) served as instruments for employment share of services. It is obvious that the other three regressors, medium and high-tech manufacturing value added to GDP (medhighmanuf), share of agricultural land (agriland_perc), and mobile cellular take-up (mobcell) can be deemed as exogenous as they are not affected by the corresponsiveness of unemployment and output over time or the business cycle.
- Finally, inasmuch as the time-varying features of Okun coefficients in Schlicht's VC framework are facilitated by simple random walk models, the estimated Okun coefficients naturally display non-stationary behaviour. Also some other variables participating in the analysis are suspect for non-stationarity or exhibit it clearly. In order to justify the regression with a mixture of stationary and non-stationary variables, the criterion of validity proposed by Baffes [114] was employed that requires that both residuals and fitted values of the estimated regression models come out as stationary in order to make the regression results valid for the analysis.

The ensuing regression models fitted for the difference version, the gap version with the smoothing parameter $\lambda = 6.25$, and the gap version with the smoothing parameter $\lambda = 100$ in a simplistic fashion by WLS and in an instrumented fashion by WSLS are presented in Table 3. A caveat must be made that these explain *average* patterns in the estimated time-varying Okun coefficients for the heterogeneous block of ASEAN economies, in consequence of which these factors need not apply to each ASEAN economy or their influence may be in fact varied across the economies. As a matter of fact, the non-instrumented estimates are reported for informative purposes, and the relevant estimates here are those yielded by the instrumented models presented in the last three columns. Only two explanatory variables survived the conventional correction for endogeneity bias, that is, medium and high-tech manufacturing value added to GDP (medhighmanuf) and share of agricultural land (agriland_perc). Nonetheless, the utilization of instrumental variables necessitated a reduction of the data sample from 309 observations (non-instrumented models) to 229 observations (instrumented models), but each ASEAN economy remained represented with at least 17 years of observations. Several R-squared metrics are presented and they indicate that fixed effects generally contribute to the explanation of the variability in Okun coefficients, and that in the instrumented models the factors reported in the table have an explanatory power of about 21% for the difference version, and about 29% for the two gap variants. Finally, the test results reported in the last rows of Table 3 vindicate the reasonability of the estimated regression models as at a 0.05 level of significance they pass the criterion formulated by [114].

On the one hand, Table 3 admits some role to two factors in shaping the unemployment-output sensitivity over time (the difference Okun's law) or across the business cycle (the gap Okun's law): medhighmanuf and agriland_perc. Here medhighmanuf is used to proxy the relative size of the digital economy and is found to pronounce the magnitude of the unemployment-output cosensitivity, whereas this exposure is somewhat higher in the course of the business cycle (–0.003, or –0.004) than in terms of year-on-year changes (–0.001). The share of the agricultural land is in some respect a partial indicator that describes the tendency towards agricultural production, but misses differentiated agricultural productivity, which is what it departs from the share of agriculture in the entire economy. Whereas agriland_perc dampens the unemployment-output sensitivity from year to year (+0.001), it amplifies it somewhat over the business cycle (–0.001 or –0.004). On the other hand, the coefficients presented in Table 3, be them statistically significant or not, are all small in substantive terms. For instance, a higher level of medium and high-tech manufacturing by 10% points strengthens on average the Okun coefficient in ASEAN countries by 0.01–0.04, whereas a higher share of agricultural land by 10% points modifies the Okun coefficient from 0.01 upwards to 0.04 downwards. That these considerations are marginal is also seen from Table 13 in the Appendix as medhighmanuf varied for the ASEAN economies over the investigated period between 0.01 and 23.74% (in the 1990s beginning at low levels) and agriland_perc ranged from 0.92 to 46.00%. Unfortunately, mobcell in the instrumented models was found insignificant, hence it is not possible to claim that digital adoption in ASEAN economies could have an impact upon Okun's law.

Table 3 Panel data regressions explaining the temporal variation in the estimated Okun coefficients

Version of Okun's law	Diff		Gap (6.25)		Gap (100)	
Non-instrumented models (WLS)						
medhighmanuf	–0.003***	(0.000)	–0.004***	(0.001)	–0.004***	(0.001)
agriland_perc	0.001**	(0.000)	–0.003**	(0.001)	–0.005***	(0.001)
unemp_t	–0.006**	(0.002)	–0.013***	(0.002)	–0.024***	(0.003)
emp_serv_t	0.000ns	(0.002)	0.002***	(0.001)	0.003***	(0.001)
mobcell	0.001***	(0.000)	0.001***	(0.000)	0.001***	(0.000)
R^2 (with fixed effects)†)	0.240		0.646		0.607	
R^2 (w/o fixed effects)†)	0.180		0.386		0.355	
R^2 (quasi-demeaned)†)	0.059		0.063		0.054	
# observations	309		319		319	
# cross-sections	10		10		10	
# of years	30–31		31–32		31–32	
P-test *p*-value (fitted)‡)	0.024		0.006		0.002	
P-test *p*-value (residuals)‡)	0.007		0.000		0.021	
Z-test *p*-value (fitted)‡)	0.027		0.006		0.001	
Z-test *p*-value (residuals)‡)	0.006		0.000		0.023	
Instrumented models (W2SLS)						
medhighmanuf	–0.001**	(0.001)	–0.003***	(0.001)	–0.004**	(0.002)
agriland_perc	0.001**	(0.000)	–0.001**	(0.000)	–0.004***	(0.002)
unemp_t	–0.007ns	(0.007)	–0.011ns	(0.007)	0.090ns	(0.072)
emp_serv_t	–0.001ns	(0.002)	0.001ns	(0.001)	0.001ns	(0.002)
mobcell	0.001ns	(0.000)	0.001ns	(0.000)	0.001ns	(0.001)
R^2 (with fixed effects)†)	0.208		0.587		0.539	
R^2 (w/o fixed effects)†)	0.209		0.286		0.288	
R^2 (quasi-demeaned)†)	0.144		0.036		0.070	
# observations	229		229		229	
# cross-sections	10		10		10	
# of years	17–24		17–24		17–24	
P-test *p*-value (fitted)‡)	0.000		0.000		0.000	
P-test *p*-value (residuals)‡)	0.021		0.022		0.000	

(continued)

Table 3 (continued)

Version of Okun's law	Diff	Gap (6.25)	Gap (100)
Z-test *p*-value (fitted)‡)	0.000	0.010	0.000
Z-test *p*-value (residuals)‡)	0.007	0.000	0.000

Notes As recommended by Saxonhouse [80], these estimates ensue from using the weighted variant of ordinary least squares (OLS) and two-stage least squares (2SLS) with weights set equal to the (square) reciprocals of standard errors associated with the estimated time-varying Okun coefficients acting here as regressands. †) Each R^2 metric is defined as the ratio of the explained sum of squares to the total sum of squares, but the difference is the definition of predictions (as well as residuals). Whilst "R^2 (with fixed effects)" is derived from predictions that include the contribution of both regressors and fixed effects, "R^2 (w/o fixed effects)" is only partial in the sense that it only accounts for the contribution of regressors. "R^2 (quasi-demeaned)" is a metric that answers to the quasi-demeaned regression that arises in the process of estimating regression parameters and that does not include fixed effects. ‡) In order to justify the estimated regressions in tune with the conditions outlined by Baffes [114], the results of two unit root tests are reported here as developed by Choi [115] for a panel data context, each having unit-root non-stationarity in the null hypothesis. The P-test is the inverse chi-squared test attributable to Maddala and Wu and the Z-test is the inverse normal test attributable originally to Stouffer et al. For both the residuals and fitted values, each implementation applied individual intercepts as exogenous variables with lags for the augmented Dickey-Fuller regression determined by the Bayesian/Schwarz information criterion (BIC) with a limit of two years

Legend Figures in parentheses positioned on the right of the estimated coefficients are conventional standard errors. Significance labels displayed at computed regression coefficients and statistics convey the following meaning: *** for p-values ≤ 0.001, ** for p-values ≤ 0.01, * for p-values ≤ 0.05, $^{\bullet}$ for p-values ≤ 0.10, and ^{ns} otherwise

It is disappointing that little evidence was found that would permit explaining Okun's law in ASEAN countries through digitalization of either their economies or societies. Still, both key metrics that were found of significance in the regression analysis echo consistently in analytical documents and economic briefs in the form of structural transformation and industrialization (medhighmanuf) and the relative importance of agriculture and its issues (agriland_perc).

Albeit the nominal share of agricultural output in GDP trended down over the investigated period in each ASEAN economy [116], there persist notable differences as to the extent with which agriculture contributes to total output creation and employment. For Singapore and Brunei Darussalam, agriculture is currently marginal, contributing to GDP by 0% or 1.3%, whereas for Myanmar and Cambodia takes a lion's share on GDP, contributing by 22.7% or 22.8%, respectively. These numbers were reported for 2021, and when they are combined with agriculture's shares on total employment [117, p. 4] or agriculture's shares on total land, they prove critical disparities in agricultural labour and land productivity. Two ASEAN economies were already commented earlier in this respect: that is, Thailand for its faltering agricultural productivity, and the agriculturally trapped Philippines for its inability to pursue modern agro-industrial policies. Whatever is at the core of difficulties to deliver reasonable productivity of agricultural production—as shared not only by these two ASEAN economies, but possibly others—, this also signals their

Table 4 Fixed broadband take-up (subscriptions per 100 people) in ASEAN and G7 countries

Time frame	1991–1997	1998–2002	2003–2007	2008–2012	2013–2017	2018–2022
Country	7 years	5 years	5 years	5 years	5 years	5 years
ASEAN countries						
Brunei	na	0.69	2.15	5.18	8.01	15.36
Cambodia	na	0.00	0.02	0.18	0.53	1.70
Indonesia	na	0.01	0.10	0.88	1.68	4.06
Laos	na	na	0.01	0.09	0.27	1.56
Malaysia	na	0.05	2.00	7.22	9.28	10.01
Myanmar	na	na	0.01	0.04	0.15	1.25
Philippines	na	0.02	0.23	1.73	2.80	6.48
Singapore	na	2.58	15.31	25.54	27.30	27.02
Thailand	na	0.00	0.88	4.76	9.15	15.57
Viet Nam	na	0.00	0.50	4.00	8.40	17.33
G7 countries						
Canada	na	5.44	20.92	31.52	36.13	40.70
Germany	na	0.84	15.54	33.33	41.28	46.40
France	na	2.26	13.95	31.89	37.52	43.01
Italy	na	0.79	11.05	21.20	24.92	30.05
Japan	na	2.25	17.59	26.37	30.46	34.77
United States	na	0.98	15.68	30.65	37.60	40.57
United Kingdom	na	3.04	16.71	27.14	31.26	35.42

Note These averages are computed from all observations available in the World Development Indicators subset of the World Bank's database (https://databank.worldbank.org) for the time frame announced in the header. If all observations were absent, this is announced by "na"

inability to automatize, robotize, and optimize agricultural production to meet current technological frontiers. Yet, there are tendencies amongst ASEAN farmers to shift from subsistence-oriented to market-based agriculture, whereas economic stimuli and support from authorities may differ from zero interest, as in Laos, to a high degree of political involvement, as in Viet Nam [118].

Once again, albeit the shares of agriculture on national GDP in the ASEAN territory have been on the gradual decline over the past few decades, the trajectories of agricultural land shares have been constantly growing (save in Singapore), which indicates that indeed agriculture is rising in importance as is claimed in official documents (e.g. [117], pp. 1–2). Simultaneously, individual ASEAN economies are known to be not uniform in their digitalization efforts or capabilities, which entails that their agricultural productivity per unit of land may rise, but they still may require a significant portion of their labour force retained in employment for the agricultural sector instead of being released for other sectors of the economy. The constant increase of agricultural land shares may indicate that employment is demanded and conserved for agriculture, and in turn Okunian sensitivity in ASEAN economies is dampened in terms of year-on-year changes. This is especially true for countries like Indonesia, the Philippines, Thailand or Viet Nam that have employment rates in agriculture above 25% [117, p. 4]. However, the trending agricultural land shares in ASEAN economies signal the growing significance of the agricultural sector at the expense of other sectors, mostly those with a higher contribution to value added creation, which explains the heightened Okunian sensitivity across the business cycle.

6 Digitalization Trends in ASEAN and G7 Economies

The results on the factors shaping the unemployment-output relationship in the ASEAN region presented and commented in the previous section are much an outcome of how fast ASEAN countries were capable during the period 1991–2022 of embracing digital technologies in their societies and commit them into economic production. Hence, in order to arrive at a fuller picture of these developmental trajectories, this section takes a comparative perspective and examines trends in digitalization for the ten ASEAN economies (Brunei Darussalam, Cambodia, Indonesia, Laos, Malaysia, Myanmar, the Philippines, Singapore, Thailand, and Viet Nam) juxtaposed with those for the seven G7 countries (Canada, Germany, France, Italy, Japan, the United States, and the United Kingdom). The comparison is effected in a tabular format and breaks down into six tables, out of which each table presents one indicator of digitalization employed in the analysis. For the period 1991–2022, five metrics of the societal dimension of digitalization and one metric of the economic dimension are assessed: fixed broadband take-up (Table 4), fixed telephone take-up (Table 5), mobile cellular take-up (Table 6), proportion of the population using the internet (Table 7), secure internet servers (Table 8), and the share of medium and high-tech manufacturing upon value added creation (Table 9).

The period of 32 years under scrutiny was divided into 6 subperiods of 7 years at the beginning (1991–1997) and 5 years elsewhere till the end (1998–2002, 2003–2007, 2008–2012, 2013–2017, 2018–2022). For each subperiod, the average value of the indicator was calculated. In some cases, there was no data point available during the entire period, which is displayed as "na" (i.e. non-available). Even though the first subperiod lasts 7 years, this is actually no disparity in comparison to the other 5-year subperiods since at the beginning of the three-decade long period, in the early 1990s, digitalization was only taking off, and its trends could only be detectible in the later years.

As all the data employed in the preceding analysis, also the material for Tables 4, 5, 6, 7, 8 and 9 comes from the World Bank's database (https://databank.worldbank.org). The description for four of these indicators displayed under the entry "development relevance" in the metadata section is fairly instructive and proves that their choice for the analysis was highly apposite. This entry states for fixed broadband take-up, fixed telephone take-up, mobile cellular take-up, and secure internet servers: "The quality of an economy's infrastructure, including power and communications, is an

Table 5 Fixed telephone take-up (subscriptions per 100 people) in ASEAN and G7 countries

Time frame	1991–1997	1998–2002	2003–2007	2008–2012	2013–2017	2018–2022
Country	7 years	5 years	5 years	5 years	5 years	5 years
ASEAN countries						
Brunei	21.31	24.96	22.80	20.04	18.00	22.68
Cambodia	0.09	0.24	0.24	2.13	1.80	0.34
Indonesia	1.41	3.07	5.71	14.80	6.85	3.29
Laos	0.30	0.80	1.44	2.78	14.56	18.88
Malaysia	14.60	19.84	17.10	15.92	15.86	23.04
Myanmar	0.33	0.61	0.98	1.01	1.04	0.99
Philippines	1.74	3.78	4.03	3.86	3.30	4.05
Singapore	39.31	46.98	43.20	38.72	36.18	34.44
Thailand	5.05	9.03	10.48	10.06	8.99	7.12
Viet Nam	0.85	3.46	10.48	15.22	6.65	3.41
G7 countries						
Canada	58.93	66.16	59.10	53.48	43.66	33.82
Germany	54.41	57.20	55.26	62.04	59.98	57.54
France	48.30	61.28	66.22	63.84	56.26	46.92
Italy	42.76	46.82	43.44	36.96	33.98	33.16
Japan	48.54	48.58	44.60	48.36	50.16	49.16
United States	49.24	58.02	56.26	53.20	50.24	47.16
United Kingdom	58.34	66.96	58.32	48.08	38.32	30.10

Note These averages are computed from all observations available in the World Development Indicators subset of the World Bank's database (https://databank.worldbank.org) for the time frame announced in the header. If all observations were absent, this is announced by "na"

important element in investment decisions for both domestic and foreign investors. Government effort alone is not enough to meet the need for investments in modern infrastructure; public–private partnerships, especially those involving local providers and financiers, are critical for lowering costs and delivering value for money. In telecommunications, competition in the marketplace, along with sound regulation, is lowering costs, improving quality, and easing access to services around the globe."

Tables 4, 5 and 6 document trends in the access to telecommunication services in the ASEAN region and in the G7 countries. This access has risen on an unprecedented scale since the beginning of the twenty-first century, which is seen in the last four 5-year subperiods. By comparing the trends in Tables 4 and 5 with those in Table 6, it becomes apparent that this growth is discernible chiefly in cellular technologies.

Fixed broadband subscriptions studied in Table 4 refer to subscriptions of high-speed access to the public internet (such as through cable modem, DSL, fiber-optic

Table 6 Mobile cellular take-up (subscriptions per 100 people) in ASEAN and G7 countries

Time frame	1991–1997	1998–2002	2003–2007	2008–2012	2013–2017	2018–2022
Country	7 years	5 years	5 years	5 years	5 years	5 years
ASEAN countries						
Brunei	7.35	30.03	69.46	109.05	116.36	127.25
Cambodia	0.11	1.44	10.03	71.00	129.92	123.64
Indonesia	0.13	2.32	21.90	85.47	138.67	125.11
Laos	0.03	0.42	11.64	59.93	60.86	61.04
Malaysia	4.04	22.93	67.53	117.97	140.35	136.09
Myanmar	0.01	0.05	0.31	2.58	67.83	130.12
Philippines	0.64	9.71	43.84	89.13	110.86	139.27
Singapore	8.80	57.82	102.55	142.10	147.75	149.64
Thailand	1.64	10.17	52.14	105.96	152.20	173.09
Viet Nam	0.05	1.13	19.31	124.05	134.79	142.60
G7 countries						
Canada	7.45	28.23	52.25	74.31	83.47	89.20
Germany	2.94	46.65	79.69	94.75	104.11	113.96
France	4.10	49.27	97.82	118.56	123.98	128.06
Italy	6.75	69.63	124.11	156.35	147.60	133.50
Japan	9.82	51.53	75.40	97.46	126.73	154.98
United States	7.59	61.38	106.93	122.24	120.99	119.36
United Kingdom	10.30	37.77	69.12	91.41	101.25	106.74

Note These averages are computed from all observations available in the World Development Indicators subset of the World Bank's database (https://databank.worldbank.org) for the time frame announced in the header. If all observations were absent, this is announced by "na"

cables, satellite and terrestrial fixed wireless broadband). Each country showed a non-trivial rise in fixed broadband take-up over the period 1998–2022 (as for the first 7-year subperiod the data were available for no country). Yet, ASEAN economies were still lagging behind G7 countries in the last subperiod 2018–2022. In contrast, their starting position was poorer than the fixed broadband take-up in the G7 countries, and the proliferation of fixed broadband was more vehement and compelling.

Whereas fixed telephone subscriptions examined in Table 5 relate especially to analogue fixed telephone lines, mobile cellular subscriptions presented in Table 6 represent a public mobile telephone service based on cellular technology. It might be argued that mobile communications were of significance in the adoption of digital technologies primarily in rural areas. Whereas Japan in 2022 had an 8.0% share of the rural population, Italy had this share at 28.3%, the other G7 countries had this share between 16.9 and 22.4%. The percentage of the rural population was rather stable for the entire period except in Japan where the rural population shrank considerably and the United States where it decreased somewhat slightly. On the contrary, in Cambodia

Table 7 Individuals using the internet (percentage of the population) in ASEAN and G7 countries

Time frame	1991–1997	1998–2002	2003–2007	2008–2012	2013–2017	2018–2022
Country	7 years	5 years	5 years	5 years	5 years	5 years
ASEAN countries						
Brunei	1.31	10.24	34.53	52.85	77.87	95.93
Cambodia	0.00	0.08	0.37	2.07	18.35	54.57
Indonesia	0.04	1.16	3.83	10.51	22.39	53.98
Laos	0.00	0.12	0.87	7.26	18.47	53.10
Malaysia	0.49	19.90	46.64	58.96	70.14	89.82
Myanmar	0.00	0.00	0.10	0.61	11.94	41.96
Philippines	0.03	2.27	5.44	20.00	37.02	55.90
Singapore	3.90	33.68	61.15	70.40	81.57	92.40
Thailand	0.09	4.06	14.44	22.17	40.71	74.91
Viet Nam	0.00	0.70	12.43	30.60	47.13	72.32
G7 countries						
Canada	4.44	46.84	69.50	80.66	89.36	93.34
Germany	1.48	17.25	46.24	75.76	80.70	84.28
France	1.95	28.28	67.34	80.54	85.32	89.58
Italy	0.61	19.45	35.20	51.44	59.32	74.56
Japan	2.35	29.98	64.14	78.04	90.66	87.88
United States	2.20	30.36	68.86	83.98	91.72	94.10
United Kingdom	8.18	43.38	67.68	72.22	78.36	90.34

Note These averages are computed from all observations available in the World Development Indicators subset of the World Bank's database (https://databank.worldbank.org) for the time frame announced in the header. If all observations were absent, this is announced by "na"

the share of the rural population in 2022 was about 74.9%, Myanmar 68.2%, Laos 62.4%, Viet Nam 61.2%, the Philippines 52.0%, Thailand 47.0%, Indonesia 42.1%, Malaysia and Brunei Darussalam around 21–22%, whilst in Singapore there was no rural population. In Malaysia, Laos, Thailand, and Indonesia, the share of the rural population diminished over the period by more than 20% points.

Whereas for the G7 countries fixed telephone subscription increased from the subperiod 1991–1997 to the subperiod 1998–2002, and then followed mostly a decreasing trend (which is striking by looking at the numbers in the first and last columns, respectively), for the ASEAN fixed telephone subscription actually increased over the entire period, although with some less regular and certainly more volatile patterns. For one half of the ASEAN economies fixed telephone take-up was the largest in the last 5-year period (Cambodia, Laos, Myanmar, the Philippines, Viet Nam), for Brunei Darussalam and Singapore this maximum was observed in the subperiod 1998–2002.

Table 8 Secure internet servers (per 1,000,000 people) in ASEAN and G7 countries

Time frame	1991–1997	1998–2002	2003–2007	2008–2012	2013–2017	2018–2022
Country	7 years	5 years	5 years	5 years	5 years	5 years
ASEAN countries						
Brunei	na	na	na	73	628	12,606
Cambodia	na	na	na	2	19	156
Indonesia	na	na	na	3	321	1900
Laos	na	na	na	1	7	92
Malaysia	na	na	na	67	1227	7094
Myanmar	na	na	na	0	7	13
Philippines	na	na	na	8	35	109
Singapore	na	na	na	1120	17,286	144,983
Thailand	na	na	na	19	172	1807
Viet Nam	na	na	na	5	337	3195
G7 countries						
Canada	na	na	na	1690	9047	38,773
Germany	na	na	na	434	5083	34,557
France	na	na	na	1572	11,212	94,227
Italy	na	na	na	207	2109	20,229
Japan	na	na	na	714	2403	21,561
United States	na	na	na	1876	8072	37,617
United Kingdom	na	na	na	3103	11,513	133,640

Note These averages are computed from all observations available in the World Development Indicators subset of the World Bank's database (https://databank.worldbank.org) for the time frame announced in the header. If all observations were absent, this is announced by "na"

Mobile cellular phones were used during the investigated period more intensively both in the G7 countries and in the ASEAN countries. The patterns of increase differed amongst the economies, but the adoption was more spectacular in the ASEAN countries, from virtually zero for Cambodia, Indonesia, Laos, Myanmar, the Philippines, and Viet Nam in the subperiod 1991–1997 to 61 subscriptions per 100 people in Laos and more than 123 subscriptions in the other five ASEAN countries in the subperiod 2018–2022. The most intensive rise in the G7 block was between the first subperiod 1991–1997 and the second subperiod 1998–2002 (between 20.78% points for Canada and 62.88% points for Italy), whilst for the ASEAN region broadly the most dynamic growth in the adoption of cellular phones was most dynamic between the subperiods 2003–2007 and 2008–2012. Save Myanmar where the rise was in fact in later subperiods, for the other countries this growth ran between 39.55% points for Singapore and 104.74% points for Viet Nam). In the last subperiod 2018–2022 the ASEAN countries in mobile cellular phones per 100 people mostly dominated the G7 countries.

Table 9 Medium and high-tech manufacturing upon value added creation (percentage of GDP) in ASEAN and G7 countries

Time frame	1991–1997	1998–2002	2003–2007	2008–2012	2013–2017	2018–2022
Country	7 years	5 years	5 years	5 years	5 years	5 years
ASEAN countries						
Brunei	0.02	0.07	0.06	0.04	0.05	0.22
Cambodia	0.07	0.04	0.05	0.05	0.06	0.07
Indonesia	6.16	7.89	8.71	9.17	7.24	6.15
Laos	0.59	0.86	1.01	0.40	0.30	0.30
Malaysia	13.05	15.30	13.68	9.77	9.60	9.88
Myanmar	2.20	2.20	2.20	3.06	7.66	21.05
Philippines	7.44	9.57	9.98	9.04	8.52	5.62
Singapore	15.41	17.19	19.90	15.05	13.67	16.30
Thailand	7.95	10.48	13.32	12.73	11.20	10.93
Viet Nam	3.84	4.05	4.49	4.15	7.76	9.28
G7 countries						
Canada	6.73	6.97	4.95	3.67	3.60	3.16
Germany	10.34	10.79	11.34	11.35	12.18	11.22
France	7.52	6.37	5.45	4.77	4.89	4.79
Italy	11.44	11.05	11.56	10.78	11.29	11.20
Japan	7.49	6.43	5.72	5.70	6.02	6.25
United States	8.17	7.35	6.36	5.59	5.33	4.81
United Kingdom	6.61	5.50	4.34	4.11	4.06	3.67

Note These averages are computed from all observations available in the World Development Indicators subset of the World Bank's database (https://databank.worldbank.org) for the time frame announced in the header. If all observations were absent, this is announced by "na"

In summary, whilst the G7 economies started off with better fixed and mobile telephone take-up, the ASEAN economies were capable of keeping up the pace with mobile cellular phones that obviously allow easier access to the internet and applications. That said, Table 7 shows that G7 countries had in the last subperiod 2018–2022 more internet users (between 74.56% in Italy and 94.10% in the United States) than ASEAN countries (under Italy's share Myanmar, the Philippines, Laos, Indonesia, Cambodia, Viet Nam, whereas the other four ASEAN economies having between 74.91% for Thailand and 95.93% for Brunei Darussalam).

Concerning the availability of secure internet servers per 1 million people there is a sizeable degree of heterogeneity not only between the two grouping of countries, but also within them. Table 8 shows that in the last subperiod 2018–2022 the leading country was Singapore with 144,983 secure internet servers per 1 million people, whereas the other ASEAN countries varied from 13 servers in Myanmar and 12,606 servers in Brunei Darussalam. In contrast, for the G7 countries, the most intensive coverage by secure internet servers per 1 million people was in the subperiod 2018–2022 in the United Kingdom at 133,640. In that subperiod, France had 94,227 servers per 1 million people and the other five G7 countries ranged between 20,229 and 38,773 servers per 1 million people.

The trends in the only economic aspect of digitalization is shown in Table 9, and it is again a case of immense heterogeneity between and inside the G7 and ASEAN group. Regarding the last subperiod 2018–2022, the leading two countries in the G7 bloc were Germany and Italy that reported an average share of medium and high-tech manufacturing upon value added creation about 11.2% and the other five G7 countries had this share between 3.16% for Canada and 6.25% in Japan. On the contrary, this sector of the economy contributed to value creation in the ASEAN bloc in the last subperiod 2018–2022 most notably in Myanmar by 21.05% and Singapore by 16.30%. Three other ASEAN countries had this contribution around 10%, that is Malaysia, Viet Nam and Thailand, whereas medium and high-tech manufacturing contributed to output in the other five ASEAN countries between 0.07% in Cambodia and 6.15% in Indonesia.

7 Conclusion

This chapter contributes to the extant literature in several respects. It is the first study that studies for all ten member ASEAN countries how unemployment and output *co*vary in time or *co*fluctuate over the business cycle and estimates Okun's law for all of them on a standalone basis. Prior studies such as [71, 74, 112] estimated Okun's law merely for some ASEAN economies and were framed as comparative studies with no special focus upon specific conditions of ASEAN economies. In this, the time-varying framework of Schlicht's VC method is adopted and Okun coefficients are realistically allowed to update with changes in the economic environment. The difference version of Okun's law is applied with two variants of the gap version simultaneously for comparative purposes. The examination of the unemployment-output relationship is extended by a second, regression stage in the hope of explaining the cross-country heterogeneity in the estimated Okun coefficients. The choice of the explanatory factors reflects the original motivation of this study, which is to discover what effects digitalization exerts upon the unemployment-output relationship in ASEAN economies across the business cycle. In addition to conventional labour market, structural and societal explanands, these factors encompass metrics of digitalization in order to capture digitalization tendencies not only in the economy, but also in society. Albeit the motivation was to prove that digitalization affects the intensity with which unemployment and output *co*move across the business cycle—since the ASEAN is open to digital transformation and endorses it—, this ambition was only of limited success.

During the explored period of three decades, 1991–2022, the ASEAN economies were mostly successful in implementing a policy-driven and carefully-implemented structural transformation, where economic growth was promoted by shifting the core axis from labour-intensive sectors to skill-intensive sectors. In this process, the labour force flew from low-productivity labour-intensive activities to higher-productivity skills-intensive activities, and employment was diverted from agriculture to industry and services. Yet, relatively recently it has taken the form of digital transformation.

In spite of the differentiated experience with transformation and structural transition, the ASEAN economies were capable of attaining secular growth rates with only small variation in unemployment throughout the entire period 1991–2022. In consequence, only in three countries Okun's law showed regular patterns for the entire period with each variant; that is, in Cambodia Okun's law was found not to hold, and for Malaysia and the Philippines it was found to be constant. For the other seven ASEAN economies it displayed less stable patterns and frequent insignificance for most of the period, interrupted by consecutive or intermittent years of significant Okun coefficients. At any rate, the estimated values of Okun coefficients are not only unstable in terms of significance, they are also small in comparison to typical estimates for advanced economies, which implies that unemployment and output variations or fluctuations are not *co*sensitive for ASEAN economies or this sensitivity is insubstantial. In consequence, economic growth and output fluctuations are not virtually reflected in unemployment, which is driven by other forces. For policy purposes this may not be an ideal finding since economic policy cannot rely on Okun's law as a standard tool of policy making and planning, but it also means that ASEAN labour markets are fairly stable in terms of their unemployment outcomes. However, this stability is frequently earned through the larger size of the informal sector.

It is rather unfortunate that digitalization of society—proxied through metrics of access to, and use of, elements of the infrastructure transmitting digital services––was not ascertained to be relevant for the explanation of the differences in the estimated Okun coefficients. In the cornucopia of other explanatory factors, two were found relevant and both turn out to be associated with digital transition. One was employed directly as a metric of digitalization of the economy, viz., the share of medium and high-tech manufacturing in GDP. On average, across the ASEAN region, with a higher contribution of medium and high-tech manufacturing to value added Okun's law becomes somewhat stronger and unemployment *co*moves with output more sizably. This indicator captures the size of a segment of the economy that is exposed to intensive structural transformative efforts and digitalization pressures. Another factor is the agricultural land share whose importance in this context stems from the gradual reorientation of agriculture with the aim of higher productivity and sustainability. In all ASEAN countries except Singapore, the share of agricultural land was on the rise during the period, whereas the share of agricultural output to GDP trended downward, with a high proportion of employment still entrenched in the agricultural sector. On average, in the ASEAN region, this increase of agricultural land was found associated with pronouncing the strength of unemployment and output *co*fluctuations across the business cycle, albeit year-on-year mutual changes were found actually feebler. That said, even though the magnitude of these effects is actually low, digitalization of the economy, manifested through its several channels, was found of effect upon the unemployment-output relationship in ASEAN economies.

8 Summary

Integration of digital technologies is a characteristic feature of current economic and social life. Among the countries or groups of countries that significantly benefit from the rapid development of digitalization is the Association of Southeast Asian Nations (ASEAN) consisting of Brunei Darussalam, Cambodia, Indonesia, Laos, Malaysia, Myanmar, Philippines, Singapore, Thailand, and Viet Nam.

Digitalization as a process has manifold impacts upon society and the economy. It changes labour market conditions by affecting labour demand, the organization of labour requirements placed upon employees and even the availability of jobs or their offerings. It also affects total economic production by enforcing sectoral changes and introducing new technical and organizational modes of enterprise operations. In order to understand better the impact of digitalization on the economy, it is necessary to examine its impact on the relationship between the labour markets and the real economy. These components of economic life are isolated here into unemployment and total output, and their relationship (the so-called Okun's law) is studied and matched against digitalization and its trends. Despite much attention being paid to ASEAN countries, research on the relationship between unemployment and output for this region of the world and how it is affected by digitalisation remains limited.

This study addresses the said research gap by focusing on Okun's law across the ten ASEAN countries from 1991–2022 and explaining it, *inter alia*, by digitalization tendencies. It spite of the fact that Okun's law is merely an empirical relationship between unemployment and output changes, it is in fact a standard and globally recognized analytical framework for examining the stimuli going between economic performance and unemployment. A two-stage procedure is applied to that end. First, the time-varying framework of Schlicht's VC method is adopted and Okun coefficients are realistically allowed to update with changes in the economic environment. Then, the examination of the unemployment-output relationship is extended by a second, regression stage in the hope of explaining the cross-country heterogeneity in the estimated Okun coefficients.

Albeit the motivation was to prove that digitalization affects the intensity with which unemployment and output *co*move across the business cycle, this ambition was only of limited success, at least for the 10 ASEAN economies over 1991–2022.

The results indicate that only in three ASEAN countries Okun's law showed regular patterns for the entire period with both the difference and gap version; that is, in Cambodia Okun's law was found not to hold, and for Malaysia and the Philippines it was found to be constant. For the other seven ASEAN economies it displayed less stable patterns and frequent insignificance for most of the period, interrupted by consecutive or intermittent years of significant Okun coefficients. At any rate, the estimated values of Okun coefficients are not only unstable in terms of significance, they are also small in comparison to typical estimates for advanced economies, which implies that unemployment and output variations or fluctuations are not *co*sensitive for ASEAN economies or this sensitivity is insubstantial. It means that economic growth and output fluctuations are not virtually reflected in unemployment, which is driven by other forces.

On the one hand, the results of the regression analysis reveal that digitalization of society—proxied through metrics of access to, and use of, elements of the infrastructure transmitting digital services—was not ascertained to be relevant for the explanation of the differences in the estimated Okun coefficients. On the other hand, two factors associated with digital transition were found relevant, namely the share of medium and high-tech manufacturing in GDP and agricultural land share. On average, across the ASEAN region, with a higher contribution of medium and high-tech manufacturing to value added, Okun's law becomes somewhat stronger and unemployment *co*moves with output more sizably. In all ASEAN countries except Singapore, the share of agricultural land was on the rise during the period, whereas the share of agricultural output to GDP trended downward, with a high proportion of employment still entrenched in the agricultural sector. On average, in the ASEAN region, this increase of agricultural land was found associated with pronouncing the strength of unemployment and output *co*fluctuations across the business cycle, albeit year-on-year mutual changes were found feebler. That said, even though the magnitude of these effects is actually low, digitalization of the economy, manifested through its several channels, was found of effect upon the unemployment-output relationship in ASEAN economies.

Acknowledgements The paper arose in partial fulfilment of tasks linked with VEGA projects 1/0346/25 and 1/0343/23.

Appendix

See Tables 10, 11, 12 and 13.

Table 10 Declaration of key variables and their definition

Notation or code	Definition and units of measurement
g_y †)	Real GDP annual growth rate (%, in constant 2015 US$ prices)
$gap_{y(6.25)}$, $gap_{y(100)}$ †)	Output gap estimated by the Hodrick-Prescott filter as described in the text
Δu †)	Annual unemployment change (% points, to total labour force)
$gap_{u(6.25)}$, $gap_{u(100)}$ †)	Unemployment gap estimated by the Hodrick-Prescott filter as described in the text
medhighmanuf †)	Medium and high-tech manufacturing value added (% of GDP)
agriland_perc	Agricultural land (% of land area)
unemp_t	Unemployment rate (%, to total labour force) ‡)
emp_serv_t	Employment rate in the services sector (%, to employed labour force) ‡)
mobcell	Mobile cellular subscriptions (per 100 people)
fdi	Foreign direct investment (% of GDP, net inflows)
polstab ‡)	(WGI) *political stability and absence of violence/terrorism* index
ruleoflaw ‡)	(WGI) *rule of law* index
popdens	Population density (persons per square kilometre, total population)

Notes The data for the analysis were downloaded in February 2025 from the World Development Indicators fraction of the World Bank's DataBank (https://databank.worldbank.org). Save the few variables identified in the superscript by †), the variables were all applied in directly available measurement units. Amongst the computed variables, medhighmanuf was derived from medium and high-tech manufacturing value added expressed relative to manufacturing value added, and manufacturing value added expressed relative to GDP. The real GDP growth rate g_y was defined in logarithmic terms, the annual unemployment change was computed from unemp_t in absolute terms. The output gaps, $gap_{y(6.25)}$, $gap_{y(100)}$, as well as the unemployment gaps, $gap_{u(6.25)}$, $gap_{u(100)}$, were estimated as cyclical components of the logarithmized real GDP in constant 2015 US$ prices and the unemployment rate expressed relative to total labour force, respectively. The Hodrick-Prescott filter [119] was employed towards that end with backcasting and forecasting the underlying series by 4 years to amend for end-point distortions by adhering to the procedure outlined by Kaiser and Maravall [120]. As indicated by the subscript, two values of the smoothing parameter were used, $\lambda = 6.25$ or $\lambda = 100$. Whereas the latter is conventional as set-up in [97], the former is recommended elsewhere [98]. ‡) WGI stands for the Worldwide Governance Indicators, a project of the World Bank aiming to provide a basis for comparing the quality of governance across the world. Both metrics, polstab and ruleoflaw, range from –2.50 to 2.50 with a higher value suggesting higher stability or better rule of law

Table 11 Descriptive statistics of the unemployment and output variables for the estimation of Okun coefficients

Country	Variable	# observations	Mean	Standard deviation	Minimum	Maximum
BRN	g_y	32	0.444	2.959	–6.889	6.151
	$gap_{y(6.25)}$	32	0.030	1.880	–3.847	4.345
	$gap_{y(100)}$	32	–0.035	2.547	–5.250	5.460
	Δu	31	0.016	0.685	–2.504	0.864
	$gap_{u(6.25)}$	32	0.001	0.381	–1.002	1.061
	$gap_{u(100)}$	32	0.028	0.606	–1.291	1.952
KHM	g_y	32	5.434	6.515	–18.054	12.490
	$gap_{y(6.25)}$	32	–0.102	4.127	–12.602	12.555
	$gap_{y(100)}$	32	0.290	7.339	–16.749	18.875
	Δu	31	–0.018	0.201	–0.575	0.323
	$gap_{u(6.25)}$	32	0.000	0.121	–0.213	0.320
	$gap_{u(100)}$	32	–0.004	0.156	–0.212	0.376
IDN	g_y	32	4.453	3.846	–14.072	7.899
	$gap_{y(6.25)}$	32	–0.283	2.655	–5.624	8.514
	$gap_{y(100)}$	32	–0.047	4.074	–6.715	12.308
	Δu	31	0.027	0.603	–1.102	1.584
	$gap_{u(6.25)}$	32	–0.010	0.321	–0.581	0.682
	$gap_{u(100)}$	32	–0.035	0.520	–0.938	1.289
LAO	g_y	32	6.148	1.768	0.502	8.268
	$gap_{y(6.25)}$	32	–0.264	1.210	–4.639	1.640
	$gap_{y(100)}$	32	–0.102	1.559	–2.910	3.432
	Δu	31	–0.042	0.278	–0.857	0.387
	$gap_{u(6.25)}$	32	0.022	0.113	–0.258	0.384
	$gap_{u(100)}$	32	0.076	0.351	–0.703	0.988
MMR	g_y	32	6.974	5.678	–12.802	12.966
	$gap_{y(6.25)}$	32	–0.035	2.571	–7.135	9.339
	$gap_{y(100)}$	32	0.026	4.321	–8.718	13.892
	Δu	31	0.072	0.619	–1.283	2.864
	$gap_{u(6.25)}$	32	–0.010	0.398	–1.241	1.560
	$gap_{u(100)}$	32	0.008	0.542	–1.309	2.334
MYS	g_y	32	5.149	3.942	–7.644	9.534
	$gap_{y(6.25)}$	32	–0.367	2.709	–6.974	6.100
	$gap_{y(100)}$	32	–0.268	3.606	–6.996	9.390
	Δu	31	0.007	0.404	–0.710	1.280
	$gap_{u(6.25)}$	32	0.020	0.274	–0.502	0.626

(continued)

Table 11 (continued)

Country	Variable	# observations	Mean	Standard deviation	Minimum	Maximum
	$gap_{u(100)}$	32	0.035	0.381	–0.792	0.943
PHL	g_y	32	4.177	3.349	–10.002	7.307
	$gap_{y(6.25)}$	32	–0.271	2.369	–7.790	5.230
	$gap_{y(100)}$	32	–0.189	2.818	–7.208	6.778
	Δu	31	–0.036	0.300	–0.800	0.876
	$gap_{u(6.25)}$	32	0.005	0.188	–0.299	0.716
	$gap_{u(100)}$	32	–0.001	0.241	–0.487	0.747
SGP	g_y	32	5.271	3.899	–3.947	13.558
	$gap_{y(6.25)}$	32	–0.365	2.549	–7.300	3.967
	$gap_{y(100)}$	32	–0.350	3.637	–9.000	7.136
	Δu	31	0.046	0.845	–1.740	1.900
	$gap_{u(6.25)}$	32	–0.005	0.541	–0.898	1.402
	$gap_{u(100)}$	32	–0.026	0.640	–1.140	1.345
THA	g_y	32	3.560	3.878	–7.941	8.212
	$gap_{y(6.25)}$	32	–0.225	2.611	–6.221	6.290
	$gap_{y(100)}$	32	–0.008	4.179	–7.336	10.433
	Δu	31	–0.055	0.598	–1.280	2.531
	$gap_{u(6.25)}$	32	–0.002	0.364	–0.943	1.210
	$gap_{u(100)}$	32	–0.020	0.507	–1.009	1.475
VNM	g_y	32	6.488	1.470	2.522	9.112
	$gap_{y(6.25)}$	32	–0.435	1.676	–6.516	1.816
	$gap_{y(100)}$	32	–0.422	1.682	–4.713	3.466
	Δu	31	–0.020	0.391	–0.862	0.940
	$gap_{u(6.25)}$	32	0.001	0.241	–0.563	0.559
	$gap_{u(100)}$	32	0.003	0.296	–0.499	0.605

Legend The meaning of codes adopted for the variables follows from Table 10

Table 12 Descriptive statistics of the estimated time-varying Okun coefficients

Country	Variable	# observations	Mean	Standard deviation	Minimum	Maximum
BRN	β_{diff}	31	–0.038	0.065	–0.212	0.065
	$sig\beta_{diff}$	2	–0.196	0.023	–0.212	–0.180
	$\beta_{gap(6.25)}$	32	–0.054	0.061	–0.191	0.001
	$sig\beta_{gap(6.25)}$	6	–0.158	0.028	–0.191	–0.117
	$\beta_{gap(100)}$	32	–0.098	0.172	–0.661	0.058
	$sig\beta_{gap(100)}$	6	–0.376	0.198	–0.661	–0.170
KHM	β_{diff}	31	–0.016	0.000	–0.016	–0.016
	$sig\beta_{diff}$	0	NA	NA	NA	NA
	$\beta_{gap(6.25)}$	32	–0.007	0.000	–0.007	–0.007
	$sig\beta_{gap(6.25)}$	0	NA	NA	NA	NA
	$\beta_{gap(100)}$	32	0.000	0.000	0.000	0.000
	$sig\beta_{gap(100)}$	0	NA	NA	NA	NA
IDN	β_{diff}	31	–0.045	0.056	–0.130	0.076
	$sig\beta_{diff}$	11	–0.093	0.057	–0.130	0.076
	$\beta_{gap(6.25)}$	32	0.061	0.085	–0.068	0.191
	$sig\beta_{gap(6.25)}$	1	0.177	NA	0.177	0.177
	$\beta_{gap(100)}$	32	–0.106	0.147	–0.361	0.178
	$sig\beta_{gap(100)}$	8	–0.295	0.053	–0.361	–0.225
LAO	β_{diff}	31	–0.025	0.003	–0.028	–0.021
	$sig\beta_{diff}$	0	NA	NA	NA	NA
	$\beta_{gap(6.25)}$	32	–0.029	0.000	–0.029	–0.029
	$sig\beta_{gap(6.25)}$	0	NA	NA	NA	NA
	$\beta_{gap(100)}$	32	0.111	0.148	–0.117	0.319
	$sig\beta_{gap(100)}$	12	0.280	0.055	0.137	0.319
MMR	β_{diff}	31	–0.041	0.045	–0.177	–0.014
	$sig\beta_{diff}$	6	–0.121	0.046	–0.177	–0.060
	$\beta_{gap(6.25)}$	32	–0.047	0.069	–0.226	–0.005
	$sig\beta_{gap(6.25)}$	5	–0.188	0.047	–0.226	–0.136
	$\beta_{gap(100)}$	32	–0.034	0.062	–0.228	0.006
	$sig\beta_{gap(100)}$	5	–0.158	0.068	–0.228	–0.081
MYS	β_{diff}	31	–0.050	0.000	–0.050	–0.050
	$sig\beta_{diff}$	31	–0.050	0.000	–0.050	–0.050
	$\beta_{gap(6.25)}$	32	–0.084	0.000	–0.084	–0.084
	$sig\beta_{gap(6.25)}$	32	–0.084	0.000	–0.084	–0.084
	$\beta_{gap(100)}$	32	–0.087	0.009	–0.102	–0.076
	$sig\beta_{gap(100)}$	32	–0.087	0.009	–0.102	–0.076
PHL	β_{diff}	31	–0.042	0.000	–0.042	–0.042

(continued)

Table 12 (continued)

Country	Variable	# observations	Mean	Standard deviation	Minimum	Maximum
	$sig\beta_{diff}$	31	–0.042	0.000	–0.042	–0.042
	$\beta_{gap(6.25)}$	32	–0.041	0.000	–0.041	–0.041
	$sig\beta_{gap(6.25)}$	32	–0.041	0.000	–0.041	–0.041
	$\beta_{gap(100)}$	32	–0.053	0.000	–0.053	–0.053
	$sig\beta_{gap(100)}$	32	–0.053	0.000	–0.053	–0.053
SGP	β_{diff}	31	–0.044	0.102	–0.215	0.271
	$sig\beta_{diff}$	9	–0.065	0.158	–0.215	0.271
	$\beta_{gap(6.25)}$	32	–0.072	0.142	–0.353	0.284
	$sig\beta_{gap(6.25)}$	11	–0.128	0.197	–0.353	0.284
	$\beta_{gap(100)}$	32	–0.089	0.128	–0.344	0.181
	$sig\beta_{gap(100)}$	11	–0.190	0.139	–0.344	0.181
THA	β_{diff}	31	–0.082	0.035	–0.186	–0.042
	$sig\beta_{diff}$	11	–0.117	0.035	–0.186	–0.067
	$\beta_{gap(6.25)}$	32	–0.090	0.039	–0.161	–0.022
	$sig\beta_{gap(6.25)}$	19	–0.116	0.022	–0.161	–0.085
	$\beta_{gap(100)}$	32	–0.080	0.029	–0.136	–0.041
	$sig\beta_{gap(100)}$	19	–0.100	0.020	–0.136	–0.072
VNM	β_{diff}	31	–0.028	0.000	–0.028	–0.028
	$sig\beta_{diff}$	31	–0.028	0.000	–0.028	–0.028
	$\beta_{gap(6.25)}$	32	–0.007	0.045	–0.088	0.066
	$sig\beta_{gap(6.25)}$	0	NA	NA	NA	NA
	$\beta_{gap(100)}$	32	–0.005	0.032	–0.071	0.029
	$sig\beta_{gap(100)}$	0	NA	NA	NA	NA

Legend Whereas "β_{diff}", "$\beta_{gap(6.25)}$", and "$\beta_{gap(100)}$" denote Okun coefficients for the difference version and the gap versions with the smoothing parameter either $\lambda = 6.25$ or $\lambda = 100$, respectively, the prefix "sig" refers to only Okun coefficients significant at an approximate 0.05 level of significance. In this, statistically insignificant portions of the estimated trajectories were disregarded, in consequence of which, *inter alia*, the number of observations at "$sig\beta_{diff}$", "$sig\beta_{gap(6.25)}$", and "$sig\beta_{gap(100)}$" states the number of statistically significant Okun coefficients. The symbol "NA" relates to cases when during the entire period the estimated trajectory was found completely statistically insignificant

Table 13 Descriptive statistics of the regressands, regressors and instruments in the panel data regressions used in the explanation of Okun coefficients

Variable	# observations	Mean	Standard deviation	Minimum	Maximum
Estimation by WLS with no heed given to possible endogeneity issues					
β_{diff}	309	–0.043	0.078	–0.449	0.743
$\beta_{gap(6.25)}$	309	–0.038	0.121	–0.776	0.570
$\beta_{gap(100)}$	309	–0.047	0.117	–0.752	0.329
medhighmanuf	309	6.781	5.866	0.016	23.735
agriland_perc	309	22.510	14.256	0.919	45.998
unemp_t	309	2.907	1.939	0.119	9.316
emp_serv_t	309	43.968	20.604	10.512	85.447
mobcell	309	64.210	58.543	0.000	181.767
Estimation by W2SLS in an instrumented manner					
β_{diff}	229	–0.046	0.085	–0.449	0.743
$\beta_{gap(6.25)}$	229	–0.041	0.126	–0.776	0.570
$\beta_{gap(100)}$	229	–0.049	0.130	–0.752	0.329
medhighmanuf	229	7.030	5.702	0.030	20.838
agriland_perc	229	23.864	14.385	0.919	45.998
unemp_t	229	2.842	2.013	0.119	9.316
emp_serv_t	229	45.071	19.703	11.506	85.447
mobcell	229	81.101	55.830	0.017	181.767
ruleoflaw	229	–0.225	0.883	–1.736	1.838
polstab	229	–0.171	0.909	–2.095	1.599
fdi	229	5.327	6.012	–2.757	31.621
popdens	229	862.194	2137.841	21.862	7965.878

Legend The meaning of codes adopted for the variables follows from Table 10. In addition, "β_{diff}", "$\beta_{gap(6.25)}$", and "$\beta_{gap(100)}$" denote Okun coefficients for the difference version and the gap versions with the smoothing parameter either $\lambda = 6.25$ or $\lambda = 100$, respectively

References

1. ASEAN key figures 2024 (2024) Association of Southeast Asian Nations Secretariat. https://asean.org/serial/asean-key-figures-2024
2. Sefrina M (2024) An inclusive digital economy in the ASEAN region. Eria discussion paper series no. 505. Economic Research Institute for ASEAN and East Asia. https://www.eria.org/uploads/An-Inclusive-Digital-Economy-in-the-ASEAN-Region.pdf
3. Google, Temasek, & Bain & Company (2023) e-conomy SEA 2023. https://services.google.com/fh/files/misc/e_conomy_sea_2023_report.pdf
4. O'Grady M (2024) The global digital economy will reach $16.5 trillion and capture 17% of global GDP by 2028. https://www.forrester.com/blogs/the-global-digital-economy-will-reach-16-5-trillion-and-capture-17-of-global-gdp-by-2028
5. Yang ATH (2023) ASEAN's splintering digital economy governance. East Asia Forum. https://eastasiaforum.org/2023/04/20/aseans-splintering-digital-economy-governance
6. APEC (2017) Apec internet and digital economy roadmap. https://www.apec.org/docs/default-source/groups/ecsg/17_csom_006.pdf
7. Hanif A, Abd Rashid A, Roselan MH (2025) Digital development among top 5 ASEAN countries: is there convergence or divergence? Open Res Eur 5(2):1–20. https://doi.org/10.12688/openreseurope.19106.1
8. World Economic Forum (2022) ASEAN digital generation report: digital financial inclusion. World Economic Forum. https://www3.weforum.org/docs/WEF_ASEAN_Digital_Generation_Report_2022.pdf
9. United Nations Development Programme (2024) Structural transformation and labour markets in Asia and the Pacific. United Nations Development Programme. https://www.undp.org/sites/g/files/zskgke326/files/2024-02/Structural%20Transformation%20in%20APAC%20Study-D5.pdf
10. Pradhan RP, Arvin MB, Mittal J, Bahmani S (2016) Relationships between telecommunications infrastructure, capital formation, and economic growth. Int J Technol Manage 70(2/3):157–176. https://doi.org/10.1504/IJTM.2016.075158
11. Pradhan RP, Arvin MB, Hall JH, Bennett SE (2018) Mobile telephony, economic growth, financial development, foreign direct investment, and imports of ICT goods: the case of the G-20 countries. Economia e Politica Industriale 45(2):279–310. https://doi.org/10.1007/s40812-017-0084-7
12. Fossen F, Sorgner A (2019) Mapping the future of occupations: transformative and destructive effects of new digital technologies on jobs. Foresight STI Gov **13**(2):10–18. https://doi.org/10.17323/2500-2597.2019.2.10.18
13. FrontierView (2018) The four key drivers of economic growth in emerging ASEAN. https://frontierview.com/insights/four-key-drivers-economic-growth-emerging-asean/
14. ASEANFocus (2023) No 41 (September 2023 issue), ASEAN Studies Centre, ISEAS-Yusof Ishak Institute
15. Rasjid A (2023) The ASEAN region is the world's economic dark horse. Here's why. World Economic Forum. https://www.weforum.org/stories/2023/08/asean-economic-growth
16. Schlicht E (1985) Isolation and aggregation in economics. Springer
17. Schlicht E (2021) VC: A method for estimating time-varying coefficients in linear models. J Korean Stat Soc 50(4):1164–1196. https://doi.org/10.1007/s42952-021-00110-y
18. Melguizo C (2017) An analysis of Okun's law for the Spanish provinces. Rev Regn Res 37(1):59–90. https://doi.org/10.1007/s10037-016-0110-7
19. Maza A (2022) Regional differences in Okun's law and explanatory factors: some insights from Europe. Int Reg Sci Rev 45(5):555–580. https://doi.org/10.1177/01600176221082309
20. Asia-Pacific Economic Cooperation (2019) APEC economic policy report 2019: Structural reform and the digital economy. https://www.apec.org/publications/2019/12/2019-apec-economic-policy-report

21. Organisation for Economic Co-operation and Development (2020) A roadmap toward a common framework for measuring the digital economy. Report for the G20 digital economy task force. https://www.itu.int/en/ITU-D/Statistics/Documents/publications/OECDRoadmap Digital Economy2020.pdf
22. Barefoot K, Curtis D, Jolliff W, Nicholson JR, Omohundro R (2018) Defining and measuring the digital economy. Bur Econ Anal (BEA). https://www.bea.gov/sites/default/files/papers/defining-and-measuring-the-digital-economy.pdf
23. International Monetary Fund (2021) Singapore: selected issues. IMF Country Reports 21/159. https://www.elibrary.imf.org/view/journals/002/2021/159/002.2021.issue-159-en.xml
24. European Commission (2022) Digital economy and society index (DESI) 2022. Thematic chapters. https://ec.europa.eu/newsroom/dae/redirection/document/88557
25. Vinayak HV, Thompson F, Tonby O (2014) Understanding ASEAN: seven things you need to know. Asia Insur Rev, 1–8
26. Brynjolfsson E, Kahin B (2000) Understanding the digital economy: data, tools, and research. MIT Press
27. van Deursen AJ, van Dijk JA (2014) The digital divide shifts to differences in usage. New Media Soc 16(3):507–526. https://doi.org/10.1177/1461444813487959
28. McKinsey Global Institute (2016) Digital globalization: the new era of global flows. McKinsey & Company. https://www.mckinsey.com/capabilities/mckinsey-digital/our-insights/digital-globalization-the-new-era-of-global-flows
29. Mesenbourg TL, Atrostic BK (2001) Measuring the digital economy: theory and practice. In: Proceedings of the 53rd session, August 22–29, 2001, Seoul. International Statistical Institute.
30. Turban E, King D, Lee JK, Liang TP, Turban DC (2015) Electronic commerce: a managerial and social networks perspective. Springer
31. Organisation for Economic Co-operation and Development (2019) Going digital: shaping policies, improving lives. OECD Publishing. https://www.oecd.org/content/dam/oecd/en/publications/reports/2019/03/going-digital-shaping-policies-improving-lives_g1g9f091/9789264312012-en.pdf
32. Vaňová A, Vitálišová K, Sýkorová K, Rojíková D, Škvareninová D (2025) The digital economy in Slovakia. In: Vinod Kumar TM (ed) Indo-Pacific core and peripheral digital economic communities. Springer, pp 333–375
33. Shome P (2021) Taxation history, theory, law and administration. Springer
34. Organisation for Economic Co-operation and Development (2011) OECD guide to measuring the information society 2011. OECD Publishing. https://doi.org/10.1787/9789264113541-en
35. Eurostat (2021) Glossary: e-commerce—statistics explained. https://ec.europa.eu/eurostat/statistics-explained/index.php?title
36. Dalman C, Mealy S, Wermelinger M (2016) Harnessing the digital economy for developing countries. OECD Development Centre Working Paper No. 334. https://www.oecd.org/en/publications/harnessing-the-digital-economy-for-developing-countries_4adffb24-en.html
37. Lazović V, Duričković T (2014) The digital economy in developing countries: challenges and opportunities. In: 37th International convention on information and communication technology, electronics and microelectronics (MIPRO), Opatija, Croatia, 2014, pp 1580–1585
38. Boccia F, Leonardi R (2016) The challenge of the digital economy. Palgrave Macmillan
39. Xia L, Baghaie S, Sajadi SM (2024) The digital economy: challenges and opportunities in the new era of technology and electronic communications. Ain Shams Eng J 15(2):102411. https://doi.org/10.1016/j.asej.2023.102411
40. Lau LJ (2023) The benefits and potential costs of a digital economy. Telecommun Policy 47(8):102594. https://doi.org/10.1016/j.telpol.2023.102594
41. Raihan A (2024) A review of the potential opportunities and challenges of the digital economy for sustainability. Innov Green Dev 3(4):100174. https://doi.org/10.1016/j.igd.2024.100174
42. United Nations (2019) Digital development: opportunities and challenges. United Nations Conference on Trade and Development. https://unctad.org/system/files/official-document/tdb66_d5_en.pdf

43. Bukht R, Heeks R (2017) Defining, conceptualising and measuring the digital economy. Manchester Centre for Development Informatics Working Paper 68. https://diode.network/wp-content/uploads/2017/08/diwkppr68-diode.pdf
44. Department for Culture, Media and Sport (2016) Digital sector economic estimates. January 2016 statistical release (revised). https://assets.publishing.service.gov.uk/media/5a75a374ed915d6faf2b489a/Digital_Sector_Economic_Estimates_-_January_2016_Revised.pdf
45. Corfe S, Dupont J (2024) State of the UK digital economy. Computer & Communications Industry Association (CCIA). https://ccianet.org/research/reports/uk-digital-economy
46. Okun AM (1970) The political economy of prosperity. The Brookings Institution
47. State Special Committee on Unemployment Problems (1965) The costs of unemployment. In: Okun A (ed) The battle against unemployment. W. W. Norton, pp 7–8
48. Rees A (1965) Dimensions of the employment problem. In: Okun A (ed) The battle against unemployment. W. W. Norton, pp 23–31
49. Okun AM (1962) Potential GNP, its measurement and significance. In: Proceedings of the business and economics. Statistics Section of the American Statistical Association, pp 98–104
50. Fontanari C, Palumbo A, Salvatori C (2020) Potential output in theory and practice: a revision and update of Okun's original method. Struct Chang Econ Dyn 54:247–266. https://doi.org/10.1016/j.strueco.2020.04.008
51. Mankiw NG (2019) Macroeconomics, 10th ed. Worth Publishers/Macmillan Learning
52. Dadkhah K (2009) The evolution of macroeconomic theory and policy. Springer
53. Hoover KD (1988) The new classical macroeconomics. Basil Blackwell
54. Priesol R (2021) Structural macroeconomic model of Slovakia. Econ Anal 56 (Institute of Financial Policy). https://www.mfsr.sk/files/archiv/50/Manual.pdf
55. Hasenzagl T, Pellegrino F, Reichlin L, Ricco G (2022) A model of the Fed's view on inflation. Rev Econ Stat 104(4):686–704. https://doi.org/10.1162/rest_a_00974
56. Hasenzagl T, Pellegrino F, Reichlin L, Ricco G (2022) Monitoring the economy in real time: trends and gaps in real activity and prices. CEPR Discussion Paper No. DP17111. Centre for Economic Policy Research
57. Ball L, Jalles JT, Loungani P (2015) Do forecasters believe in Okun's law? An assessment of unemployment and output forecasts. Int J Forecast 31(1):176–184. https://doi.org/10.1016/j.ijforecast.2014.03.009
58. Lee J (2000) The robustness of Okun's law: evidence from OECD countries. J Macroecon 22(2):331–356. https://doi.org/10.1016/S0164-0704(00)00135-X
59. Sögner L, Stiassny A (2002) An analysis on the structural stability of Okun's law—a cross-country study. Appl Econ 34(14):1775–1787. https://doi.org/10.1080/00036840210124180
60. Knotek ES (2007) How useful is Okun's law? Econ Rev 92(Q IV):73–103
61. Daly MC, Fernald JG, Jordà Ò, Nechio FF (2014) Interpreting deviations from Okun's law. FRBSF Economic Letter 2014–12. Federal Reserve Bank of San Francisco. https://www.frbsf.org/research-and-insights/publications/economic-letter/2014/04/okun-law-deviation-unemployment-recession/
62. Foroni C, Furlanetto F (2022) Explaining deviations from Okun's law. ECB Working Paper No. 2699. European Central Bank. https://www.ecb.europa.eu/pub/pdf/scpwps/ecb.wp2699~a1985f19ed.en.pdf
63. Boďa M (2025) Metodologický príspevok k významu a interpretácii Okunovho zákona [Methodological contribution on the meaning and interpretation of Okun's law]. Slovenská štatistika a demografia [Slovak Statistics and Demography] 35(1):24–40
64. Kiley MT (2013) Output gaps. J Macroecon 37:1–18. https://doi.org/10.1016/j.jmacro.2013.04.002
65. Boďa M, Považanová M (2023) How credible are Okun coefficients? The gap version of Okun's law for G7 economies. Econ Chang Restruct 56(3):1467–1514. https://doi.org/10.1007/s10644-022-09438-9
66. Boďa M, Považanová M (2025) A quarter century of Okun's law in scholarly literature. J Knowl Econ Forthcom Online First. https://doi.org/10.1007/s13132-024-02425-7

67. de Masi P (1997) IMF estimates of potential output: theory and practice. In: Staff studies for the world economic outlook, 1997. International Monetary Fund, pp 40–46
68. Office for Budget Responsibility (2022) Forecasting potential output—the supply side of the economy. Office for Budget Responsibility briefing paper No. 8. https://obr.uk/docs/dlm_uploads/BriefingPaperNo8.pdf
69. Owyang MT, Sekhposyan T (2012) Okun's law over the business cycle: was the great recession all that different? Fed Reserv Bank St Louis Rev 94(5):399–418. https://doi.org/10.20955/r.94.399-418
70. Ball L, Leigh D, Loungani P (2017) Okun's law: Fit at 50? J Money Credit Bank 49(7):1413–1441. https://doi.org/10.1111/jmcb.12420
71. Ball L, Furceri D, Leigh D, Loungani P (2019) Does one law fit all? Cross-country evidence on Okun's law. Open Econ Rev 30(5):841–874. https://doi.org/10.1007/s11079-019-09549-3
72. Boumans M (2019) Econometrics: the Keynes-Tinbergen controversy. In: Dimand RW, Hagemann H (eds) The Elgar Companion to John Maynard Keynes. Edgar Elgar, pp 283–289
73. Gandolfo G (2010) Economic dynamics, 4th ed. Springer
74. Furceri D, Jalles JT, Loungani P (2020) On the determinants of the Okun's law: new evidence from time-varying estimates. Comp Econ Stud 62(4):661–700. https://doi.org/10.1057/s41294-019-00111-1
75. Huang H-C, Lin S-C (2006) A flexible nonlinear inference to Okun's relationship. Appl Econ Lett 13(5):325–331. https://doi.org/10.1080/13504850500398625
76. Grant AL (2018) The great recession and Okun's law. Econ Model 69:291–300. https://doi.org/10.1016/j.econmod.2017.10.002
77. Aguiar-Conraria L, Martins MMF, Soares MJ (2020) Okun's law across time and frequencies. J Econ Dyn Control 116:103897. https://doi.org/10.1016/j.jedc.2020.103897
78. Jalles JT (2019) On the time-varying relationship between unemployment and output: what shapes it? Scott J Polit Econ 66(5):605–630. https://doi.org/10.1111/sjpe.12200
79. Lucchetti RJ, Valentini F (2024) Linear models with time-varying parameters: a comparison of different approaches. Comput Statistics 39(7):3523–3545. https://doi.org/10.1007/s00180-023-01452-3
80. Saxonhouse GR (1976) Estimated parameters as dependent variables. Am Econ Rev 66(1):178–183
81. Lewis JB, Linzer DA (2005) Estimating regression models in which the dependent variable is based on estimates. Polit Anal 13(4):345–364. https://doi.org/10.1093/pan/mpi026
82. Blackley PR (1991) The measurement and determination of Okun's Law: evidence from state economies. J Macroecon 13(4):641–656. https://doi.org/10.1016/S0164-0704(05)80017-5
83. Hilbert M (2020) Digital technology and social change: the digital transformation of society from a historical perspective. Dialogues Clin Neurosci 22(2):189–194. https://doi.org/10.31887/DCNS.2020.22.2/mhilbert
84. Spinola D (2023) Instability constraints and development traps: an empirical analysis of growth cycles and economic volatility in Latin America. CEPAL Rev 2023(139):7–27. https://doi.org/10.18356/16840348-2023-139-1
85. Bührer C, Hagist C (2017) The effect of digitalization on the labor market. In: Ellermann H, Kreutter P, Messner W (eds) The Palgrave handbook of managing continuous business transformation. Palgrave Macmillan, pp 115–137
86. Jaimovich N, Siu HE (2020) Job polarization and jobless recoveries. Rev Econ Stat 102(1):129–147. https://doi.org/10.1162/rest_a_00875
87. Mathy G (2024) Are jobless recoveries history? Okun's law, insufficient stimulus, and slow recoveries. Rev Keynes Econ 12(4):435–452. https://doi.org/10.4337/roke.2024.0002
88. World Economic Forum, & The Boston Consulting Group (2015) Expanding participation and boosting growth: the infrastructure needs of the digital economy. World Economic Forum. https://www3.weforum.org/docs/WEFUSA_DigitalInfrastructure_Report%202015.pdf
89. Pohjola M (2001) Information technology and economic growth: international evidence and implications for economic development. Oxford University Press

90. Hassan AFMK, Islam MR (2007) Temporal causality and dynamics of financial development, trade openness, and economic growth in vector auto regression (VAR) for Bangladesh, 1974–2003: implication for poverty reduction. J Nepal Bus Stud 2(1):1–12. https://doi.org/10.3126/jnbs.v2i1.50
91. Cheng C-Y, Chien M-S, Lee C-C (2021) ICT diffusion, financial development, and economic growth: an international cross-country analysis. Econ Model 94:662–671. https://doi.org/10.1016/j.econmod.2020.02.008
92. Veeramacheneni B, Ekanayake EM, Vogel R (2007) Information technology and economic growth: a causal analysis. Southwest Econ Rev 34(1):75–88
93. Pradhan RP, Arvin MB, Nair M, Bennett SE, Bahmani S (2019) Short-term and long-term dynamics of venture capital and economic growth in a digital economy: a study of European countries. Technol Soc 57:125–134. https://doi.org/10.1016/j.techsoc.2018.11.002
94. Oloyede AA, Faruk N, Noma N, Tebepah E, Nwaulune AK (2023) Measuring the impact of the digital economy in developing countries: a systematic review and meta- analysis. Heliyon 9(7):e17654. https://doi.org/10.1016/j.heliyon.2023.e17654
95. Ehlgen J (1998) Distortionary effects of the optimal Hodrick-Prescott filter. Econ Lett 61(3):345–349. https://doi.org/10.1016/S0165-1765(98)00190-6
96. Flaig G (2015) Why we should use high values for the smoothing parameter of the Hodrick-Prescott filter. Jahrbücher für Nationalökonomie und Statistik 235(6):518–538. https://doi.org/10.1515/jbnst-2015-0602
97. Ravn MO, Uhlig H (2002) On adjusting the Hodrick-Prescott filter for the frequency of observations. Rev Econ Stat 84(2):371–376. https://doi.org/10.1162/003465302317411604
98. Backus DK, Kehoe PJ, Kydland FE (1992) International real business cycles. J Polit Econ 100(4):745–775. https://doi.org/10.1086/261838
99. International Monetary Fund (2023) World economic outlook 2023: a rocky recovery. International Monetary Fund. https://www.imf.org/en/Publications/WEO/Issues/2023/04/11/world-economic-outlook-april-2023
100. Serrano MR (2014) The rise of non-standard employment in selected ASEAN countries: between flexibility and security. ASEAN Services Employees Trade Unions Council (ASETUC)
101. International Labour Organization (2024) Employment challenges and policies in ASEAN (executive summary). Research Brief, September 2024. https://labordoc.ilo.org/permalink/41ILO_INST/j3q9on/alma995382993502676
102. International Monetary Fund (2024) Regional economic outlook notes. Asia and Pacific: Asia-Pacific's structural transformation: the past and prospects. International Monetary Fund. https://www.imf.org/-/media/Files/Publications/REO/APD/2018/apd-reo-full-report.ashx
103. International Monetary Fund (2018) Regional economic outlook. Asia and Pacific: Asia at the forefront: Growth challenges for the next decade and beyond. IMF Country Reports 21/159. https://www.elibrary.imf.org/view/journals/002/2021/159/002.2021.issue-159-en.xml
104. Rahman AA, Schmillen A (2023) From farms to factories and firms—structural transformation and labor productivity growth in Malaysia. Asian Econ Pap 22(1):82–111. https://doi.org/10.1162/asep_a_00862
105. Yi S, Simler K, Loh WS, Tan Wei K (2020) Who is keeping score? Estimating the number of foreign workers in Malaysia. World Bank. http://documents.worldbank.org/curated/en/892721588859396364
106. Dartanto T, Yuan EZW, Sofiyandi Y (2017) Two decades of structural transformation and dynamics of income equality in Indonesia. ADBI Working Paper 783. Asian Development Bank Institute. https://www.adb.org/publications/two-decades-structural-transformation-and-dynamics-income-equality-indonesia
107. Lanzona LA (2022) Philippine structural transformation in the context of technological change. Discussion Paper Series No. 2022–09. Philippine Institute for Development Studies. https://pidswebs.pids.gov.ph/CDN/PUBLICATIONS/pidsdps2209.pdf
108. Briones RM (2025) Is the Philippines in an agricultural trap? Structural change, agricultural productivity and drivers of transformation. Millenn Asia, Forthcoming, Online First. https://doi.org/10.1177/09763996251313868

109. Huang H-C (River), Yeh C-C (2013) Okun's law in panels of countries and states. Appl Econ 45(2):191–199. https://doi.org/10.1080/00036846.2011.597725
110. Bartolucci F, Choudhry MT, Marelli E, Signorelli M (2018) GDP dynamics and unemployment changes in developed and developing countries. Appl Econ 50(31):3338–3356. https://doi.org/10.1080/00036846.2017.1420894
111. An Z, Ghazi T, Gonzalez Prieto N, Ibourk A (2019) Growth and jobs in developing economies: trends and cycles. Open Econ Rev 30(5):875–893. https://doi.org/10.1007/s11079-019-09551-9
112. Widarjono A (2020) Asymmetric Okun's law in an emerging market: ASEAN-3 case. Econ Sociol 13(3):203–217. https://doi.org/10.14254/2071-789X.2020/13-3/12
113. Kim J, Yoon JC, Jei SY (2020) An empirical analysis of Okun's laws in ASEAN using time-varying parameter model. Physica A 540:123068. https://doi.org/10.1016/j.physa.2019.123068
114. Baffes J (1997) Explaining stationary variables with non-stationary regressors. Appl Econ Lett 4(1):69–75. https://doi.org/10.1080/758521836
115. Choi I (2001) Unit root tests for panel data. J Int Money Financ 20(2):249–272. https://doi.org/10.1016/S0261-5606(00)00048-6
116. Astriana F, Tobing AL, Chandra AC (2017) Agriculture in ASEAN: trade and investment guidebook. The Habibie Center
117. EU-ASEAN Business Council (2023) Towards achieving UN SDGs in ASEAN: the role of agriculture. Agriculture paper 2023. https://www.eu-asean.eu/wp-content/uploads/2023/11/2024-Report-Towards-Achieving-UN-SDGs-in-ASEAN_The-Role-of-Agriculture.pdf
118. Burra DD, Parker L, Than NT, Phengsavanh P, Long CTM, Ritzema RS, Sagemueller F, Douxchamps S (2021) Drivers of land use complexity along an agricultural transition gradient in Southeast Asia. Ecol Ind 124:107402. https://doi.org/10.1016/j.ecolind.2021.107402
119. Hodrick RJ, Prescott EC (1997) Postwar U.S. business cycles: an empirical investigation. J Money Credit Bank 29(1):1–16. https://doi.org/10.2307/2953682
120. Kaiser R, Maravall A (1997) Estimation of the business cycle: a modified Hodrick-Prescott filter. SpanEconRev 1(2):175–206. https://doi.org/10.1007/s101080050008
121. Warr P, Suphannachart W (2022) Benign growth: structural transformation and inclusive growth in Thailand. In: Alisjahbana AS, Sen K, Sumner A, Yusuf A (eds) The developer's dilemma. Oxford University Press, pp 88–114

Martin Boďa The scientific and professional expertise of Martin Boďa falls upon intersections of three main knowledge areas, viz. (i) macroeconomics and finance, (ii) applied statistics and econometrics, and (iii) operations research. On the one hand, he works as an academic worker at Matej Bel University in Banská Bystrica, Slovakia (Faculty of Economics) and at Ján Evangelista Purkyně in Ústí nad Labem, Czechia (Faculty of Social and Economic Sciences, Faculty of Natural Sciences). On the other hand, he is the Scientific Secretary and the Vice President for international relations of the Slovak Statistical and Demographic Society (SSDS) as well as an Executive Committee member of the Federation of European National Statistical Societies, Luxembourg (FENStatS). Over the past few years, his research interests have been directed upon three chief fields: 1. macroeconomic empirics focused upon the modelling of Okun's law and business cycle, 2. aspects of financial stability policy with a focus upon financial intermediation, 3. selected aspects of corporate finance (financial distress prediction and dividend policy).

Mariana Považanová is an assistant professor at the Deparment of Economics of the Faculty of Economics, Matej Bel University in Banská Bystrica, Slovakia. Her research area in the field of macroeconomics focuses upon studying short-run economic fluctuations with an emphasis on the use of Okun's law and upon diverse aspects of European economic integration.

Katarína Vitálišová is an associate professor at the Department of Public Economics and Regional Development of the Faculty of Economics, Matej Bel University in Banská Bystrica, Slovakia. She orients her research on issues of public and local governance, on building and maintaining relationships and partnerships between stakeholders, strategic planning of spatial development and implementation of its new approaches, on aspects of creative and smart cities, marketing of places, which is represented by more than 180 publications. She is a member of several international expert networks in the field of regional and local development, management and marketing places.

E-Global Gateways of Digital Economic Communities as Metropolitan Strategies for Anti-fragility: Experiences from Intelligent Spatially and Institutionally Metropolitan Systems Connected by International Economic Linkages

Antonella Contin, Fioralba Vela, Alessandra Maria Pandolfi, Arxhenda Lipovica, Dicheng Yang, Hasibe Şimal Keskin, Javad Kianisadr, Tianduo Yan, Glen Carlo Canlas, Morteza Khajehahsani, Nika Haghpanah, Eugenia Cicero, Prarthana Ulaganathan, and Aafreen Saleem

Abstract The Green and Digital Transition is a significant shift in society, requiring equitable strategies and anticipating socio-economic implications. This chapter advocates for relational urban living, integrating communal existence, regenerative knowledge, and digital tools. The Quintuple Helix Innovation model is used to define the Twin Transition policy, emphasizing digital technology in metropolitan construction and the growing role of intangible components like knowledge, local culture, and

A. Contin (✉) · A. M. Pandolfi · A. Lipovica · D. Yang · J. Kianisadr · T. Yan · G. C. Canlas · M. Khajehahsani · N. Haghpanah · P. Ulaganathan · A. Saleem
Department of Architecture and Urban Studies, Politecnico di Milano, Milano, Italy
e-mail: antonella.contin@polimi.it

A. M. Pandolfi
e-mail: alessandramaria.pandolfi@polimi.it

A. Lipovica
e-mail: arxhenda.lipovica@polimi.it

D. Yang
e-mail: dicheng.yang@polimi.it

J. Kianisadr
e-mail: javad.kianisadr@mail.polimi.it

T. Yan
e-mail: tianduo.yan@mail.polimi.it

G. C. Canlas
e-mail: glencarlo.canlas@mail.polimi.it

M. Khajehahsani
e-mail: morteza.khajehahsani@mail.polimi.it

T. M. Vinod Kumar (ed.), *Indo-Pacific E-Gateways*, Advances in 21st Century Human Settlements, https://doi.org/10.1007/978-981-95-4824-8_3

labour. The Twin Transition concept involves place-based solutions, sustainability, networking opportunities, and technology transfer. The research aims to develop guidelines for sustainable habitats in vulnerable territories in metropolitan regions, focusing on data-driven innovation and strategic partnerships. An 'E-Gateway for digitisation in the metropolitan economy' strategy is proposed to enhance economic potential in Italian urban systems, including digital infrastructure development, skills enhancement, supportive policies, and collaborative partnerships. E-Global Gateways can facilitate global economic integration, enhancing local digital economy participation, and promoting cultural exchange and understanding between countries.

Keywords Twin Transition · Regenerative knowledge · Metropolitan Architecture and Cartography · Anti-fragility and Resilience · Sustainable habitats · Skills enhancement in vulnerable territories

1 Sustainable Habitats in Intelligent Metropolitan Systems Connected by International Economic Linkages

The chapter proposes a new idea of the Green and Digital Transition, which will, as we know, significantly reshape societies in the near future, necessitating equitable strategies and anticipating socio-economic implications. This chapter advocates for a shift towards relational urban living, integrating communal existence, regenerative knowledge, and digital tools. The Quintuple Helix Innovation model is essential for defining the Twin Transition policy, emphasizing the role of digital technology in metropolitan construction and the growing role of intangible components like knowledge, local culture, and labour. The Twin Transition concept involves place-based solutions that foster learning and inter-regional and international synergies, addressing key challenges, developing innovation policies, focusing on sustainability, networking

N. Haghpanah
e-mail: nika.haghpanah@mail.polimi.it

P. Ulaganathan
e-mail: prarthana.ulaganathan@mail.polimi.it

A. Saleem
e-mail: aafreen.saleem@mail.polimi.it

F. Vela
Business Department, University of Vlora "Ismail Qemali", Vlorë, Shqipëri, Albania

H. Ş. Keskin
UN-Habitat, Human Settlements Programme, Nairobi, Kenya
e-mail: hskeskin@kth.se

E. Cicero
MVRDV, Rotterdam, Netherlands
e-mail: eugenia.cicero@mvrdv.com

opportunities, providing smart and digital skills with resources, and focusing on technology transfer. The research aims to develop guidelines for analysing sustainable habitats in vulnerable territories in metropolitan regions, focusing on data-driven innovation and fostering strategic partnerships to address global challenges.

An "E-Gateway for digitisation in the metropolitan economy" is a strategy to enhance economic potential in Italian urban systems. This involves a comprehensive approach that includes digital infrastructure development, digital skills enhancement, supportive policies, and collaborative partnerships. The strategy consists of expanding high-speed internet infrastructure, investing in 5G technology, and deploying smart city technologies. It also supports the adoption of Industry 4.0 technologies, digital platforms, and cloud computing. Digital skills and talent development are also emphasized, with education and training programs, startup incubators, and university collaborations. Government policies and regulations are also crucial, with a comprehensive digital policy framework, robust data protection measures, and regulatory support. Collaborations between government, private sector companies, academia, and research institutions are also encouraged. Digital inclusion and accessibility are also crucial, with broadband access, digital literacy programs, and user-centric design. Performance metrics and feedback mechanisms are also used to measure the impact of digital initiatives.

E-Global Gateways could be digital platforms that facilitate global economic integration, enhancing local digital economy participation. They could provide direct market access, streamline trade processes, serve as innovation hubs, and contribute to digital infrastructure. These platforms can also advocate for policies supporting digital economic integration, influence regulatory frameworks, and develop digital skills for participation in global economic activities. They also promote cultural exchange and understanding between countries, fostering a more interconnected global community. By leveraging these platforms, metropolitan systems can strengthen its position as a competitive player in the worldwide marketplace.

The green and digital transition represents an epochal transformation that calls for strategies capable of addressing social and economic inequalities. This chapter proposes a new vision of sustainable living in *Intelligent Metropolises*, focusing on the role of metropolitan cities as strategic spaces in which to root federated digital infrastructures and build inclusive and resilient development models. The E-Gateway concept, in this perspective, is not reduced to a technical infrastructure, but becomes a relational tool connecting territories, urban practices, and markets through what we call metropolitan anthropology: a situated knowledge of communities, local knowledge, and new forms of urban coexistence.

Throughout the different chapters, the text explores sustainable housing systems (Chap. 1), European joint digitization strategies in the Indo-Pacific corridors (Chap. 2), the construction of digital economic communities (Chap. 3), the adaptation of smart specialization strategies in the Western Balkans, especially in Albania (Chap. 4), and a series of international case studies (Chaps. 5, 6) showing how local projects can contribute to the construction of global economic and digital corridors.

The proposed vision values the metropolitan scale as the key to integrating multilevel governance, civic innovation, sustainability, and digital sovereignty. The E-Gateway thus becomes a cognitive infrastructure, capable of promoting not only economic but also cultural, institutional and anthropological exchanges, making Europe the protagonist of a digital transformation oriented towards the common good.

The chapter addresses, in fact, a new idea of sustainable habitats in Intelligent spatially and institutionally Metropolitan systems connected by international economic linkages through:

1. *Sustainable Habitats in Intelligent Metropolitan systems Connected by International Economic Linkages*

   Antonella Contin, Dicheng Yang

The Twin Transition will significantly impact the future, necessitating policymakers to address existing inequalities and socioeconomic effects. Strategies embracing de-risking and anti-fragility principles are implemented to improve community well-being, conserve local identity, and preserve natural heritage. Anti-fragility, a concept by philosopher Nassim Nicholas Taleb [1], promotes growth and improvement by embracing instability and uncertainty. This approach can be applied to personal development, business strategy, risk management, and policy making. By embracing Anti-fragility, metropolitan areas can transform challenges into opportunities for growth and sustainability. The paradoxical moment of stasis and acceleration, creative surplus and nostalgia, and the hybrid nature of human and artificial capabilities is crucial.

Religion and technology, particularly AI, are not separate entities. Technology is a fundamental part of human existence, as seen in the Bible and Jesus' creation of artefacts. AI profoundly impacts anthropology, social organization, truth, identity, and faith. The Church must acknowledge the human transformed by technology, recognizing its creative potential. AI can be used to order, divide, and exercise power, but it must also seek the other, generate relationships, and protect humans.

2. *E-Gateway Projects and Strategies in Europe for the Joint Development Strategies in the Indo-Pacific Corridors*

   Antonella Contin, Alessandra Pandolfi

The European Union Global Gateway strategy aims to enhance sustainable connectivity between Europe and the Indo-Pacific region, promoting democratic values, transparency, and environmental sustainability. Launched in 2021, it focuses on digital infrastructure, energy, and transportation sectors. The EU is raising up to 300 billion EUR in investments for infrastructure development between 2021 and 2027, aligning with the UN Sustainable Development Goals and Agenda 2030. The strategy aims to build resilient partnerships, support green initiatives, and enhance connectivity and infrastructure development. The EU is also funding multilateral projects in Asia, including the ASEAN Green Initiative and the ASEAN Team Europe Initiative on Sustainable Connectivity. However, concerns have been raised about the EU commitment to sustainable and transparent investment and its reliance on Chinese supply chains and markets.

Digital Economic Global Enterprise Ecosystem Designs and Digital Economic Global Governance aim to transform economic communities into digital ones. International agencies and regional groups promote this, focusing on strengthening global digital infrastructure and encouraging values-based investment. Local community projects and embracing digital tax and anti-corruption are also crucial. Addressing social conflict and inclusion in metropolises is essential.

Metropolises are not just commercial spaces but aggregation centres, determining complex interconnected markets on different scales. The real market value is linked to the need to anchor itself to each city local dimension, representing its complex market. The proposal is to understand how to build a network of markets at different scales, from micro to macro, and compare them based on their social, economic, and environmental impacts on the lives of citizens. The EU must take a central role in defining the international metropolitan community, in collaboration with UN-Habitat. To build an ambitious urban agenda, the EU should give the urban and metropolitan dimension a prominent role in cohesion programmes, define the next EU budget, and outline a European framework for metropolitan governance.

3. *Digital Economic Communities for the E-Gateway Strategies in Europe*

Antonella Contin, Fioralba Vela, Alessandra Pandolfi

Intelligent Metropolitan systems should adopt new strategies to support diversity and coexistence, such as establishing green and sustainable spatial connections, integrating historical and modern levels in urban plans, and improving the quality of spaces. Digital economic communities, which involve networks of businesses, institutions, cities, and regional networks, can collaborate through digital platforms to create, exchange, and scale economic value across borders. An E-Gateway Strategy envisions Europe as a global digital logistics and innovation hub, acting as a gateway for goods, knowledge, services, technologies, and digital products. This approach could accelerate smart, sustainable trade, foster innovation ecosystems across regions, ensure green and inclusive growth, increase Europe global competitiveness, and promote strategic autonomy.

Federated digital infrastructures and distributed platforms for logistics, data sharing, innovation exchange, and smart contracts are essential for this approach. City-region digital twins and data hubs can optimize transport, energy, and environmental management, while cross-border innovation networks linking startups, universities, and innovation hubs across different cities into thematic communities can be implemented. Multimodal smart logistics corridors and digital Green Deals for local economies can also be implemented to achieve the Green Deal and Digital Europe Programme objectives.

Digital Social Hubs can help Intelligent Metropolitan systems develop proper digital economic communities, involving younger generations in the process of creating community identity and shared values. Technology transfer initiatives supporting research as the primary engine of long-term, sustainable economic growth can contribute to the spread of sustainable solutions and increase the resilience and competitiveness of regions.

4. *Territorialization of Smart Specialization Strategies in Albania*

Antonella Contin, Fioralba Vela, Alessandra Pandolfi, Arxhenda Lipovica

The European Commission has adopted the Smart Specialization Strategy (S3) framework to identify regions with significant growth potential in research and innovation. The strategy focuses on investments in areas where a region can achieve knowledge-driven growth and gain a competitive edge. Key components include a place-based approach, interactive priority-setting, a broad view of innovation, and a monitoring and evaluation system. The West Balkan Region has identified technology in education, sustainable city solutions, and smart communities as key areas for smart specialization initiatives. Albania is preparing for the S3 process, with priority areas identified through a qualitative mapping exercise and quantitative mapping of nation potential. Smart specialization involves using intelligent functions and decision-supporting simulations to adapt to changing systems and improve prosperity, community, and quality of life.

5-6. *Case Studies in Italy, Albania, China, Philippines, Canada, Argentina*

Simal Keskin, Javad Kianisadr, Tianduo Yan, Glen Canlas, Morteza Khajehahsani, Nika Haghpanah, Eugenia Cicero, Prarthana Ulaganathan, Aafreen Saleem

Indo-pacific policies are crucial for the economic corridor and the EU, as they will change the balance in world economic sectors. Case studies from Italy, Italy, and the Philippines highlight the challenges faced by agricultural communities in rapidly urbanizing areas. The Melzo AgriHub project aims to integrate green and digital innovations into a unified strategy, focusing on ecological farming techniques and digital platforms. Piacenza, Italy, proposes a holistic landscape strategy through Sustainable Urban Drainage Systems to redefine the urban-river relationship. Resilient landscape design approaches, such as those used in China and India, can help reduce the impact of unstable water levels on ecological sites and create a multi-dimensional landscape corridor. In the Philippines, "May Nilad" aims to redefine resiliency for coastal communities through aqua silviculture using an E-gateway approach.

Several Americas countries, including Canada, the USA, and South America, have developed their own national Indo-Pacific policies to enhance international collaboration with the region. The Indo-Pacific region is a global hub of economic dynamism, impacting Canadian national security, economic prosperity, and human rights. Canada Indo-Pacific Strategy (2022) aims to deepen relationships with the region and promote peace, resilience, and security. The region makes up over one-third of global economic activity and will account for over half of the global economy by 2040. Latin America faces the challenge of maintaining diplomatic balance between China and the US, and a Latin American vision for the Indo-Pacific could prompt states to rethink regional schemes of integration and cooperation. Argentina has relaunched in 2016 its relationship with the Asia–Pacific region.

1. **Twin Transition and Anti-fragility**

The Twin Transition will significantly impact the future; thus, policymakers must recognize existing inequalities and their socioeconomic effects, starting from existing policies that might worsen existing disparities or lead to new ones. Strategies that embrace de-risking and anti-fragility principles are being implemented to combat this. These policies seek to improve community well-being, conserve local identity, and save natural heritage. Additionally, the study explores how the Twin Transition intersects with existing disparities in the Far East, Mediterranean, and Latino-American metropolitan areas. This topic will be investigated using cutting-edge mapping methods, and research findings will be shared with the public and decision-makers.

Nassim Nicholas Taleb is a philosopher and author specializing in studying randomness, probability, and uncertainty. He believes that society needs to understand and develop strategies to navigate uncertainty, which is central to his concept of 'anti-fragility' [2]. It refers to systems, organizations, or entities that not only tolerate stress, chaos, and volatility, but also benefit and thrive from such challenges. Unlike resilience, which is the ability to recover from adversity, anti-fragility thrives and improves because of it.

Practical applications of the concept of antifragility start from personal development, which is related to accepting challenges and learning from setbacks. They can help you grow and become more resilient. In terms of business strategy, companies can promote innovation and adaptation by promoting experimentation and learning from market disruptions. Passing to risk management, implementing strong methods in this sense, such as redundancy, diversification, and flexible response tactics, can help a company become more resilient. Other possible applications include policymaking, thanks to which governments and organizations can create policies that enable flexibility and learning from societal and economic changes, encouraging long-term resilience and growth. Anti-fragility can also be considered for its benefits, starting from increased resilience, as anti-fragile systems recover from shocks and emerge stronger, resulting in increased overall resilience. Plus, by thriving in the face of volatility and change, antifragile systems are better suited for long-term sustainability. In terms of continual progress, regular exposure to problems and the ability to adapt result in continual improvement and innovation.

Anti-fragility is about using instability and uncertainty to promote growth and improvement. It promotes a proactive attitude to coping with problems, converting potential weaknesses into strengths. Applying the concept of Antifragility to metropolitan areas involves creating urban environments that not only withstand but also thrive amid challenges such as economic shocks, natural disasters, and social changes. By embracing anti-fragility, metropolitan areas can transform potential challenges into opportunities for growth, ensuring they remain vibrant, resilient, and sustainable in the face of future uncertainties.

2. Human, AI and the Need for a Structured Thought

Looking at artificial intelligence from a cultural point of view means, first, not taking the centrality of the human being for granted. We must not start with the idea that our role is guaranteed but instead ask ourselves what new horizons are opening. AI is not a future hypo-project: it is already inside our lives much more than we often imagine. Accepting this fact also means abandoning comfort and placing ourselves in a productive discomfort capable of stimulating critical thinking.

We are in a paradoxical moment: stasis and acceleration, creative surplus and nostalgia, progress and collapse. In this context, it no longer makes sense to rigidly distinguish between human and artificial: the availabilities on the table are hybrid, and the challenge is to understand how to use them, integrate them, and co-inhabit them.

Even Pope Francis, in his speech at the G7, recalled that the human being is the creator of tools; he is pushed towards the beyond, towards the other. But it is precisely this tension that risks disappearing if organizations and institutions do not immerse themselves in the changes of the present—a present that accelerates, fragments, and puts in crisis. And it is precisely in that fracture that we should decide to live.

Considering religion, technology, and the illusion of separateness, at first glance, religion and new technologies—especially AI—seem to belong to separate worlds. Yet this is not the case.

Technology is a structural part of humans; it has always been in the Bible. Salvation comes with Noah's Ark, a machine. Today, AI has a profound impact on anthropology, social organization, truth, identity, and faith.

For this reason, the Church, which believes in a God made man, cannot ignore the human transformed by technology. The first approach is often one of amazement: "Look what these tools can do!". But immediately afterwards the defensive reaction is triggered: "We humans are different, we are creative". And yet, this is where the cultural stakes come into play: what does it mean to admit forms of creativity and agency in machines? Does it belittle the human or, perhaps, relocate it in a more honest way?

It could be a healthy exercise in radical humility—humility not as submission, but as the ability to generate the other. We are on the edge of the cosmos, peripheral, yet we are the ones who think about it. As Paul VI said when man landed on the moon, we have accomplished something great because we desire the beyond.

3. The Myth of Technological Neutrality

Do data lie? Or, worse, do they turn us into means rather than ends? In the face of these technologies, we oscillate between wonder and defence. But accepting the challenge means rethinking our position in the world.

Like any technology, artificial intelligence is not neutral. It is a techno-socio-political product, that expresses the values of the society that creates it. Whenever a technology presents itself as aseptic, it is highly political.

Paul VI again: "The greatness of human beings lies in thinking." Not in power, but in thought as a practice of fragility, as openness to the other. The fundamental difference between humans and machine is not intelligence but thought: the ability to remain in one's own vulnerability and make it the engine of otherness.

Today, AI is often used to order, divide, and exercise power. However, it can also seek the other, generate relationships, and protect the humanity. Ultimately, it depends on us. No technology is self-justifying; all must be governed with transparency, accountability and a democratic vision.

Considering exploration as a human horizon, the narrative of space is emblematic: a mission is accomplished not when you land but when you return alive, conscious, and happy. This is how AI should be: not just efficiency but visionary.

Yes, every artefact solves a glitch. But it must also protect humans. It must be oriented towards another horizon, high and distant: a prophecy, not a shortcut.

Darwin wrote in The Origin of Species that life comprises "infinite beautiful forms". A fragile beauty: 99% of species have become extinct. This vision is not easy to accept. There is no finality, no centre; we are not special.

And yet, we are the ones who observe the universe, who discover laws, who question ourselves. The hypo-project of the Red Queen of Carroll teaches us that everything evolves because everything co-evolves; it is a relentless race. The other changes us while we change. This is the meaning of evolution, of the human, of thought.

4. A New Possible Approach to Promote the Diffusion of E-Gateways Communities Starting from Real Experiences of Green and Digital Transition Management Projects

The chapter proposes a Green and Digital Transition, a shift towards relational urban living that integrates communal existence, regenerative knowledge, and digital tools. The Twin Transition policy emphasizes digital technology in metropolitan construction and the growing role of intangible components like knowledge, local culture, and labor. The research aims to develop guidelines for sustainable habitats in vulnerable territories in metropolitan regions, focusing on data-driven innovation and strategic partnerships. An "E-Gateway for digitisation in the metropolitan economy" strategy is proposed to enhance economic potential in Italian urban systems, including digital infrastructure development, skills enhancement, supportive policies, and collaborative partnerships. E-Global Gateways are digital platforms that facilitate global economic integration, enhancing local digital economy participation and strengthening metropolitan systems' competitiveness in the global marketplace.

The proposed vision for Europe future focuses on integrating multi-level governance, civic innovation, sustainability, and digital sovereignty at the metropolitan scale. The E-Gateway will serve as a cognitive infrastructure, promoting economic, cultural, institutional, and anthropological exchanges. The Twin Transition will significantly impact the future, necessitating policymakers to address existing inequalities and socioeconomic effects. Strategies embracing de-risking and anti-fragility principles are being implemented to improve community well-being, conserve local identity, and preserve natural heritage. The study will investigate the intersection of the Twin Transition with disparities in the Far East, Mediterranean, and Latino-American metropolitan areas. Anti-fragility, a concept by Nassim Nicholas

Taleb, refers to systems that thrive from stress and uncertainty, promoting growth and improvement. By applying anti-fragility to metropolitan areas, cities can transform potential challenges into opportunities for growth, ensuring their vibrant, resilient, and sustainable future.

The concept of developing an AI-powered cartographic tool model to map territorial assets, focusing on regional cohesion and sustainable development, is emerging from case studies. This multidisciplinary approach integrates historical, political, economic, environmental, and social data, providing a place-based approach for regional planning and policy-making decision-making. The approach uses Metropolitan Cartography, Development and E-gateway Strategies, and the Quintuple Helix Model to generate synthetic maps, offering a holistic view of territorial interactions.

The multidisciplinary nature of the approach fosters cross-sectoral insights, benefiting public authorities, urban and regional planners, researchers, academic institutions, and local communities. Potential beneficiaries include public authorities and policymakers, urban and regional planners, researchers, and local communities, who benefit from improved governance and sustainability planning.

This approach not only enhances territorial knowledge and decision-making but also serves as a model for future applications in regional policies, addressing a complex, multidisciplinary challenge with high innovation potential. It could deliver innovative, multidisciplinary results that align with the objectives of EU territorial cohesion, sustainable development, and data-driven policymaking.

The multi-scale approach, focused on sustainability and the enhancement of cultural and environmental capital, represents a replicable model for other metropolitan contexts, as it goes far beyond mapping and is a proactive system, integrating the physical dimension with the virtual one. This becomes even more important in the light of climate change, which affects economic communities and landscape architects are tasked with proposing interventions to improve city conditions.

The European Digital Public Goods represent shared, scalable, and democratically compliant infrastructures that can drive digitization in partner countries in an inclusive and sustainable way. The role of SMEs, support of public institutions, and integration of local communities are indispensable elements to make the EU not only an exporter of technologies but a builder of collaborative and resilient platforms.

The European Union faces the strategic challenge of building a single, inclusive, and interoperable digital identity, with metropolitan cities emerging as central hubs to implement a federated vision of the European digital infrastructure. A federated metropolitan network would promote a culture of interoperability and facilitate the secure circulation of digital identities between territories, avoiding duplication, enhancing transparency, and reducing administrative costs.

Multi-level and participatory metropolitan governance, based on the quintuple helix of innovation (environment, citizens, research, industry, and institutions), finds the space to be practiced in metropolitan cities, becoming a laboratory and platform for a polycentric digital ecosystem capable of integrating public and private resources, technological infrastructures, and social capital in the construction of an equitable, open, and resilient European digital identity. In this scenario, digital diplomacy, a distinctive feature of Europe, emerges as a fundamental pillar for science diplomacy.

2 E-Gateway Projects and Strategies in Europe for the Joint Development Strategies in the Indo-Pacific Corridors

The European Union Global Gateway strategy is a comprehensive initiative to enhance sustainable, secure, and rules-based connectivity between Europe and the Indo-Pacific region. Launched in 2021, it counterbalances other global infrastructure initiatives by promoting democratic values, transparency, and environmental sustainability [3]. In addition to bolstering health, education, and research institutions, it focuses on projects in essential industries including digital infrastructure, energy, and transportation. With a more open and sustainable approach to infrastructure development, this project seeks to serve as a global substitute for the Chinese Belt and Road Initiative.

On October 25–26, 2023, the EU hosted its inaugural Global Gateway Forum in Brussels, drawing 40 leaders, ministers, and officials from more than 20 partner nations. With a focus on international infrastructure investment, the forum signed renewable energy finance agreements with Asian nations such as Bangladesh, the Philippines, and Vietnam. The EU Global Gateway policy seeks to improve smart, clean, and secure links in the digital, energy, and transportation sectors to strengthen global health, education, research, and innovation. The approach, which focuses on creating long-lasting relationships, has been expanded to encompass Latin America and Africa [3]. Through its Team Europe strategy, the EU is raising up to 300 billion EUR in investments for infrastructure development between 2021 and 2027. The Global Gateway supports principles that align with the Sustainable Development Goals and Agenda 2030 of the UN, including human rights, the rule of law, and international norms and standards. The forum allowed leaders to talk about potential collaborations on global infrastructure projects and promote tangible action on the ground. Additionally, the EU is funding multilateral projects in Asia, including the ASEAN Green Initiative and the ASEAN Team Europe Initiative on Sustainable Connectivity.

The strategic objectives of the Global Gateway strategy align with the EU broader Indo-Pacific Strategy, focusing on:

- Promoting sustainable and inclusive prosperity,
- Advancing the green transition and digital governance,
- Enhancing connectivity and infrastructure development,
- Strengthening security and defence cooperation,
- Fostering human security and resilience.

Through these initiatives, the EU aims to build resilient partnerships that support peace, prosperity, and sustainable development in the Indo-Pacific region. i.e. The EU is expanding its network of digital partnerships with Indo-Pacific nations, including potential Digital Partnership Agreements with countries like Japan, South Korea, and Singapore. Additionally, the EU is supporting green initiatives such as the Just Energy Transition Partnership (JETP) with Vietnam, providing a €500 million credit facility to aid in the country decarbonization efforts.

In December 2022, the EU and ASEAN launched the Sustainable Connectivity Team Europe Initiative (SCOPE TEI; see Fig. 1), a €60 million program focusing on sustainable transport, energy, and digital connections within Southeast Asia. Notable projects include:

- Upgrading roads across Laos, Thailand, and Vietnam,
- Co-funding a new metro line in Jakarta alongside Japan,
- Establishing the first national Copernicus data centre in the Philippines to enhance environmental and disaster monitoring.

The Chinese government introduced the Belt and Road Initiative (BRI), a global infrastructure and economic development plan, in 2013. The Silk Road Economic Belt and the 21st Century Maritime Silk Road are its two constituent parts. As of September 2023, the initiative had expanded quickly to include 154 countries. However, issues with transparency, geopolitical ramifications, environmental impact, and debt-trap diplomacy have led to criticism.

Because of the EU dedication to sustainable and transparent investment, the Global Gateway initiative, which seeks to offer high-quality investments without compromising national sovereignty, is regarded as challenging. Additionally, the EU is working to lessen its excessive reliance on Chinese supply chains and markets.

With an additional 3 billion EUR in new agreements, the Global Gateway Forum 2023 sought to foster communication between stakeholders and government leaders about infrastructure investment. Concerns were expressed, meanwhile, about the EU countries' dedication to the forum as well as the dearth of data and policies pertaining to the actual operation and execution of the EU side.

The EU should consider strengthening ties with Japan, a longtime supporter of regional infrastructure development, to better carry out its infrastructure initiatives. Japanese considerable significance in this field is shown by its substantial commitment, which is demonstrated by the 259 billion USD in ongoing projects throughout Southeast Asia. To coordinate their overlapping endeavours in Bangladesh, the Philippines, and Vietnam in areas like digital infrastructure and the transition to green energy, the EU and Japan should strengthen their governmental cooperation channels.

In brief, the EU should advance and strengthen the efficiency of its Global Gateway projects in the Indo-Pacific, including by coordinating initiatives with other partners and enhancing private sector involvement. This becomes even more crucial since several of the top economies in the world have created national Indo-Pacific policies to interact with the most economically active area of the coming century. Whether

Fig. 1 EU-Indo-Pacific flagship projects. *Source* eeas.europa.eu[1]

their nation is inside or outside the Indo-Pacific, these strategies aim to boost their respective economies. Therefore, encouraging the designs of International Corridors and E-Global Gateways of Digital Economic Communities for Indo-Pacific Smart Metropolitan Systems could contribute to the success of the entire system. Digital economic communities' e-global gateways are cutting-edge electronic platforms that promote international economic integration. Any specialized service platform, as

[1] See https://www.eeas.europa.eu/sites/default/files/documents/2024/EU-indo-pacific-flag-proj-2023-24-v09-no-borders-no-highlightsV03.pdf.

well as comparable services for logistics or small and medium-sized businesses, can connect to this networked platform.

Digital Economic Global Enterprise Ecosystem Designs and Digital Economic Global Governance are two notable accomplishments of a digital community approach, that serve as the foundation for this gateway. By utilizing access to digital infrastructure and public goods, the architecture of the digital economic global enterprise ecosystem focuses on transforming economic communities into digital economic communities. International agencies like the World Trade Organization and regional groups like ASEAN promote this. To increase economic opportunities, the plan entails locating and funding itself the gaps in the physical and digital public infrastructure. Strengthening global digital infrastructure and encouraging values-based investment are the goals of the 2030 Digital Compass, Africa-Europe Investment Package, Digital Sovereignty and Human-Centric Approach, Global Gateway Initiative, and other E-Global Gateways projects. Adoption of national or subnational/federal state laws such as the Digital Services Act, Digital Markets Act, Data Act, Artificial Intelligence Act, and Media Freedom Act is part of the global governance design. Local community projects like I2U2 and QUAD can be carried out locally. Digitally embracing tax and anti-corruption, encouraging fair competition, and combating tax evasion and corruption constitute the fourth pillar of the Indo-Pacific Economic framework.

1. **Towards a Metropolitan Urban Agenda: Cohesion, Inclusion and Development**

Recognising the metropolitan scale is not only imperative for the localisation of the SDGs, as established in the resolution of June 2023, but it is also an essential condition for promoting territorial cohesion based on an ambitious urban agenda. This must identify and support those who cannot sustain the transition, encourage inclusion, and contribute to the definition of the EU strategic objectives, which must align with the new geopolitical situation.

Considering social conflict and inclusion, metropolises are characterised by social conflict, which can be addressed through a multiethnic society founded on shared principles and values, developed through institutions that guarantee the equitable distribution of social services. The academy can contribute by providing education for the metropolitan common good, teaching a common language between disciplines and administrative sectors, and building a collective space of cohesion based on a cooperative rather than a confrontational vision.

The EU 'fifth freedom', aimed at strengthening research, innovation, and education in the single market, must incorporate the drivers of research and innovation into metropolitan strategies, enhancing science diplomacy.

Metropolises are not just commercial spaces but centres of concrete and symbolic aggregation. Their local dimension determines complex interconnected markets on different scales, from micro to macro to hyper-macro. The metropolis is the meeting point between global and local, without one dominating the other.

The intrinsically multiscalar nature of the metropolis requires us to go beyond the abstract approach and adopt methodologies that consider topography and social dynamics.

The Letta's Report emphasises the market role not only as a commercial space but, above all, as a centre of aggregation, concrete rather than abstract.

The real market value is linked to the need to anchor itself to each city local dimension, representing its complex market. Metropolitan cities constitute a network of different markets, bringing the question of scale to the forefront. In fact, metropolis is subject to scale effects ranging from micro to macro to hyper-macro. Today, we consider it the 'showcase' of the micro and country systems.

Therefore, the concrete proposal is to understand how, in the frame of a metropolitan network, to build a network of markets at different scales, from micro to macro, and compare them based on their social, economic, and environmental impacts on the lives of citizens at different scales rather than in an abstract way. The metro scale represents a different territorial, social and economic level of cohesion.

Regarding territorial cohesion and overcoming local interests, it is possible to develop integrated models for managing resources, such as water, and to promote the integration of administrative languages only on a metropolitan scale. Territorial cohesion must be based on a revision of the European regional organisation, breaking up super-local territorial interests in favour of metropolitan districts comprised of societies with diversified territorial interests. Science Diplomacy is fundamental for the management of territorial conflicts.

Developing scenarios based on critical maps allows us to identify and address metropolitan issues, adopting historical and spatial approaches to understand the complexity of identity and geopolitical and socio-economic factors.

This is why we propose constant participation in round tables with European and international partners to define a resolution on the metropolitan question within a year and a half. These events will highlight territorial claims and the need for metropolitan governance to manage and resolve conflicts.

In collaboration with UN-Habitat, the EU must play a central role in defining the international metropolitan community. A systematic analysis of the solutions adopted by metropolitan cities for the localisation of the SDGs through the Smart Specialisation Strategies and the Quintuple Helix Innovation Model [4] will enable verification of how their application varies according to geographical, historical and social attributes.

2. An Ambitious Urban Agenda: The Key Points

The current context makes the situation even more complex. The recent announcement by the European Commission that it will allocate part of the funds to the defence industry at the expense of cohesion makes the message clear. On the one hand, the issue of sustainability and the Green Deal is losing centrality; on the other hand, resources for territorial cohesion will be increasingly limited. This means that efforts must be concentrated on a targeted and incisive strategy.

The EU already has numerous documents on the metropolitan dimension and territorial cohesion, but the actual set of policies is still more focused on the administrative level than on the urban and metropolitan approach proposed internationally.

If we want to build an ambitious Urban Agenda, we need to start from the EU strategic objectives:

1. *Give the urban and metropolitan dimension a prominent role in cohesion programmes.*
   - It has been stated that the term 'urban' indicates any sub-regional level for the EU and the Eurostat definitions are still very generic.
   - It has been said that the JRC deals with the metropolitan scale. However, in reality, it still only defines it as conurbation, without adopting an approach that allows monitoring the localisation of the SDGs locally.
   - The metropolitan dimension is the closest to the concept of bioregion, which makes it essential to address the issue of climate neutrality. However, without data below the national scale, any study on environmental vulnerability remains random.
2. *Define the next EU budget.*

With cohesion funds progressively decreasing, it is necessary to demonstrate that a part of the budget should be allocated to metropolitan governance and local regions.

3. *Structure a European framework for metropolitan governance.*

Given the territorial diversity within the EU, it is necessary to develop a flexible but effective model of metropolitan governance.

With the European Commission recent shift towards increasing armament funds at the expense of cohesion, it is crucial to demonstrate the need for a European metropolitan strategy. The REGIO Directorate must negotiate the inclusion of the metropolitan dimension in European policies, demonstrating its role in territorial cohesion and competitiveness.

There could be some strategic actions to be taken:

1. Increasing the urban dimension in cohesion programmes: redefining the metropolitan scale so that it allows for effective monitoring of the SDGs localisation, especially on climate neutrality, and since it is the scale that is well attained at the prevention approach.
2. Defining the future European budget: ensuring that a share of cohesion funds is allocated to metropolitan governance and local regions.
3. Delineating a European framework for metropolitan governance: identify a model suitable for the territorial diversity of Europe.
4. Outlining Metropolises and Market Networks: A Model for Territorial Cohesion.

Metropolises are not only spaces for economic exchange, but dynamic ecosystems in which markets play a key role both as micro showcases and as structures that favour economic dynamics from micro to hyper-macro. In Europe, where markets are hybrid

Fig. 2 Strategic objectives of the Global Gateway Strategy. *Source* eeas.europa.eu[2]

and complex, the metropolitan dimension becomes the strategic level for articulating territorial cohesion based on the market network (see Fig. 2).

3. A Model of Territorial Cohesion Through Markets

Metropolitan areas are not just administrative entities but the critical interface between local and global dynamics, shaping economic, social, and environmental transformations. However, despite their strategic role, the metropolitan dimension remains politically and institutionally underrepresented within European governance frameworks.

[2] Infographics generated through AI.

We can build a new level of territorial cohesion by integrating:

- Localisation of the SDGs through Smart Specialisation Strategies (S3) and the Quintuple Helix Innovation model, which allow urban policies to be adapted to specific territorial characteristics.
- Connection between territories and markets at different scales, using the metropolitan dimension as a real and constructed hub to integrate local economic ecosystems into a larger network.
- Synergistic networks between markets and cities by building maps of conflicts (including environmental ones) and identifying specific problems to be addressed with collaborative and multiscalar approaches.

The objective is to transform metropolises into platforms for innovation and cohesion, where market networks not only respond to the SDGs but also act as territorial infrastructures capable of balanced and sustainable connecting of the local with the global in a balanced and sustainable way.

In short, a Metropolitan Area Constituency must receive adequate funding based on its role in European territorial cohesion. This would allow it to offer a better metropolitan strategy different from current approaches.

The current EU approach to urban policies largely treats 'urban' as synonymous with any sub-regional level, neglecting the challenges and opportunities metropolises face. This lack of recognition weakens Europe ability to address pressing issues such as:

- Territorial cohesion: The ability to balance regional disparities and promote integrated development.
- Climate resilience: Implementing effective prevention, and not only mitigation, and adaptation strategies at the scale where environmental vulnerabilities are most evident.
- Economic transformation: Supporting market networks that function across multiple scales, from micro to hyper-macro.

A Metropolitan Area Constituency is essential to bridge this gap, ensuring that metropolises receive dedicated funding, governance structures, and strategic planning frameworks aligned with their specific role in European cohesion policy.

4. Towards a Metropolitan Area Constituency: A New Framework for Territorial Cohesion

By influencing social, economic, and environmental changes, metropolitan regions play a critical role in both local and worldwide developments. They do not, however, have enough representation in European governing structures. The objective is to turn metropolises into hubs for innovation and unity, allowing market networks to sustainably link the local and the global. To close the gap in existing methods, a Metropolitan Area Constituency is required to supply specific funding, governance frameworks, and strategic planning frameworks for these areas.

A Metropolitan Area Constituency could, then, achieve

1. *Strengthening Territorial Cohesion*

The metropolitan scale is the missing link in the EU regional framework. Unlike traditional administrative regions, metropolises transcend local boundaries, integrating different socio-economic ecosystems and fostering cooperation across jurisdictions.

A Metropolitan Area Constituency would:

- Facilitate multi-level governance, reducing territorial fragmentation.
- Promote synergistic networks of cities and markets that function beyond administrative borders.
- Ensure fair distribution of EU cohesion funds, avoiding inefficiencies tied to rigid local or regional allocations.

2. *Advancing the Green and Digital Transitions*

Without metropolitan-scale policies, efforts to localize the Sustainable Development Goals (SDGs) remain fragmented. The Green Deal shift away from a central EU priority further highlights the need for localized yet integrated strategies.

A Metropolitan Area Constituency would:

- Align Smart Specialisation Strategies (S3) and the 5 Helix Innovation Model with the metropolitan scale.
- Enhance climate neutrality efforts, ensuring that data and policies go beyond national-level statistics to reflect real environmental vulnerabilities.
- Develop inventive urban models that integrate sustainability, resilience, and digital transformation.

3. *Reframing Markets as Engines of Territorial Cohesion*

European markets are increasingly hybrid and complex, with metropolitan areas acting as both local showcases and hubs for global economic flows. Current urban policies do not adequately address the multi-scalar nature of market networks.

A Metropolitan Area Constituency would:

- Recognize the economic role of metropolitan areas in structuring national and transnational market systems.
- Support market integration strategies connecting local, metropolitan, and global scales, ensuring economic resilience and inclusive growth.
- Facilitate multi-sector collaboration, integrating research, education, and business within a Science Diplomacy framework.

A Metropolitan Area Constituency can be, then, implemented through:

1. *Political and Institutional Recognition*

The EU must formally acknowledge metropolitan governance as a distinct level of territorial management. This requires:

- Negotiating a dedicated metropolitan budget within the EU cohesion policy.

- Establishing a European framework for metropolitan governance that respects territorial diversity while ensuring policy coherence.
- Strengthening international collaboration, positioning the EU as a leader in metropolitan strategies within global institutions like UN-Habitat.

2. *Evidence-Based Metropolitan Strategies*

Existing European research, including ESPON studies, provides fragmented insights into metropolitan dynamics. However, there is no systematic metropolitan-scale approach to:

- Monitoring the localisation of SDGs at a meaningful scale.
- Assessing environmental risks and vulnerabilities beyond national averages.
- Evaluating the socio-economic impact of metropolitan policies on different territorial scales.

A Metropolitan Area Constituency would ensure that these gaps are addressed through comprehensive data collection, territorial mapping, and evidence-based policy design.

3. *Institutional Alliances and Advocacy*

To make this vision a reality, Europe must:

- Organize regular roundtables with European and international partners to define a Metropolitan Resolution within 18 months.
- Develop a strategic alliance between REGIO, UN-Habitat, and metropolitan networks to strengthen policy influence.
- Advocate for a dedicated metropolitan funding mechanism within the next EU budget cycle.

With cohesion funds shrinking and the EU focus shifting away from territorial policies, now is the time to make the case for a Metropolitan Area Constituency. This is not just a technical adjustment—it is a political necessity to ensure that Europe urban and economic future remains cohesive, competitive, and resilient.

A Metropolitan Area Constituency would provide the institutional framework, resources, and strategic direction needed to:

- Reinforce territorial cohesion through metropolitan networks.
- Drive innovation and climate neutrality at the most effective scale.
- Ensure that markets and cities function as interconnected systems rather than isolated entities.

The EU cannot afford to overlook the metropolitan scale any longer. The time to act is now.

3 Digital Economic Communities for the E-Gateway Strategies in Europe

In order to support diversity and coexistence, Intelligent Metropolitan systems should take into account new strategies, such as establishing new, green, and sustainable spatial connections between disjointed neighbourhoods and parts of cities through a new classification of fast and slow connections; integrating historical and modern levels in a single urban strategy; defining macro and micro scales for the urban design and reinventing the economic side of open spaces; improving the quality of spaces; developing a new smart city through the renovation of its spaces; easing a sustainable transition in light of recent droughts; driving investments, including some new landscape strategies.

Digital economic communities refer to networks of businesses, institutions, cities, and regions that collaborate through digital platforms to create, exchange, and scale economic value across borders. An E-Gateway Strategy envisions Europe as a global digital logistics and innovation hub—acting as a gateway not just for goods (like traditional ports and airports) but also for knowledge, services, technologies, and digital products. When combined, such an approach proposes building integrated, digital, territorial ecosystems across Europe that connect cities, regions, and markets through smart platforms, AI-driven logistics, blockchain-trusted supply chains, and innovation clusters—all aligned with Smart Specialisation and SDGs. It could be connected to SDG 8 (Decent Work), SDG 9 (Innovation & Infrastructure), SDG 10 (Reduced Inequalities), SDG 11 (Sustainable Cities), SDG 13 (Climate Action), and SDG 17 (Partnerships).

The objectives of such an approach could be accelerating smart, sustainable trade, fostering innovation ecosystems across regions, ensuring green and inclusive growth, increasing Europe global competitiveness, and promoting strategic autonomy and open strategic autonomy goals.

Such an approach will imply the creation of federated digital infrastructures and. distributed platforms (not just centralized tech giants) for logistics, data sharing, innovation exchange, and smart contracts, such as European Blockchain Services Infrastructure (EBSI) and Gaia-X for data sovereignty. Similarly, city-region digital twins and data hubs could boost the role of Metropolitan areas in developing digital skills and networks (virtual copies of their physical systems) to optimize transport, energy, and environmental management, which could be interconnected across Europe for smart routing of goods, people, capital, and knowledge. Cross-border innovation networks, linking startups, universities, innovation hubs across different cities into thematic communities (e.g., CleanTech, AgriFoodTech, Circular Economy) could be implemented. They should be supported by platforms, like the Smart Specialisation Platform (S3P) and the European Innovation Council ecosystems.

Developing on the last decades of European policies, multimodal smart logistics corridors could be implemented. E-Gateways are not just ports: they're digitally enhanced supply chains. Smart corridors (e.g., the Rhine-Alpine, Mediterranean

Corridor) should then integrate rail, road, air, and sea with real-time digital orchestration. Similarly, digital Green Deals for local economies could establish sustainability standards (carbon footprint tracking, circularity metrics) for digital communities, to be embedded in trade and production. This would be critical for achieving Green Deal and Digital Europe Programme objectives simultaneously.

The establishment of Digital Social Hubs, where people could gather through events (a kind of open lab), is one potential future concept that can help Intelligent Metropolitan systems develop proper digital economic communities, as we will see in the following case studies. This would involve the younger generations in the process of creating the community identity and sense of belonging while also fostering shared values. This can be included into Innovation Platforms or incubators for business accelerator programs, which can serve as a community reference for social innovation and Twin Transition management in addition to scientific research.

Promoting also cultural-based solutions as drivers of improved digital economic communities, the reduction of the Twin Transition gaps should be addressed. Technology transfer initiatives that support research as the primary engine of long-term, sustainable economic growth could be another possible goal of these initiatives. Opportunities for all ecosystem participants, including university and doctoral students, researchers, creative start-ups and research spinoffs, SMEs, corporations, civil society organizations and citizens, and public administration, were discussed. These participants will be able to contribute to the spread of sustainable solutions, making the various regions more inclusive, sustainable, and appealing. There are several ways to promote research-industry collaborations: collaborative research opportunities, industrial research services, and the development of human capital will help make the industry more competitive, create new jobs and businesses, and increase the resilience and competitiveness of regions overall.

A regional framework of policies on public engagement, digital incubators that involve universities, R&D departments, and research centres and offer top-notch services to researchers, developers, innovation teams, and trainers, as well as local Technopoles that have specialized services and labs based on the unique local economic profiles, digital skill development programs for communities to match their competencies with the skills needed by markets, and a networking location in every major city should be the main pillars of the regional and local innovation ecosystems. Strictly related to Spokes activities, technological widespread demonstrators support all aspects of the technology transfer process, such as field testing and evaluation of new technology adoption or prototyping. A few spinoff accelerators and incubators collaborate closely with technological demonstrators, businesses, and venture funds.

1. **SDGs and Smart Specialisation Strategies**

Considering the localisation of the SDGs through Smart Specialisation Strategies (S3) and the Quintuple Helix Innovation model, urban policies could be efficiently adapted to specific territorial characteristics, promoting a place-based vision of territorial development. This is the first element to promote a new idea of digital economic communities, which can promote the integration of tailored digital solutions into a new dimension of local policies.

The foundation of a strong economic ecosystem is a flexible and diverse model of production specialization, whose strength lies in the arrangement of various but complementary activities, a long-standing system of relationships and cooperation, the calibre and vibrancy of subcontracting, and the strong link with services. It must be a highly inventive system with a robust and competitive manufacturing sector, a territory full of expertise, skills, services, and infrastructures, and the ability to create research and development-based collaborative networks with a range of organizations.

Investment in research and development must be encouraged by local policy to advance a new knowledge-based manufacturing system. Since they are thought to be essential for bolstering the connections between innovation, product diversification, and capabilities, education, technical training, research, and welfare policies are also essential elements of a resilient and balanced territorial development. In accordance with educational and training establishments, universities, research institutes, business organizations, local authorities, corporations, and professionals, a local balanced development should strive for an innovation ecosystem.

Localising the Sustainable Development Goals (SDGs) through Smart Specialisation Strategies (S3s) and the Quintuple Helix Innovation Model enables urban and regional policymakers to tailor global goals to local realities, considering specific territorial, environmental, economic, and social contexts. The principles that guide this localisation process should focus on Place-Based Policy Design activities, as S3s emphasize building on the unique assets, strengths, and potentials of regions, rather than applying 'one-size-fits-all' strategies. It encourages local stakeholders to identify and prioritize development areas where they have a comparative advantage. Urban policies must, then, be tailored to local needs, environmental conditions, social dynamics, and economic structures.

A Participatory Governance through the Quintuple Helix Innovation Model should, then, be promoted, involving Academia, industry, governments, civil society and a strong environmental view. This framework will ensure broad stakeholder engagement, bringing diverse knowledge systems and values into the policy-making process. It will help foster innovation that is both inclusive and ecologically responsible, aligning with SDGs like SDG 9 (Industry, Innovation), SDG 11 (Sustainable Cities), and SDG 13 (Climate Action).

Evidence-Based and Innovation-Driven Planning can also be implied, addressing S3s towards data-aware analysis (e.g., regional economic data, innovation potential, sustainability indicators) to guide investment decisions. This aligns SDG localisation with real, measurable capacities of cities and regions, and ensures policy coherence.

Alignment of Global and Local Agendas is also crucial, as the S3 approach should become the translation of global SDG targets into regional priorities, ensuring relevance and ownership. e.g., a coastal city might prioritize SDG 14 (Life Below Water) through marine biotech innovation, while an inland industrial region might focus on SDG 12 (Responsible Consumption and Production).

The environmental component in the Quintuple Helix Innovation Model makes ecological sustainability a central concern in innovation and urban planning, promoting Sustainability as a Cross-Cutting Priority. S3 strategies should integrate sustainability across sectors, aligning development with long-term ecological goals, promoting anti-fragile approaches to territorial development.

An effective Continuous Learning and Adaptation process can support S3s, as they should be conceived as evolving strategies, based on monitoring, feedback, and evaluation. Likewise, the localisation of the SDGs is not a static approach: it requires iterative adjustment as cities evolve and as new challenges and opportunities emerge.

Territorial Cooperation and Knowledge Sharing can also encourage collaboration between regions, fostering peer learning and trans-regional innovation ecosystems. It can support SDG 17 (Partnerships for the Goals) through interregional cooperation platforms.

2. **Multiscalar Connections to Markets**

Connection between territories and markets at different scales should be promoted, using the metropolitan dimension as a real and constructed hub to integrate local economic ecosystems into a larger network. Within programs for creative and scientific research activities, governance models and cooperative urban forms of governance are designed to manage the twin transition while considering potential social and territorial differences.

The metropolitan dimension plays a strategic role as both a real and constructed hub to connect local economic ecosystems with wider territorial and market networks. This aligns with the logic of Smart Specialisation and SDG localisation, helping to scale up innovation and sustainability outcomes. This is because the Metropolitan dimension is promoting sustainable city solutions, smart communities, and educational technology are pertinent subjects for smart specialization initiatives. To establish the prerequisites for sustainable development, multiscale planning is required, as is collaboration and consistent advancement across many forms of intelligent specialization.

In terms of the priority production systems deemed strategic by Regions to address the primary economic, social, and territorial challenges—including those previously identified from the policy objectives of the new cohesion policy—thematic areas of specialization should be distinguished by a high degree of inter-sectorality. They specifically helped to identify the difficulties associated with sustainability and the ecological and green transition.

The relationship between the local communities and the territories in the research and innovation processes: they are meant to be both the carriers of difficulties and the beneficiaries of the proposal. Plus, the function of the research system as a catalyst for the regions revolutionary innovation processes.

Metropolitan Areas should, then, become Connective Hubs, working as intermediaries between local (municipalities, rural areas) and global (international markets, global supply chains) levels. They should serve as nodes in multi-scalar networks of production, innovation, logistics, knowledge, and governance. Their infrastructure (transport, digital, institutional) will enable mobility of goods, people, and ideas.

Integrating Local Economies into Global Value Chains can help S3 strategies to promote regional innovation ecosystems that can plug into global value chains via metropolitan networks. E.g., a rural agri-tech cluster can gain access to export markets or research facilities through its proximity to a metropolitan innovation hub.

Polycentric Development and Functional Urban Areas (FUAs) will promote views of cities and towns as interlinked, not isolated. Functional urban areas, often centred on a core city but including surrounding territories, can allow coordinated strategies that leverage complementarities, reduce territorial disparities, and align investments across scales.

Governance and Collaboration Platforms, at the base of the digital economy community principle, can help creating new constructed spaces. The 'constructed' dimension can refer to institutional coordination frameworks, such as Metropolitan governance bodies, Inter-municipal cooperation, and Innovation alliances (e.g., regional S3 platforms). These enable joint visioning, planning, and investment, key to achieving SDGs across diverse jurisdictions.

This new vision of multiscalar markets can also promote Smart Specialisation Platforms as Scalable Connectors, as S3 platforms and European Territorial Cooperation (ETC) initiatives (e.g., Interreg, Vanguard Initiative) should foster cross-regional innovation alliances. These will allow smaller or peripheral territories to benefit from metropolitan resources and visibility while contributing their unique strengths.

Plus, in terms of Market Access and Branding, Metropolitan regions can act as branding platforms for local products (e.g., food, design, clean tech), giving them international visibility. This will support SDG 8 (Decent Work and Economic Growth) and SDG 10 (Reduced Inequalities) by elevating lesser-known regions through shared metropolitan identities.

Intelligent Metropolitan systems will face numerous hindrances in the next years in terms of Digital Innovation for the economic communities, including but not limited to Regional and socioeconomic (and policy-making ability) differences and regulations gaps. A lack of recommendations for the establishment or support of transitional clusters with a distribution of local and foreign actors at various levels of the value chain; a need for fact-based cluster policies that address the framework conditions for innovation and competitiveness in a more reflective manner.

The creation of a new 'smart city' model that is applied to the whole region and the shift in economic activity toward a circular economy with a focus on the blue growth, tourist, and agricultural sectors can help address these issues in some way.

3. **Intelligent Metropolitan systems and Markets as Synergistic Networks**

Creating a new digital economic community through E-Gateways should start from synergistic networks between markets and cities by building maps of conflicts (including environmental ones) and identifying specific problems to be addressed with collaborative and multiscalar approaches. Defining synergistic networks between markets and cities through the mapping of conflicts and challenges, especially environmental ones, could create a foundation for collaborative, multiscalar, and strategic responses, which is key to effective localisation of the SDGs and the Smart Specialization strategies.

The first step would be, then, mapping conflicts and frictions, using territorial diagnostics to identify conflict zones, such as land use tensions (e.g. urban sprawl vs. agricultural land), environmental degradation (e.g. air/water pollution from industrial hubs), social inequalities (e.g. gentrification, spatial segregation), infrastructure bottlenecks (e.g. limited mobility or digital divide), and similar other factors that Metropolitan Cartography can systematically represent in static and dynamic protocol maps. These maps could reveal where and why certain SDGs are not being achieved, creating a territorialised problem map.

Identifying strategic problems, instead of generic goals, could lead such an approach to focus on pivotal factors that require collective action, such as poor intermodal transport between rural producers and metropolitan markets (SDG 9, SDG 11), climate-related vulnerabilities in coastal cities (SDG 13, SDG 14), or water resource competition across industrial, urban, and agricultural uses (SDG 6, SDG 12), a crucial issue of the Third Millennium, mostly in South America and Africa.

Multiscalar governance models will be needed, defining frameworks that could encourage shared responsibility across different levels of policy and geography. Similarly synergistic market-city networks could leverage Smart Specialisation to connect local production systems with market opportunities, while addressing environmental limits (circular economy, green logistics), social inclusion (local jobs, affordability), and digital access (platforms, digital marketplaces). E.g., a regional circular bioeconomy strategy could be introduced to connect agricultural regions with urban innovation labs and global agri-tech markets.

Embedding the Process in S3s and SDGs localisation can help defining Smart Specialisation processes that could provide the framework to prioritize problems and invest in solutions that have both economic and sustainability impacts. SDGs can give the normative direction, ensuring that synergy doesn't come at the cost of equity or ecology.

A similar approach will require visualization and decision-making tools, using geospatial tools and participatory mapping (GIS, stakeholder workshops) to visualize hotspots and flows (pollution, economic value chains, access gaps) and co-create solution pathways based on shared evidence.

Smart Specialization Regional Territorialization goals, methodologies, and activities could be used for regional research and the industrial ecosystem overall, promoting its structural strengthening, improvement of ongoing or completed projects, services, and initiatives, encouraging the leap in scale and structural consolidation, are some possible uses for regional territorialization goals, methodologies, and activities.

Facilitation of technology transfer and speeding up the digital transformation of business production; processes aimed at achieving economic and environmental sustainability; strengthening of ties and reaffirming the role of local communities; and gradually allowing the public administration to participate in research and innovation processes as both challenge owners and recipients of interventions, while also enhancing the role of the research system as a facilitator of the local communities' path of transformative innovation.

4. A New Idea of Market Based on a European E-Gateway

We live in an era in which artificial intelligence and sustainability are no longer separate themes, but horizons that intertwine. Looking at AI from a cultural, anthropological and territorial point of view means recognizing that we are no longer at the centre of a static world: we are in a profound transition, in which the human being is called upon to redefine itself in dialogue with intelligent tools, new development paradigms and new forms of co-habitation with the planet. In this scenario, a European E-Gateway can only be based on an integrated and inter-disciplinary vision, capable of holding together critical thinking, technology, sustainability and territorial innovation.

AI is already in our lives, often invisibly. Accepting this presence also means assuming a state of 'productive discomfort', capable of activating new critical thinking. The human-artificial dichotomy is obsolete today: the challenges are hybrid, and invite us to explore new ways of integration, agency, and creativity. Even the religious dimension—as Pope Francis reminds us—cannot ignore the transformation of the human being brought about by technology.

The real challenge is not intelligence, but thought: the ability to inhabit fragility, interdependence, relationships. Every technology is a political and cultural act, and as such it must be guided, regulated, imagined in new forms.

At the same time, European territories—and in particular metropolitan ones—are called upon to respond to environmental, social and economic urgencies. Research projects must strengthen the capacity of local institutions to implement the three vectors of the National Strategy for Sustainable Development (SNSvS): integration of development dimensions, promotion of ecological transition, monitoring of the SDGs.

In the context of Metropolitan cities, TTMT (Territorial Tools for Metropolitan Transition) could become fundamental to promote integrated inter-municipal projects, with solutions based on Nature-Based Solutions (NBS) for climate adaptation and the enhancement of green and blue infrastructure. However, critical issues emerge related to the lack of resources for project support and monitoring, essential elements for integrating Ecosystem Services into municipal plans.

A TTMT focused on innovation in production and service spaces can be linked to the issues of antifragility and territorial equalization, which are fundamental for equitably distributing the benefits and impacts of infrastructure choices. In this context, a reflection on service innovation is proposed through the lens of the Twin Transition—digital and green—imagining sustainable supra-municipal poles, hotspot networks, and new public spaces capable of hosting dematerialized but accessible services, supported by digital technologies and green architectural solutions.

To build a credible and innovative European E-Gateway, it is necessary to overcome the fragmentation between cultural, technological and environmental issues. Artificial intelligence, territorial sustainability and service innovation must be treated as parts of a complex, co-evolutionary ecosystem, in which every design choice is also a political, ethical and cultural choice.

Projects such as those developed by MSLab and Polimi MetroHUB between Segrate, Melzo and Pioltello already offer a concrete direction, exploring sensitivity indicators, logistical infrastructures and integrated services. The goal is clear: to restore to the human a centrality that is not arrogant, but conscious, capable of thinking and inhabiting the future—together with machines, in the territories, within a new idea of Europe.

Urban planners and architects must consider these shifts as they design future cities. If public spaces are changing, how should public spaces be designed to accommodate the needs of all? What kinds of new spaces will emerge as a result of the interaction between different digital and cultural norms? Will there be a blending of public and private spaces, or will entirely new space typologies emerge?

5. The European Union as a Digital and Logistics Hub for the Indo-Pacific: A Systemic and Propositional Model for the Exportability of Public Digital Infrastructures

In the current geopolitical context, the Indo-Pacific region emerges as a strategic area for the European Union (EU), both commercially, politically, and technologically. In response to the challenges of global digitization, the European Commission has promoted a series of initiatives aimed at transforming the EU into a sovereign, resilient digital platform capable of projecting its infrastructure also in foreign markets. This calls for a systemic and proactive approach to understanding how the EU can effectively act as a digital and logistical hub for the Indo-Pacific, enhancing the exportability and scalability of digital public goods (DPGs), with a focus on the role of European SMEs.

The concept of a hub is not only to be understood in terms of infrastructure or connectivity, but as a digital platform based on open standards, interoperability and multi-level governance. The EU, through the Digital for Development Hub (D4D Hub https://capacity4dev.europa.eu/resources/team-europe-tracker/partner-countries/global/digital-4-development-hub-d4d-hub_en), the Global Gateway and the Digital Public Goods Stack initiative (DPG Stack, https://www.govstack.global/news/digital-public-infrastructure-building-blocks-and-their-relation-to-digital-public-goods/), is structuring itself as an international reference point capable of offering secure, scalable and reusable digital infrastructures.

The European added value lies in the human-centric approach: digital infrastructure must serve citizens, respect fundamental rights and promote democracy [5]. Moreover, European digital projects are also strongly in line with European values on digital public spaces and civic participation.

The European experience in DPGs represents a model that is exportable to the Indo-Pacific. Solutions such as Decidim (https://decidim.org/), OpenFisca (https://github.com/openfisca), X-Road (https://x-road.eu/), and eIDAS 2.0 (https://www.eid.gov.it/it/nodo-eidas) have demonstrated an inherent ability to adapt to different cultural, regulatory, and administrative contexts. Technical modularity, open-source licensing, participatory governance, and compatibility with European standards make these tools replicable in third countries, offering an alternative to proprietary technology models.

Projects such as MultiPO, Multi-level Participatory and Deliberative Democracy (https://multipod-project.eu/) and Civic Assistant (https://civicassist.org.au/) not only embody the principles of open innovation but also those of digital diplomacy. They are instruments that strengthen European soft power, promoting transnational cooperation in public data management, civic participation, and transparent decision-making.

DPGs are increasingly mentioned in European policy initiatives. The vision is to support European entrepreneurship and competitiveness (a diverse ecosystem of businesses, SMEs, start-ups), create resilience, protect autonomy and sovereignty in an unstable world and empower European people and businesses.

AI is not the only way to close the competitiveness gap. European citizens and businesses need sovereign technology infrastructures, obtained through industrial leadership, to support all our digital experiences. The EU can enhance this process by federating existing resources, coordinating them with public and private investment, and establishing stakeholder governance to protect intermediary infrastructures from data capture.

SMEs are the beating heart of the European economy and a key player in the digital projection towards the Indo-Pacific. Programmes such as EU Gateway, Enterprise Europe Network (EEN) https://www.een-italia.eu/, and the D4D Hub have already facilitated SMEs access to emerging markets in Asia and the Pacific. However, there is a need to move beyond a transactional model to build genuine local digital ecosystems.

European SMEs can offer flexible, adaptable and interoperable solutions, contributing to the localization of DPGs and skills training in partner countries.

European public instruments, such as OSPOs, Open Source Programme Offices (https://github.com/todogroup/ospodefinition.org), can support these companies in technology transfer and in the management of Europe Digital Public Goods compliance.

In Europe, digital infrastructure—whether for participatory democracy, public budgeting, public health, or other purposes—is primarily dedicated to the participation of European public institutions and civic technology communities. Moreover, many open-source innovations developed within cities, such as **Decidim** (https://decidim.org/), are gaining recognition as digital public goods (DPGs).

The Decidim Free Software Association is a democratic organization dedicated to the governance of the Decidim community.

Decidim, born in Barcelona, fosters participation by providing online spaces for participatory budgeting, participatory texts, functions as a tool to collect feedback and opinions on projects and public policies, offering citizens the possibility to participate directly in decision-making, and is a platform designed to be accessible to all, regardless of their level of education or digital skills, with translations in different languages and accessibility options. The platform design and project development are produced by Metadecidim, the community that collaborates to create new features and to report bugs for the continuous improvement of the platform.

The Next Generation Internet project (https://ngi.eu/) is a European Commission initiative launched in 2018 with the aim of being at the service of people through the implementation of technologies that address people needs in terms of privacy, trust, transparency, security and inclusiveness. In the Horizon research projects, it was part of Pillar II,

"Global Challenges and the Competitiveness of European Industry", Pole 4: "Digital, Industry and Space" and focused on the definition of a dynamic community of innovators and stakeholders in the Internet sector, capable of setting the course for the evolution of the Internet by adopting a people-centered approach; on understanding and broadly supporting the NGI initiative vision of an Internet at the service of people; and on enhanced cooperation between the Union and the United States and between the Union and Canada in the development of technologies, services and standards for the next generation Internet.

The Open-Source Programme Office (OSPO, https://github.com/todogroup/ospodefinition.org) is the umbrella structure within an organization in charge of the strategic and coordinated management of all open source-related activities. Its main task is to define, implement, and oversee policies and strategies governing the use, selection, distribution, review, and contribution to open-source software, while ensuring legal compliance, technical efficiency, and community involvement.

OSPOs also serve an educational and support function, providing training to internal and external employees and collaborators, promoting the sustainable use of existing open-source components, and encouraging contributions to external projects whenever possible. In addition, it can accompany teams in publishing their software in open source, help build an internal culture geared towards transparency and collaboration and ensure that engineering practices meet high-quality and regulatory standards.

Closely related to OSPOs is the InnerSource model (https://innersourcecommons.net; https://innersourcecommons.net/learning/; https://github.com/InnerSourceCommons) that adopts open-source practices for internal collaboration between corporate teams, often coordinated by the open source office itself.

OSPOs act as a bridge between the organization and the global open-source community. Externally, it reinforces the organization credibility in the open-source landscape, facilitates its integration with communities, and mitigates legal and reputational risks. Internally, it provides a single point of reference for all open-source-related activities, integrating legal, technical, economic, and social expertise.

Each OSPO is customized according to the reality of the host organization, its products, culture and objectives. There are also public resources to learn more about the global OSPO landscape, such as landscape.todogroup.org, and guidelines on how to create one.

Artificial intelligence and large language models are increasingly entering public sector applications, but with continuing concerns about security, bias and transparency. The Consul Democracy Foundation (https://consuldemocracy.org) is taking a different approach. Their new Civic Assistant is an open-source assistant based on LLM (a language model trained on vast amounts of text to understand, generate, and translate natural language), explicitly designed for public participation platforms. The platform is based on Consul Democracy, a registered DPG, to experiment with the use of artificial intelligence tools that help implement Consul and make it useful for local Public Safety Organizations (PSOs).

Unlike proprietary solutions, this AI aims to be based on public values: transparency, explainability and inclusiveness. By incorporating it into the workflows of public consultations, the Consul Democracy Foundation aims to provide a first glimpse of how open AI infrastructure could become part of the DPG landscape, provided it is carefully governed and equipped with appropriate safeguards. The project will run from 2025 to 2027 and is currently supported by three technical providers, including the Slovenian non-profit service provider Danes je nov dan, the Scottish municipal association COSLA, and the German service provider Mehr Demokratie.

In the context of the global digital transformation, the European Union has initiated the development of a Digital Public Goods Stack (DPG Stack, digitalpublicgoods.net), an open, modular and interoperable technological framework designed to support the creation of reliable and inclusive digital public services, both within the Union and in partnerships with third countries.

The concept of digital public goods is based on the idea that specific tools, such as open-source software, standardized protocols, digital platforms, or data, can be made available to all, guaranteeing fair access, transparency, reusability, and respect for fundamental rights. The EU Digital Public Goods Stack embodies this vision, serving as a common infrastructure available to EU Member States and partner countries.

The adoption and dissemination of the DPG Stack responds to precise political and strategic aims:

- Promoting European digital sovereignty, reducing dependence on large external private providers and ensuring greater control over public data.
- Support international cooperation by offering partner countries, particularly in Africa, Latin America and the Indo-Pacific, digital tools that reflect European values: inclusiveness, equity, transparency and protection of rights.
- Strengthening digital resilience and security by fostering shared and controlled solutions based on trusted infrastructures.
- Accelerating the digital transformation of public administration through ready-made and standardized tools.

The EU Digital Public Goods Stack aligns with the broader framework of the Global Gateway (https://commission.europa.eu/strategy-and-policy/priorities-2019-2024/stronger-europe-world/global-gateway_it), an EU initiative aimed at promoting sustainable investments and securing digital infrastructures in partner countries. Through the DPG Stack, Europe not only strengthens its position as a key player in global digital development but also offers a concrete alternative to undemocratic technology models, promoting digital technologies as common goods rather than instruments of control or dependence.

Moreover, the European approach supports other international initiatives on digital public goods, such as those promoted by the United Nations, the Digital Public

Goods Alliance, and the World Bank, offering regulatory, technical, and institutional contributions in line with multilateral standards.

EuroStack (https://euro-stack.eu/) is a digital industrial infrastructure aimed at promoting European technological sovereignty; it represents a European Union industrial policy initiative that integrates technology, governance, and investment for the development and adoption of a cohesive European digital infrastructure. The objective is to create strategic assets that encompass connectivity, cloud computing, artificial intelligence, and digital platforms, thereby enhancing the continent technological autonomy.

Underlying the project is an ambitious vision: to support European entrepreneurship and competitiveness, particularly through a diverse ecosystem of companies, SMEs, and start-ups, by building resilience and reducing technological dependence on non-European players. In an unstable world, EuroStack aims to ensure digital sovereignty and power of choice for citizens and businesses.

More than simply bridging a competitiveness gap, the objective is to provide European territories with sovereign, reliable and shared digital infrastructures capable of supporting all dimensions of the digital experience. To achieve this, it is crucial to leverage existing resources by federating and coordinating them through public and private investment and establishing a multi-level and participatory governance framework that prevents the concentration and capture of intermediary infrastructures.

National digital strategies, e.g. Germany digital strategy, or France Tech.gouv (https://ingroupe.com/insights/tech-gouv-programme-digital-transformation-public-services/) increasingly refer to open source, interoperability and public goods. There has been a noticeable increase in European open-source infrastructure efforts (e.g. the OSPOs in governments, or projects such as X-Road and eIDAS 2.0), many of which are linked to DPG communities.

Europeans have tremendous confidence in their government to lead the digital transformation of public services. They tend to welcome this simplification of procedures, which they see as an effective tool for combating fraud (87%), simplifying daily life (79%) and reducing environmental impact (66%). The Tech.Gov programme, promoted by DINUM (Interministerial Directorate for Digital Services) and launched in 2019, is a three-year action plan (2019–2021). The government therefore articulates its strategy around 6 themes, 8 missions and 35 priority actions. The aim is to seize the opportunities offered by digital to bring public services closer to citizens, businesses and their employees in order to offer the best possible quality of service.

Eight are the missions of the Tech.Gov:

- LABEL: Labelling the digital solutions used by public services to create a coherent set of tools.
- DIGID: Creating a simplified and unique identification service that offers the user a more convenient experience.
- DATA: Exploiting the wealth of data to maximize the potential of public policies.
- INFRA: Pooling the digital infrastructure of public services to enable greater interoperability between them.

- PILOT: Establish a dedicated management for the digital transformation of public administration, with precise and measurable targets every year.
- TALENT: Successfully attracting top talent in the sector by enhancing the digital attractiveness of public services.
- CREATE: Promoting innovation through the creation of state and regional start-ups capable of designing new systems for tomorrow public services.
- TRANSFO: Using digital intelligence to support transformation projects led by ministries and public services.

Special mention must be made of the Estonian e-Residency, a model digital gateway of the European Union with significant implications for the future.

The European Union is progressively strengthening its position as a global digital hub through the adoption of policies and instruments promoting economic inclusion, connectivity and sustainability. In this context, the e-Residency developed by Estonia is an emblematic example of a European public digital good. It is not only an innovative solution for digital entrepreneurship but also a replicable model for facilitating cross-border access to European markets while strengthening economic and digital ties between the EU and the Indo-Pacific region. As we have seen, there are now many European e-gateway projects. Still, the e-residency project is a step forward even from eIDAS 2.0 itself, as it is not only a platform for fast access to online services of public administrations or private individuals, but could be the driver for a European residency, independent of state borders but linked to the federation of the United States of Europe. This project could start with the federation of the metropolitan regions and support their network of intercity markets.

Launched in 2014, the Estonian e-Residency programme enables non-residents to obtain an official digital identity, which is beneficial for setting up and running a business registered in Estonia (and thus within the European Union), signing documents of legal value, and accessing banking and tax services remotely. The project was created to overcome geographical and bureaucratic obstacles that limit international business activity, offering a secure and simplified platform that positions Estonia as an advanced laboratory of digital citizenship. This model can be understood as an archetypal European gateway, facilitating the extension of digitization benefits to other countries.

By extending the e-Residency logic to the continental level, the EU would have the opportunity to build a true ecosystem of shared digital public goods. A federated European infrastructure could include interoperable digital identities, uniform access to public and commercial services, tools for digital commerce such as blockchain, smart contracts, and e-invoicing, as well as a harmonized regulatory framework for data governance. Such a structure would create a unified digital platform for the Union, capable of attracting entrepreneurs and innovators from around the world, particularly from the Indo-Pacific region.

In this scenario, e-Residency presents itself as a highly relevant tool: it enables companies and individuals to access the European market without the need to establish a physical presence, while providing legal, tax, and technological support through fully digital channels. This approach is particularly beneficial for SMEs and start-ups from countries such as India, Indonesia, Vietnam, the Philippines, or Sri Lanka, as well as for island micro-states that view European digital platforms as a concrete opportunity for integration into global economic flows.

Already today, thousands of Indo-Pacific entrepreneurs utilize the Estonian e-Residency programme, but its impact could be significantly expanded if replicated at the EU level. A pan-European version of the project, based on a common digital license for foreign companies, could be complemented with bilateral or multilateral cooperation initiatives with governments in the Indo-Pacific region. The aim is to ensure interoperability, mutual recognition of digital securities, and the joint use of advanced digital infrastructures, particularly public blockchains, for the transparency of economic and contractual flows.

The Estonian experience clearly shows that digital citizenship is not a utopia, but a concrete reality capable of generating public and private value. In an increasingly interconnected world, where physical borders count less than digital networks, tools such as e-Residency can become key elements for the construction of Euro-Indo-Pacific E-Gateways. To fully realize this potential, the European Union will have to address three strategic challenges.

First, there is a need to harmonize the digital regulatory framework to eliminate inconsistencies and overlaps between the laws of different Member States. Common regulations on data protection, cybersecurity, artificial intelligence and cryptocurrencies would foster a more stable environment for innovation and international cooperation.

Secondly, the EU will need to invest in interoperable digital infrastructures, developing networks, databases, and software platforms that can communicate with each other, including across different countries and administrations. Indeed, interoperability is the key to overcoming technological silos and making the European digital ecosystem truly integrated. Finally, it will be crucial to adopt a strategic and coordinated approach to international digital cooperation, rather than limiting ourselves to occasional projects, and building structured alliances with countries in the Indo-Pacific region. This involves jointly developing technical standards and common projects in cybersecurity and physical infrastructure, such as submarine cables or satellite networks, to strengthen European digital sovereignty and foster balanced and inclusive development.

The replication of the Estonian model in a broader European context thus represents an extraordinary opportunity for the Union to position itself as a leader in the global digital transformation, through a model that combines openness, security, sustainability and scalability.

The development of these digital projects is linked to the eIDAS 2.0 platform project (https://www.eid.gov.it/it/nodo-eidas). eIDAS is the first step towards a digitally integrated Europe, as it facilitates citizens' and businesses' access to digital public services across different Member States through simple eID authentication.

The Regulation promotes cross-border cooperation and interoperability of national electronic identification systems by requiring European public administrations to make their services accessible across all Member States' digital identities. It also provides a common regulatory basis for secure electronic interactions between citizens, businesses, and public authorities, thereby enhancing the security and effectiveness of e-services, e-business, and e-commerce transactions within the EU.

Building on this foundation, eIDAS 2.0 was introduced, further enhancing the digital infrastructure and efficiency of the system. It provides individuals and businesses with access to secure and user-friendly digital identity solutions that work seamlessly across all EU Member States, improving data security and promoting a level playing field for qualified trust services throughout the EU. It refines the rules for qualified trust service providers by ensuring clearer and more harmonized regulations, introducing electronic attestation of attributes, website authentication certificates, management of qualified devices for remote electronic signature (or electronic seal) creation, as well as electronic storage and electronic records. eIDAS 2.0 also introduced the European Digital Identity Wallet (EUDI), https://digital-strategy.ec.europa.eu/it/factpages/european-digital-identity-wallet?utm_source=chatgpt.com, which allows users to store, manage, and share their data, credentials, and identity attributes, making them accessible both online and offline.

The Italian eIDAS node enables the use of a Member State digital identity to access online services of public administrations or private individuals in EU Member States. By activating the eIDAS Login, Administrations that make their online services available at the national level also extend the use of these services to EU citizens through their own digital identity of their home country. At the same time, other EU states, through their eIDAS nodes, allow Italian citizens to access their services using a digital identity.

In particular, AgentConnect was a European pilot project developed in the framework of the eIDAS (electronic IDentification, Authentication and trust Services) regulation, with the aim of testing and facilitating the interoperability of digital identity systems between EU Member States. Initiated to demonstrate the possibility of secure and standardized cross-border access to public digital services, AgentConnect was the first concrete step towards the construction of an integrated European digital identity.

Through this project, European citizens were able to access online public services in other Member States using their existing national digital identity, such as SPID in Italy, via national eIDAS nodes. The project demonstrated real use cases in areas such as education, social security, taxation and digital health, where cross-border authentication is crucial for European integration.

The initiative involved several public authorities and national technologies, creating the operational and technical conditions for implementing interoperable digital services. This has contributed significantly to the development of the new generation of tools envisioned by eIDAS 2.0, particularly the European Digital Identity Wallet (EUDI Wallet), designed to offer citizens a simple, secure, and controllable way to identify themselves digitally across the EU.

The AgentConnect project was coordinated under the Connecting Europe Facility (CEF) programme, and the results and technical resources are available via the official website (https://ec.europa.eu/cefdigital/wiki/display/CEFDIGITAL/AgentC onnect). AgentConnect has provided a solid experimental and methodological basis for implementing a reliable, interoperable, and privacy-compliant European digital identity, paving the way for an increasingly integrated Union in digital public services.

The European Community is increasingly moving in this direction; as of 2024, the new European Digital Identity Regulation (EUDI) enables the development of a secure, interoperable, and accessible digital wallet throughout the EU (https://dig ital-strategy.ec.europa.eu/it/policies/eudi-wallet-implementation).

The European Digital Identity Regulation (EU) 2024/1183 represents a significant shift in the way citizens and businesses manage their digital identities within the European Union. Building on the experience of the 2014 eIDAS Regulation, the new legislative framework introduces a European digital identity wallet, known as the EUDI Wallet, designed to be secure, trusted, and universally accepted.

The previous eIDAS regulation allowed Member States to voluntarily notify their electronic identification systems, which then had to be recognized by other EU countries. However, there was no obligation to develop a national electronic identity and interoperability between systems was based on a fragile technical superstructure, with limited effects, especially for private digital services. The new EUDI regulation overcomes these limitations, making it compulsory for all Member States to offer citizens a digital wallet through which other certified personal attributes, such as driving licenses, diplomas, bank accounts or health cards, can be associated with their digital identity. These wallets can be provided by both public bodies and recognized private entities, according to common security and interoperability standards.

One of the central elements of the new system is user control: people will be able to decide what information to share, with whom and in what context, thus ensuring the protection of their personal data and complete transparency of digital transactions. The digital wallet will be usable across the EU to access public and private services, simplifying online identification and increasing security in authentication processes.

To ensure uniform adoption, the regulation provides for a common technical architecture, cybersecurity requirements, and an obligation for service providers (both public and private) to accept the wallet where unique identification is required. In addition, the system will be open source, ensuring maximum transparency, auditability and innovation possibilities for developers and governments.

The impact of the EUDI portfolio extends to all levels:

- For citizens, it will mean an easy-to-use, interoperable and free tool for non-professional electronic signatures, with the possibility of accessing numerous online services without having to provide the same information several times.
- For governments, it will provide a secure channel to simplify access to public services, strengthen fraud prevention and improve data security.
- For digital enterprises, it will reduce the costs associated with authentication processes and limit dependence on third-party commercial platforms that do not offer sufficient data protection guarantees.

- For the company, it will be a driver for economic growth and digital innovation, stimulating new business models and making online transactions more straightforward and more secure.

The new regulation officially entered into force in 2024, and member states will have 24 months to implement the system and make at least one compliant digital wallet operational. Five implementing acts, adopted in November 2024, set the technical criteria and conditions for the certification and operation of wallets.

Meanwhile, large-scale pilot projects are underway in several European countries, involving public authorities and private partners to test the wallet infrastructure in the most critical areas: from government services to opening bank accounts to secure digital payments.

The introduction of the EUDI Wallet marks a milestone in the European digitization process. Combining security, accessibility and digital sovereignty, the EU finally has a unified tool to provide citizens with a portable, secure and fully recognized digital identity across Europe.

The EU increasingly supports digital cooperation and infrastructure initiatives in partner countries. This has opened new avenues for European actors and institutions to contribute with DPGs to international platforms, participate in multilateral efforts on DPGs and collaborate in global digital infrastructure initiatives.

Europe supports DPGs but does so in a more fragmented and 'political' way than other countries that have a more operational and centralized management of public digital infrastructures. This approach has advantages (diversity, adaptability), but also limitations in terms of speed and consistency. Nevertheless, the convergence of political will, institutional capacity, and international collaboration on open source and infrastructure suggests that Europe could become a significant player in the global DPG ecosystem.

The European Union has already launched several initiatives and platforms like E-Gateways to facilitate access to international markets for small and medium-sized enterprises (SMEs), including those in the Indo-Pacific region. These include the EU Gateway Business Avenues programme, which runs until 2020 and has supported over 2000 European SMEs through sectoral trade missions to countries such as Japan, South Korea, Singapore and Vietnam.

EU Business Avenues in Southeast Asia is an EU-funded initiative that helps European companies establish lasting business partnerships in Singapore, Vietnam, Malaysia, Thailand, Indonesia and the Philippines. It is a branch of the broader EU Gateway Business Avenues, a flagship EU programme that also targets Korea and China. The programme offers market information, coaching and support to selected European companies participating in a one-week trade mission to South-East Asia.

At a more structural level, the Enterprise Europe Network (EEN) https://www.een-italia.eu/ continues to provide support for the internationalization of SMEs, offering advisory services, regulatory assistance and partnership opportunities in over 60 countries. In addition, the EU Global Gateway initiative, launched in 2021, aims to promote sustainable and digital infrastructure in partner countries, creating opportunities for European companies as well.

Finally, tools such as the Digital for Development (D4D) Hub https://www.google.com/search?q=Digital+for+Development+(D4D)+Hub&oq=Digital+for+Development+(D4D)+Hub+&gs_lcrp=EgZjaHJvbWUyBggAEEUYOdIBCTE4ODRqMGoxNagCCLACAQ&sourceid=chrome&ie=UTF-8&sei=g_5CaLKgHob_7_UP6f_MyAI and digital trade agreements with Indo-Pacific countries (e.g. Singapore) strengthen international digital cooperation, facilitating the entry of SMEs into foreign markets through shared infrastructure and rules.

The EU is already actively engaged in providing SMEs with tools similar to E-Gateways, with a strategic and multilateral approach to strengthen their global competitiveness. https://capacity4dev.europa.eu/resources/team-europe-tracker/partner-countries/global/digital-4-development-hub-d4d-hub_en.

The Digital for Development Hub (D4D Hub) is a Team Europe proposal to promote equitable, inclusive and sustainable digital transformation in partner countries, following a human-centered model. Its aim is to facilitate, through dialogue, the identification of new initiatives in line with the Team Europe approach to increase investment and foster innovation.

The platform is open to all EU Member States and promotes coordination and synergies between the EU, Member States, their development agencies and international financial institutions, while ensuring the involvement of actors from the private sector, civil society and academia through advisory groups. A Multi-Partner Contribution Agreement (MPCA) supports the shared operation of activities.

The D4D Hub is a multi-level coordination structure operating through a Global Council and regional and local branches in Africa, Latin America and the Caribbean, Asia and the Pacific. It is supported by nine fully operational thematic groups, each co-led by the Directorate-General for International Partnerships (DG INTPA)-currently under the leadership of Czech Commissioner Jozef Síkela- and one or two Member States, and focused on: digital transition and green, entrepreneurship and innovation, digital governance, digital literacy and skills, connectivity, digital rights, gender and digital, cybersecurity, and digital health.

In order to make the EU an effective digital platform in the Indo-Pacific, a systemic approach is needed, with four main strands:

- DPG standardization and certification: promote a European standard for DPGs certifying their compliance with technical and regulatory criteria (e.g. GDPR, eIDAS, ISA[2]), making them immediately adoptable in partner countries.
- Local adaptation-oriented public–private partnerships: co-create projects with local authorities and SMEs in the Global South, adapting instruments to local socio-institutional needs.
- Financing and capacity building: establish dedicated funds for the implementation of DPGs in the Indo-Pacific, including local training, technical support and sustainability over time.
- Multi-level and inclusive governance: encourage the integration of civic communities, universities, NGOs and businesses in a collaborative governance model, as already experimented with Decidim and the OpenFisca Association.

4 Territorialization of the Smart Specialization Strategies in Albania

According to the European Commission, EU regions and member states adopt a Smart Specialization Strategy (S3) framework to determine their distinct strengths and sectors with significant growth potential in research and innovation.[3] It focuses on making investments in areas where a region can attain knowledge-driven growth and develop a competitive edge. By focusing on certain goals and encouraging cooperation between companies, the public sector, and knowledge institutions, S3 seeks to optimize the impact of expenditures in research and innovation, according to the European Commission.

The key components of smart specialization strategies are as follows, per the EU platform that lays out the idea and offers guidance on how to create research and innovation strategies for smart specialization: Smart specialization is a place-based approach; Priority-setting is an interactive process; The strategy should embrace a broad view of innovation, supporting technological as well as practice-based and social innovation. A good strategy must include a monitoring and evaluation system. However, to establish priorities, regions must find the agents who can produce knowledge-driven growth in the direction of change agents who can identify the transformation required to address the most pressing issues facing society and the built and natural environments.

The West Balkan Region has determined that technology in education, sustainable city solutions, and smart communities are pertinent subjects for smart specialization initiatives. It is also crucial that private investment participate in transition studies. Investments in tourism infrastructure, rural tourism, and agriculture require collaboration and consistent development among many forms of smart specialization, and multiscale planning is required to establish the prerequisites for sustainable development. With collaborative urban forms of governance and governance models handling the twin transition, the international perspective is well-focused.

A Proposed Smart Specialization Thematic Area Strategy and Country Information are included in the NSSS information. This strategy should include sections that describe the win–win situation between ecology, knowledge, and innovation, discuss whether the strategy is appropriate for addressing global warming, explain the scale dimension of smart specialization, and analyse how infrastructure and space integration patterns affect NSSS activities.

Albania is getting ready for the S3 process after joining the Smart Specialization Platform in 2017. Priority areas for the Smart Specialization Strategy have been identified through a qualitative mapping exercise and a quantitative mapping of the nation potential. The Entrepreneurial Discovery Process is anticipated to be deployed by the end of 2022, and five priority domains have been designated. In 2020, the GDP is projected to reach 1,644,077 million Albanian Lek, with the largest percentages coming from forestry, fisheries, and agriculture. Albania economy has been

[3] See https://s3platform.jrc.ec.europa.eu/ris3-guide.

severely impacted by the Covid-19 outbreak; the only industries that have grown are agriculture and real estate.

There are sixteen industries with dynamic potential and thirteen with static economic potential. The energy industry wired and satellite telecommunications, and radio and television broadcasting are all included in the Smart Specialization design. The research identifies supply chain integration and cross-fertilization difficulties, and the entrepreneurial Discovery Process might offer more information. To cut down on redundancy and boost effects, the NSSS determines long-term investment objectives.

Smart specialization includes the use of a variety of simulations and optimizations pertinent to decision support, as well as the use of intelligence functions with far wider participation in real-time decision-making and building. Decisions on planning must be telescopic in both time and space. The capacity to consider changes at various temporal and spatial scales must be improved by intelligent specialization. We could create city models that adapt to a system that is changing. How the planning and decision-making process may adapt to incorporate the capacity to view the city from the standpoint of intelligent specialization transition is the question.

Using intelligent functions with a much wider involvement in real-time construction and decision-making, as well as numerous simulations and optimizations pertinent to decision support, is known as 'smart specialization'. To consider changes at various spatial and temporal scales, planning decisions need to be telescopic in both space and time. It might make it possible to create city models that adapt to a system that is changing. How can the planning and decision-making process be altered to accommodate the capacity to view the city from the Smart Specialization transition perspective?

Intelligent behaviours have existed throughout history, as seen by the development of cutting-edge technologies and creative ideas in urban areas. We characterize intelligent behaviour as a combination of social infrastructure, communication, and knowledge quality. We connect that idea to the Quintuple Helix innovation model, which talks about Smart Specialization and explains a win–win scenario between ecology, knowledge, and innovation. We propose that smart specialization balances efficiency and equity, intelligent behaviour, and civil economy to enhance prosperity, community, and quality of life.

The topic of smart specialization is also covered, with an emphasis on expanded decision-making involvement, additional intelligent functions, and decision-supporting simulations and forecasts. We also highlight the focus on networks and the significance of considering a change at various spatial and temporal scales in Smart Specialization.

1. Territorializing Smart Specialization Strategies

Smart specialization includes the use of a variety of simulations and optimizations pertinent to decision support, as well as the use of intelligence functions with far wider participation in real-time decision-making and building. Decisions on planning must be telescopic in both time and space. The capacity to consider changes at various temporal and spatial scales must be improved by intelligent specialization. We could create city models that adapt to a system that is changing. How the planning and decision-making process may adapt to incorporate the capacity to view the city from the standpoint of intelligent specialization transition is the question.

As a 'Made in EU' idea, smart specialization is a place-based and creative approach to innovation. It has expanded outside of the EU and could contribute to the global Sustainable Development agenda. Due to its methodological features and guiding principles' compatibility with various socioeconomic and territorial settings, smart specialization has the potential to be implemented globally. By encouraging learning and creating opportunities for international and interregional synergies, it can have a multiplier effect on the EU.

Although S3s are identified at the national and regional levels in the European Union, some supranational strategies—such as the EU Macro-Regional Strategies and the Atlantic Arc—involve specific regions of the continent based on geographical groupings. At the local, regional, national, and international levels, the S3 strategy provides a strengthened setting for fostering greater cooperation and synergy among stakeholders. Building strategic alliances between nations and regions to address shared issues in a systemic manner throughout Europe requires interterritorial cooperation. The JRC divides the information display into EU Territories (Atlantic Arc, EU Macro-Regional policies, Lagging Regions) and Non-EU territories (EU Enlargement, EU Neighbourhood, Latin America, Africa, Arctic, Asia–Pacific) on the Smart Specialization Platform, presenting the key components of the territorialization of the SS policies.

Across European regions and member states, the Smart Specialization concept should be put into practice, tackling important issues and creating innovation strategies. To accomplish objectives, stakeholders have taken ownership of the concept, and success-driven elements and results are emphasized. Several hands-on experiences are offered by the Smart Specialization Platform on the following subjects: Governance and EDP, Policy tools for S3 implementation, Monitoring and Evaluation, Entrepreneurial discovery process (EDP) cycle, Good governance, Project selection: criteria and procedures, Transnational cooperation and value chains, Monitoring.

The following best practices emerge as relevant from the related case studies: encouraging sustainability and twin transition in local sectors; networking opportunities to assist innovators in transforming their ideas into profitable ventures; offering startups resources like workspace, capital, mentorship, and research opportunities; creating collaborative spaces; concentrating on technology transfer to close the gap between industry and research, resulting in new products and services; connecting with the local environment, providing access to facilities for research and development; developing artificial intelligence; integrating policies at various levels; supporting businesses to help members connect, encourage the development of ecosystems, scouting, brokerage, ecology-smart agri-food chains, and water management; offer startups customized support by developing business strategies and visions; concentrate on youth, engagement, education, and sharing; and engage in regional networking activities.

Projects with favourable socioeconomic effects should be the outcome of smart specialization strategies that work (EU, 2020). Since controlling authorities frequently struggle to allocate funds and generate observable socioeconomic effects, the S3 concept execution is vitally important. The European Commission new industrial strategy for the digital transition and green economy emphasizes the value of public funds allocated to large-scale, impactful enterprises. Although it is too soon to evaluate the socioeconomic effects of projects financed using the S3 strategy, it helps to address the difficulties that different regions experience. The upcoming S3 generation ought to be even more goal-oriented, with a stronger emphasis on ambitious results and impacts as well as high-potential projects.

Global Aims as SDGs must also be addressed in a Smart Specialization Thematic Strategy (see Fig. 3). Considering this, the approach overall goal is to pinpoint and comprehend the skills required to localize placed-based activities while applying the Smart Specialization Strategy to the instability topic. Through the telescopic dimension of space–time, the accessibility of proposed localized services, the networking required to achieve equity and effectiveness, and the necessary competencies to be introduced to achieve distributed intelligence, the approach will comprehend the Quintuple Helix innovation model as an incentive for the green and digital transition facing the global warming crisis. To create a collective structure that satisfies the demands of the Fourth Industrial Revolution, the approach will also identify and strengthen current sustainable habitats.

2. Research as Driver for Territorialization

The buildings, resources, systems, and associated services that research communities use to carry out top-tier research in their subjects are known as research infrastructures, or RIs. In collaboration with businesses and research facilities, universities play a vital role in generating information and converting it into cutting-edge goods and services. Achieving comprehensive regional strategy can be greatly aided by the effective mobilization of university resources. By encouraging the entrepreneurial spirit of their employees and students, offering guidance and services to SMEs, taking part in programs that train and place top-level graduates in innovative companies,

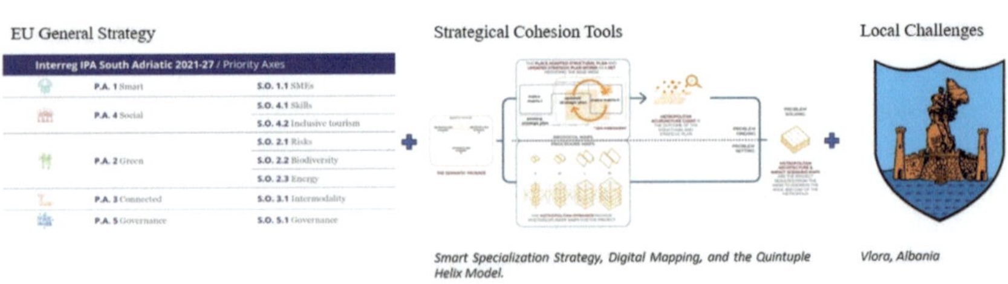

The data we consider are based on the International Strategy Plan for Cohesion like:

- SSN STRATEGY
- **The EU Strategy for the Adriatic and Ionian Region (2014)**

Fig. 3 From EU urban challenges to localized metropolitan strategies: the case of Vlora. *Source* A. Lipovica

setting up incubators for spin-offs in science and technology parks, and contributing significantly to innovative clusters and networks, universities can support regional innovation systems.

Member states are required to make sure that their research infrastructures are professionally run and capable of offering SMEs high-value financial and non-financial support services. The ability of universities to connect with the larger region may be one of the internal obstacles to enhancing their role in regional growth. Their skill in converting knowledge into innovation will be useful to their business and public partners, and universities will value the potential their areas offer for their work as 'living laboratories' that are open to worldwide connections.

Although the EU research infrastructure landscape is changing rapidly, improvements and a more equitable distribution have not yet been achieved. To build a path to excellence for research and innovation infrastructures and enterprises within their purview, regional and municipal authorities ought to collaborate. The available funds and instruments should be combined for investments in both new and existing RIs. Member States should define clear goals and provide long-term support for operational expenses and the participation of important ESFRI projects or regional RIs. The Regional Policy can offer financial assistance for the construction of RIs outlined in the ESFRI roadmap. The ERDF offers funding for the establishment and operation of research infrastructures, technology transfer centres, and centres of expertise in certain technologies.

The National Smart Specialization Strategy aims to specialize the ecosystem and identify the West Balkan Region scientific and economic vocation. The strategy partners determined several priority domains, including agriculture, food and biotechnology, ICT, biodiversity and the environment, tourism and education, renewable energy, and building and urban efficiency. It is crucial that private investment participate in the transition study. Other objectives including traffic infrastructure, environmental restoration, and property valuation can be accomplished by multiscale planning.

The relationship between territorial context and initiatives to promote stakeholder collaboration and interaction, capacity building, and local advantages is highlighted in the NSSS application. It also talks about how the initiative implementation is impacted by the local environment and what certain conditions need to be fulfilled for it to be successful. Economic, energy, agricultural, and tourist-related analysis tools can stimulate investments in rural tourism, tourism infrastructure, and agriculture. The twin transition is managed through the establishment of collaborative urban forms of governance and governance models.

The development of national and regional innovation initiatives for smart specialization should be closely coordinated with universities and other knowledge institutions (see Fig. 4). Stronger collaborations within the knowledge triangle and linking universities to regional growth are the main investment targets of the Cohesion Policy. Universities ought to play a major role in rural development by offering informational and knowledge-sharing initiatives, advisory services, research and investments pertaining to cultural and natural heritage, and collaboration between various Union actors, as well as those in the agriculture, food chain, forestry, and other sectors.

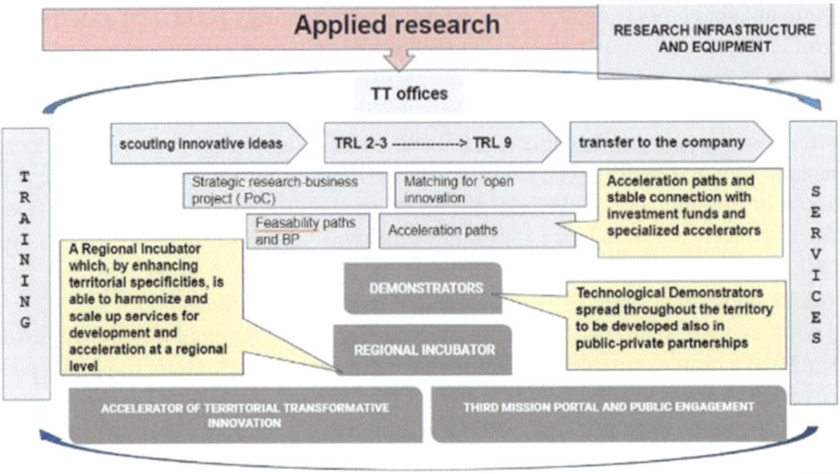

Fig. 4 The structure of S3 innovation strategies. *Source* Arter Emilia Romagna

Promoting sustainability, networking possibilities, resource provision, collaborative space creation, knowledge transfer, developing ecology-smart agri-food chains, establishing connections with the local environment, and water management are all examples of best practices in the field of smart specialization.

3. S3s and Intelligent Metropolitan Systems

The concept of smart cities, which is a combination of physical capital, or hard infrastructure, and the quantity and quality of social, communication, and knowledge infrastructure, is overlapping with smart specialization. By fostering synergies between the economy, society, and democracy, the Quintuple Helix innovation model explains a win–win scenario including ecology, knowledge, and innovation. The suggested Smart Specialization Thematic Areas Strategy needs to be appropriate for addressing global warming and connected to digital and green transformations.

A qualitative shift based on artificial intelligence (AI) represented by information technology (IT) at all scales is implied by smart specialization. We cannot overlook the fact that cities are a part of a world that is growing more interconnected and 'smart'. A hybrid top-down-bottom-up method (feedback) may be adopted because of the Fourth Industrial Revolution, in which choices are made at the individual level using data gathered at the metropolitan level. The outcomes of local applications will subsequently generate knowledge feedback that will be disseminated and transformed at the metropolitan, national, and European levels. A systemic telescopic approach is what that is.

Metropolitan systems should use S3 localization to raise awareness of the need for immediate action against climate change and global warming, territorial vulnerability, and socioeconomic inequality. Given that short-term threats are becoming more common, obvious, and urgent, it is essential to address climate change. A flexible model that operates based on knowledge and on the emergency time scale is connected to smart specialization activities to: initiate creative processes for comprehending the intricacy of metropolitan dynamics; develop new analytical tools and scientific theoretical knowledge through technology; investigate new fields of knowledge through the new dimensions of applied science; and encourage digital learning for logical action-design awareness, a model of co-evolution as a cultural phenomenon of transformation is proposed.

One of the main recommendations is to concentrate on proper positioning, which entails determining the best strategies to guarantee convenient access to markets, resources, and qualified personnel. Concentrating on local issues is a crucial component of S3 localization. This entails determining the issues that the relevant context has and creating solutions that are suited to those issues. Furthermore, establishing a specialized market that offers secure and stable conditions for growth while leveraging the local market might be essential to fostering an innovative atmosphere.

Another important point raised for the development of a correct localization process for S3s in Intelligent Metropolitan systems is cultural adaptation, which highlights the necessity of comprehending and adjusting to the cultural variations present in the relevant context. It is also advised that all stakeholders be included, which entails incorporating everyone that S3s would impact, such as economic actors, academic institutions, and the community.

For S3s to succeed in Intelligent Metropolitan systems, it is also essential to guarantee long-term systemic financing and to create a supportive atmosphere. Finally, financing investments in S3s in Intelligent Metropolitan systems can be accomplished efficiently using hybrid funding models that combine short-term and long-term funding.

All things considered, these suggestions offer a framework for developing and putting into practice S3s in Intelligent Metropolitan systems that are suited to the local community and foster an innovative atmosphere. To address the local Smart specialization areas in the regional strategies, a place-rooted proposal should therefore address coherence with the territorial professions, the definition of an Innovation Ecosystem. Transversally to all the NSSS intervention areas, an Innovation Ecosystem integrated with existing research infrastructures could address the issues that regions should face soon, as highlighted in the territorial dynamics analysis; the ecosystem should be conceived as a Digital Social Hub, as described in the previous chapters, to address the diffuse culture in the local productions of human capital. It should also serve as an incubator for specialized and advanced services for the metropolitan areas and their specializations and integrate the concept of Digital Urban Hubs, with the idea of the place-based S3s development platforms.

As cooperation should always be the primary focus of all policies related to S3s (see Fig. 5), an Innovation ecosystem should address the issue of agglomeration and attractivity towards people and other specialized functions in terms of the goals of sustainable transition connected with digital transformation and the human-centred approach. This is because there is a general lack of promotion of the territory and networking opportunities, as well as a problem of lack of integration of planning and development strategies at the various scales.

STRATEGY

South Adriatic Priority Axes	Specific Priorities	Focus	EUSAIR Pillar	Aligned SDG	Culture-Based Solutions	Quintuple Helix Contributions	Smart Specialization Strategy (S3)	Twin-Transition Focus/mapping
PA 1 Smart	SMEs	Boosting SME competitiveness with smart, green, and digital innovation.	Pillar 1: Blue Growth	SDG 8 (Economic Growth) SDG 9 (Innovation) SDG 14 (Life Below Water)	Cultural Entrepreneurship: Promoting local traditions and heritage for innovation.	**Education:** Training SMEs in digital skills and sustainability. **Economic:** Supporting SMEs to adopt green digital solutions. **Natural Environment:** Using digital technologies to monitor marine ecosystems.	**Innovation Ecosystem:** Promoting innovation through collaboration between SMEs, research centers, and government. **Smart, Green, Digital:** Encouraging the adoption of digital and green technologies.	**Green Transition:** SMEs adopting green and digital innovations. **Digital Transition:** Digital tools for smarter resource management.
PA 2 Green	Risk	Promoting climate adaptation, risk management, biodiversity, and energy efficiency.	Pillar 3: Environmental Quality	SDG 6 (Clean Water) SDG 13 (Climate Action) SDG 15 (Life on Land)	Cultural Heritage Protection: Using culture to foster environmental sustainability	**Natural Environment:** Applying smart sensors for environmental monitoring. **Civil Society:** Engaging citizens in sustainable practices. **Economic:** Promoting green technologies for resource management.	**Sustainability Innovation:** Developing green technologies for climate adaptation, energy efficiency, and biodiversity preservation.	**Green Transition:** Climate risk mitigation and biodiversity preservation. **Digital Transition:** Digital tools for environmental monitoring.
	Biodiversity			SDG 14 (Life Below Water) SDG 15 (Life on Land)				
	Energy		Pillar 2: Connecting the Region	SDG 7 (Afordable and Clean Energy), SDG 9 (Innovation), SDG 13 (Climate Action)			**Knowledge Economy:** Fostering research on green and digital transitions.	
PA 3 Connected	Intermodality	Improving connectivity and sustainable transport.	Pillar 2: Connecting the Region	SDG 9 (Infrastructure) SDG 11 (Sustainable Cities) SDG 12 (Responsible Production)	Cultural Tourism Transport: Integrating cultural heritage sites into sustainable transport networks.	**Education:** Promoting skills for digital mobility solutions. **Economic:** Leveraging digital tools for efficient transport networks. **Civil Society:** Encouraging adoption of sustainable transport.	**Smart Infrastructure:** Developing smart transport systems that enable efficient, green mobility. **Sustainable Connectivity:** Enhancing regional mobility and connectivity through digital platforms.	**Green Transition:** Green transport infrastructure. **Digital Transition:** Smart transport systems and cross-border mobility.
PA 4 Social	Skills	Promoting inclusive tourism and improving skills through cross-border actions.	Pillar 4: Sustainable and Competitive Tourism	SDG 4 (Education) SDG 8 (Economic Growth) SDG 10 (Reduced Inequality)	Cultural Heritage as Tourism: Promoting cultural tourism that enhances social inclusion.	**Education:** Providing digital and sustainable tourism training. **Economic:** Enhancing tourism economy through digital platforms. **Civil Society:** Engaging communities in sustainable tourism initiatives.	**Cultural and Tourism Innovation:** Leveraging digital tools to enhance the tourism experience and sustainability.	**Green Transition:** Sustainable tourism practices. **Digital Transition:** Enhancing tourism experience with digital tools.
	Inclusive Tourism			SDG 8 (Economic Growth) SDG 11 (Sustainable Cities), SDG 12 (Responsible Production)			**Inclusive Growth:** Promoting social inclusion through accessible tourism.	
PA 5 Governance	Administrative Efficiency	Improving governance through digitalization and efficiency.	EUSAIR Governance (Cross-cutting)	SDG 16 (Peace and Justice) SDG 17 (Partnerships)	Cultural Diplomacy in Governance: Using cultural exchanges to strengthen cross-border governance.	**Political:** Using digital tools for better policy and governance. **Civil Society:** Engaging citizens through digital governance platforms. **Media:** Using digital media to improve transparency and accountability in governance.	**Digital Governance:** Implementing e-governance systems for efficient administration and transparency. **Resilience in Policy Making:** Encouraging sustainable and transparent decision-making processes.	**Green Transition:** Sustainable governance practices. **Digital Transition:** E-governance solutions for transparency.

Fig. 5 Strategical integration of localized priorities within the eu strategical development plans. *Source* A. Lipovica

4. NSSS in Albania and Its Territorialization Process

Albania has registered in the Smart Specialisation Platform (S3P) since November 2017, with assistance from the Directorate-General for European Neighbourhood Policy and Enlargement Negotiations (DG NEAR) and the Joint Research Centre (JRC). The country has a National Strategy on Scientific Research, Technology and Innovation 2017–2022 and is associated with the EU Horizon 2020 science and innovation framework programme. In 2018, a comprehensive training on S3 focused on the Entrepreneurial Discovery Process (EDP) in Ljubljana. The Technology Transfer Capacity Building in the Western Balkans (EU4Tech) project aims to enhance technology transfer, contract research, and investor readiness. An action plan is planned to be finalized by the end of 2022, followed by a stakeholder dialogue in 2022.

Remaining challenges, as in many economies of the area, concern mainly the following fields:

- Availability of data to perform high quality quantitative analysis, especially in relation to the innovative and scientific potential;
- Ensuring the participation of stakeholders from the regional level during the EDP.

The Institute of Statistics in Albania (INSTAT) conducted an innovation activity survey from 2017 to 2019, identifying two NACE 1-digit industries with current and emerging innovation potential: J-Information and communication technologies and M-Professional, scientific, and technical activities. The quantitative analysis highlighted the importance of the ICT sector, with subsectors such as radio broadcasting, television programming, wired telecommunications, and satellite telecommunications activities. The qualitative analysis combined an online survey with 60 in-depth interviews to validate previous statistical results and identify subsectors with high potential for innovation. The analysis identified subsectors such as the blue economy, healthy food, renewable resources energy, and Extended Business Process Outsourcing as potential areas for further exploration. Albania GDP in 2020 was estimated at 1,644,077 million Albanian Lek (ALL), with agriculture, forestry, fishing, wholesale and retail trade, repair of motor vehicles and motorcycles, and construction being the largest sectors. The report comprises eight chapters, with the first chapter providing an economic outlook and methodology.

The Covid-19 pandemic has significantly impacted Albania economy, with a contraction of about 4% in 2020. The private sector has faced shortages of clients and liquidity, while trade has been affected, particularly in industrial sectors. The government has adopted two support packages for vulnerable people and businesses, allocating 45 billion ALL (2.8% of GDP). Agriculture and real estate activities are the only sectors growing, while trade, transport, and hospitality services, wholesale and retail trade, and accommodation and food service activities have declined. Tourism, which accounts for over 20% of Albania GDP, has been severely affected due to travel restrictions. The Institute of Statistics in Albania (INSTAT) conducted an innovation activity survey, identifying two NACE 1-digit industries with high potential for innovation. The analysis identified subsectors with high potential for further exploration, such as the blue economy, healthy food, renewable resources energy, and Extended Business Process Outsourcing. Albania GDP in 2020 was estimated at 1,644,077 million Albanian Lek (ALL), with GDP per capita at 579 thousand ALL (€ 4681).

Albanian agricultural sector, which accounts for 24% of its territory, has seen significant growth since 2010, with exports reaching 30 million USD in 2016 and 41 million USD in 2019. The Covid-19 pandemic has severely impacted the economy, with a contraction of about 4% in 2020. The government has adopted two support packages for vulnerable people and businesses affected by the pandemic, allocating 45 billion ALL (2.8% of GDP). Agriculture and real estate activities are the only sectors growing, while trade, transport, hospitality services, wholesale, retail trade, accommodation, and food service activities are the least.

The analysis identified 13 industries with static economic potential and 16 with dynamic potential, with two having both static and dynamic potential: J61.3 Satellite telecommunications activities and N82.2 Activities of call centres. The energy sector should be added to these industries. The Smart Specialisation design should focus on industries that could provide the largest increase in wages and employment from targeted research and innovation investment.

The Albanian BPO sector, despite high growth potential, faces competition and digital technologies. Positive assets include cultural proximity, economic fabric, and international business advisory players. Opportunities may arise from expanding BPO perimeter, exploring knowledge-intensive market segments, and integrating IT solutions.

Smart Specialization Strategies (S3) serves as an evolving role with a critical framework for territorial development, with particular attention to their implementation within metropolitan and macro-regional contexts. Originating as a place-based innovation policy within the European Union, S3 now offers a strategic lens to align regional assets, societal needs, and global challenges, notably the green and digital transitions.

The concept of territorializing smart specialization is rooted in the recognition that planning must be adaptive across temporal and spatial scales. By integrating real-time intelligence, stakeholder participation, and distributed innovation ecosystems, S3s enable cities and regions to respond to complex urban dynamics and climate vulnerabilities. The Quintuple Helix Model further enhances this approach by embedding the environment and civil society alongside academia, industry, and government in innovation governance.

Research infrastructures and universities emerge as pivotal actors in this territorial innovation process. Their ability to function as "living laboratories" facilitates the transfer of knowledge to regional stakeholders, supports applied innovation, and fosters a culture of collaborative governance. Effective mobilization of these institutions is essential to stimulate entrepreneurial ecosystems, address socioeconomic disparities, and advance SDG localization.

In the context of intelligent Metropolitan systems, S3 offer a methodology to manage the complexity of urban transformation. Through digital social hubs, innovation districts, and localized strategic platforms, S3 fosters systemic feedback loops between local experimentation and multi-level policy learning. This model supports the development of cities as adaptive, inclusive, and resilient systems capable of engaging with global networks while remaining grounded in local identity.

The case of Vlora, Albania, demonstrates both the potential and the limitations of implementing S3 in regions with emerging innovation systems. While national strategies have identified key priority sectors, including ICT, energy, tourism, and the blue economy, challenges persist in data availability, stakeholder engagement, and coordination across scales. Integrating cultural and historical dimensions into metropolitan cartography highlights the need for strategies that are not only spatially aware but also socially and culturally contextualized.

To conclude, Smart Specialization Strategy is not merely a policy tool but a strategic framework for territorial cohesion and transformation (see Fig. 6). It connects the local and the global, the economic and the ecological, and the technical and the cultural, offering a dynamic roadmap for regions to build resilient innovation ecosystems fit for the challenges of the twenty-first century.

5. The Smart Specialisation Strategy in Vlora, Albania—An Exercise in Sustainable and Collaborative Territorialisation

The exercise conducted in Vlora provided an essential opportunity for operational testing of the European Smart Specialisation Strategy (S3) concept within an emerging national and municipal context. The aim of the exercise was to illustrate, in dialogue with local government, the necessity for the Albanian state to establish a

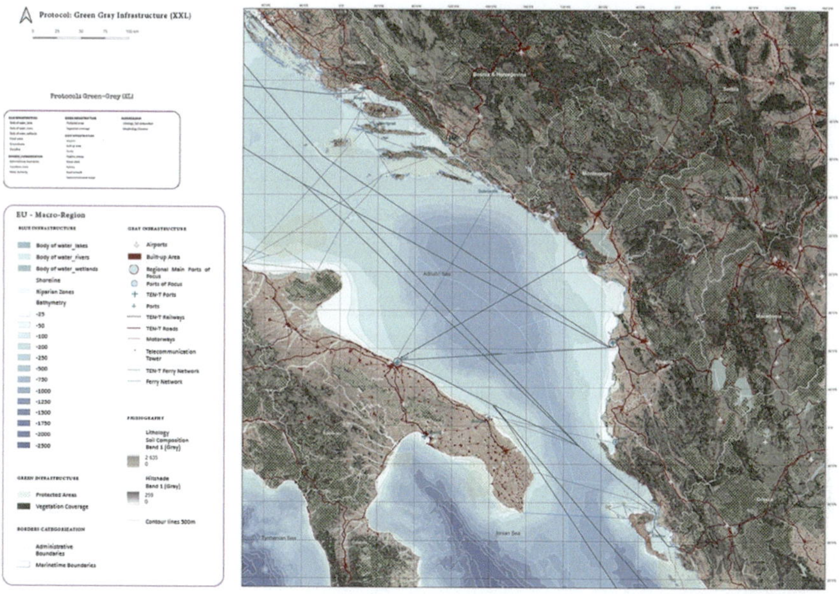

Fig. 6 XXL Green-Gray infrastructure, Balkans in relation to EU. *Source* A. Lipovica

coherent national smart specialisation strategy and, simultaneously, to encourage the adoption of such a strategy at the local level through adaptive localisation processes.

In comparison to the Municipality of Vlora, the focus was on the territorialisation of the S3 according to the model of the quintuple helix, which recognises the environment as an actor entitled to participate in defining development trajectories, alongside university, industry, government, and civil society. This approach allows for framing the environment not as a mere constraint or resource but as an active player in the governance of innovation.

During the research, the importance of considering the specialisation strategy not only as a technical tool, but also as a political, cultural, and social process rooted in the territories, was highlighted. Indeed, the strategy requires joint action between the central and local levels, where the municipality assumes an essential role in translating national objectives into concrete choices, consistent with the characteristics, vocations and vulnerabilities of the urban and metropolitan context.

The city of Vlora was analysed as a strategic urban node with potential for specialisation in sectors such as ICT, sustainable tourism, renewable energy and the blue economy. However, the difficulties that emerged in the retrieval and utilisation of reliable statistical data, as well as the lack of structured participation of local actors in the Entrepreneurial Discovery process, made evident the need to strengthen institutional capacities, multi-level governance tools and territorial information systems.

Universities and research infrastructures were identified as central elements for stimulating local entrepreneurial ecosystems and strengthening the interaction between applied research and territorial development. The adoption of an S3 on an urban and metropolitan scale was understood as an opportunity to transform the territorial system into a living lab capable of integrating digital, green and cultural transitions.

Vlora experience demonstrates how the success of the smart specialisation strategy depends on its ability to adapt to the territory and actively mobilise local resources—cognitive, institutional, and environmental—through a shared vision of development. Finally, the urgency to consider S3 not only as an economic growth agenda, but as a strategy of systemic transformation, capable of promoting cohesion, resilience and inclusiveness in territories in transition, emerged.

5 Case Studies in EU and in the Indo-Pacific Corridors

The Indo-pacific policies are crucial both for the economic corridor and for the EU, as they will completely change the actual balance in the world economic sectors. This is why some case studies are presented here, starting from the Italian territory and spreading through the Indo-pacific corridor itself.

The case study of Melzo examines the relationship between urbanization and traditional agriculture in Lombardy, Italy. The study highlights the challenges faced by agricultural communities in rapidly urbanizing areas, such as land use changes, resource competition, and economic pressures. Cascina Grande, a historic farm in Milan, serves as a representative case, illustrating both threats and opportunities posed by urban expansion into traditional farming systems. The study also highlights the importance of integrating digital innovation into traditional agricultural frameworks and strategic logistic connectivity in smart metropolitan systems. The study proposes a green and digital twin transition based on an integrated closed-loop system, linking sustainable agricultural production, smart logistics, and energy/resource cycles, to illustrate how peri-urban areas can evolve into resilient nodes within global digital economic communities. The case study also highlights the importance of balancing economic growth with environmental sustainability in urban development, with the potential for partnership and collaboration between traditional farming communities and the industrial sector.

The Melzo AgriHub project aims to integrate green and digital innovations into a unified strategy, focusing on ecological farming techniques and digital platforms. The closed-loop system at Cascina Grande uses regenerative agriculture, automated machinery, and modular facilities for efficient harvesting and processing. The project also prioritizes community engagement and social sustainability, with public green spaces and cycle paths connecting the farm to neighbouring municipalities. The AgriHub model demonstrates how modular design, regenerative practices, and digital innovation can redefine agricultural processes in the face of urban expansion and technological evolution. Further research is needed to understand the long-term impacts of these models and the evolving dynamics of farmer-industry-community collaboration.

Piacenza, a strategically located city in Italy, is facing challenges like urban sprawl and climate change. The project proposes a holistic landscape strategy using the Sustainable Urban Drainage Systems approach to redefine the urban-river relationship. The project identifies four strategic zones, each corresponding to different levels of proximity to the Po River and integrates them into a connected ecological corridor. The landscape is designed to purify water and soil, promote digital economy, and support sustainable urban development.

The project aims to link Piacenza to the Po River through a gradual ecological and urban transition. It focuses on transforming abandoned spaces into green infrastructure, addressing environmental risks, promoting biodiversity, and introducing digital amenities. The station hub, a digital and ecological gateway, will function as a multifunctional hub, integrating ecological design with digital innovation. Key features include a smart waiting park, a digital agricultural market app, urban farming spaces, interactive information kiosks, and green infrastructure. The project serves ecological, social, and economic purposes by creating multifunctional spaces and improving community engagement. Piacenza strategic location and historical significance position it as a model for mid-sized cities facing similar challenges.

Landscape design plays a crucial role in enhancing people quality of life and societal values, including economic development. Climate change is increasing the frequency and intensity of natural disasters and flooding events, necessitating innovative solutions and electronic tools for global multifunctional integration. Resilient landscape design approaches, such as those used in the Xiangtan section of the Xiangjiang River in Hunan Province, China, can help reduce the impact of unstable water levels on the ecology of the site and create a multi-dimensional landscape corridor. By living in harmony with nature and building multiple layers of defence, communities can benefit from more efficient benefits. The design should be tailored to local conditions, protect architectural ruins, and incorporate native aquatic plants, wetlands, fishponds, artificial lakes, and channels. Resilient landscape planning and design offer a way forward for communities, utilizing multi-layered protection systems with diverse, scalable elements that can fail safely in the event of a disaster.

The project on the west bank of the Xiangjiang River in Xiangtan aims to upgrade the landscape without destroying its original ecological appearance. The plan includes two small dams, 30 km of trails, 15 km swamp trails, landscape buildings, 12 landscape squares, landscape bridges, 12 artificial wetlands, 3 wetland parks, 3 beaches, 19 Diaoyutai, 2 wind farms, and 224.2 hectares of new flood control green infrastructure. The project also includes a small dam designed by Yangmei Island, which will generate electricity during the rainy season. The project also aims to improve the attractiveness of waterfront space, attract investment, and drive local tourism.

Metro Manila, the most populous metropolitan area in the Philippines, faces challenges due to rapid urbanization and ecological issues. The region dense urban environment, particularly in coastal areas like Navotas, is particularly affected by environmental disasters and climate change. A project called "May Nilad" aims to redefine resiliency for coastal communities through aqua silviculture in Navotas, using an E-gateway approach. The project focuses on circular economy principles, focusing on regenerative systems, resource productivity, and production/consumption systems. The project includes restoration of coastal ecosystems, regenerative aqua silviculture, waste-to-resource innovations, community-led local resource management, and resilient public spaces. This project demonstrates resilience beyond disasters, promoting self-sufficiency and resilience in coastal communities.

1. **Melzo, Italy**

This case study explores the dynamic interplay between urbanization and traditional agriculture in Lombardy, one of the Italian most economically and demographically significant regions. The focus progressively narrows from the regional scale to the municipalities of Liscate and Melzo, emblematic of the pressures of industrialization on rural systems, culminating in an in-depth examination of Cascina Grande, a historic farm situated in the peri-urban landscape of Milan.

In Lombardy, urban and natural environments coexist in a complex and multifaceted setting, where rapid development continuously reshapes the agricultural fabric. Through an analysis of land use changes, resource competition, and economic pressures, this study highlights the critical challenges faced by agricultural communities in rapidly urbanizing areas.

Cascina Grande serves as a representative case, illustrating both the threats and opportunities posed by urban expansion into traditional farming systems. The farm strategic location offers insights into how digital and green transitions can revitalize peri-urban agriculture and foster urban–rural continuity, crucial elements for sustainable metropolitan development.

While geographically rooted in the Milan Metropolitan Region, the dynamics explored through Cascina Grande offer broader insights relevant to the formation of E-Global Gateways in smart metropolitan systems. The integration of digital innovation into traditional agricultural frameworks, combined with strategic logistic connectivity, mirrors the challenges and opportunities faced by emerging urban systems in the Indo-Pacific region.

Through the proposal of a green and digital twin transition based on an integrated closed-loop system, linking sustainable agricultural production, smart logistics, and energy/resource cycles, this case study illustrates how peri-urban areas can evolve into resilient nodes within global digital economic communities, balancing local heritage with global connectivity.

Zooming in from the regional context to the local scale, Melzo emerges as a significant example of the interaction between urbanization and agricultural transformation. Project focus Cascina Grande stands as a good example of the rich agricultural and historical legacy of Milan. As one of the city venerable farms, it has witnessed the ebb and flow of time. Situated approximately 20 km to the east of Milan, Cascina Grande finds itself strategically positioned on the administrative boundaries of two significant municipalities: Melzo and Liscate (see Fig. 7). Melzo is about 18 km from Milan, while Liscate is roughly 22 km away.

Melzo, strategically positioned in the eastern suburbs of Milan, stands as a testament to the transformative power of infrastructure and its role in industrialization. Located at the heart of Italy most dynamic economic region, Melzo benefits from an

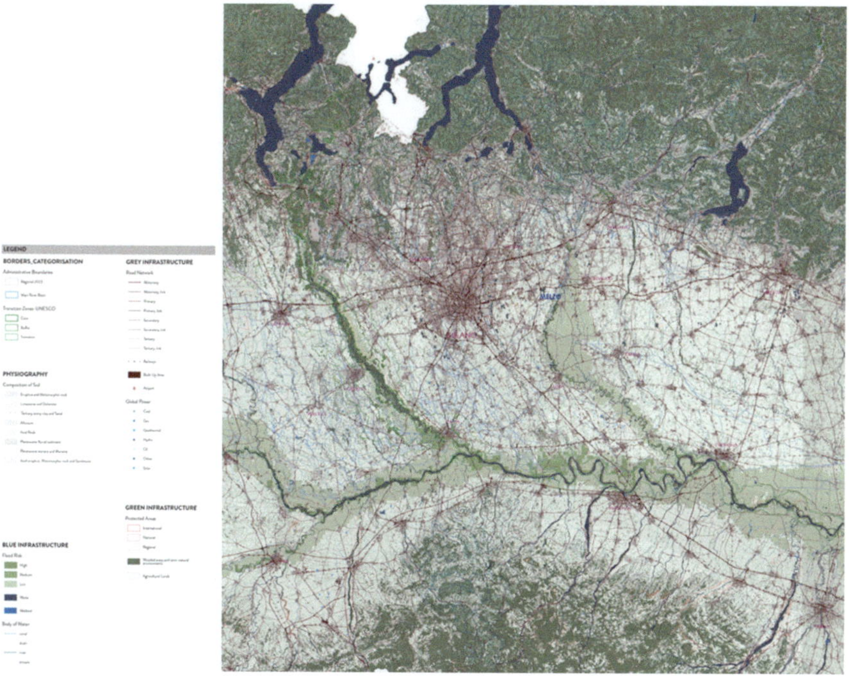

Fig. 7 Green-grey protocol map of Melzo—XL scale. *Source* S. Keskin

intricate network of highways, including the A4, A1, the recently inaugurated A35, and the Tangenziale Est Esterna di Milano (TEEM). These roads seamlessly connect it to major industrial and consumer hubs like Bergamo, Monza, and Brescia. The urban fabric in Melzo has been significantly influenced by its transportation infrastructure, resulting in a distinct spatial division. The town layout, dissected by roads and the rail network, has led to discernible patterns in its development. On the right side of SP (Strada Provinciale-Provincial Road) 13 Monza-Melzo, the residential expansion plateaued in the early 2000s. In contrast, the burgeoning industrial sector began to encroach upon agricultural lands, pushing the city boundaries towards the northern limits of Liscate. This industrial sprawl necessitated the repurposing of some agricultural lands, leading to the loss or transformation of several farms. Historical cartography, such as the Martesana maps from the 1820s, provides a snapshot of the past of Melzo, illustrating a town cradled by agricultural expanses and punctuated by primary transit routes (see Fig. 8). A closer examination of orthographic imagery from subsequent decades reveals the town evolving urban morphology. By the turn of the twenty-first century, residential and public facility zones had largely reached their current extents. However, the twentieth century marked a pivotal era for the industrial landscape in Melzo. Beginning in the early 1900s, there was a noticeable emergence and growth of industrial clusters. By the late twentieth century and onwards, this industrial cluster had expanded south to the Provincial Road 14 (SP

Fig. 8 Ecosystem services protocol map of Melzo—L scale. *Source* S. Keskin

14). The strategic introduction of SP 14 between 1954 and 1975 played a dual role: it acted as a buffer, limiting residential expansion to Liscate north, while simultaneously fostering the development of the "Zona Industriale" of Liscate in the direction of Vignate (see Fig. 9).

In essence, the urban evolution of Melzo is a testament to the interplay between transportation infrastructure and land-use decisions. Its strategic position has made it a magnet for logistics and transportation enterprises, underscoring the town significance in the broader regional industrial narrative. During this period, key industrial entities, like the Galbani Egidio production campus established in 1911, set up operations in Melzo to capitalize on its strategic and advantageous location (Heritage Galbani, https://www.galbani.com/heritage.php).

The TEEM is a significant addition to the region transportation infrastructure. As an outer eastern bypass of Milan, it further enhances connectivity in Melzo, providing faster and more efficient routes for both commercial and passenger traffic. Additionally, the 19th-century inception of the Milan–Venice railway, which was conceived to bridge the Austrian Kingdom of Lombardy-Venetia dual capitals, marked a significant turning point for Melzo (see https://it.wikipedia.org/wiki/Ferrovia_Milano-Venezia). The railway phased development, from its early segments in the mid-1800s to the introduction of a more direct route in 1878, has continually enhanced connectivity in Melzo. Today, this railway not only links Melzo to key cities like Brescia,

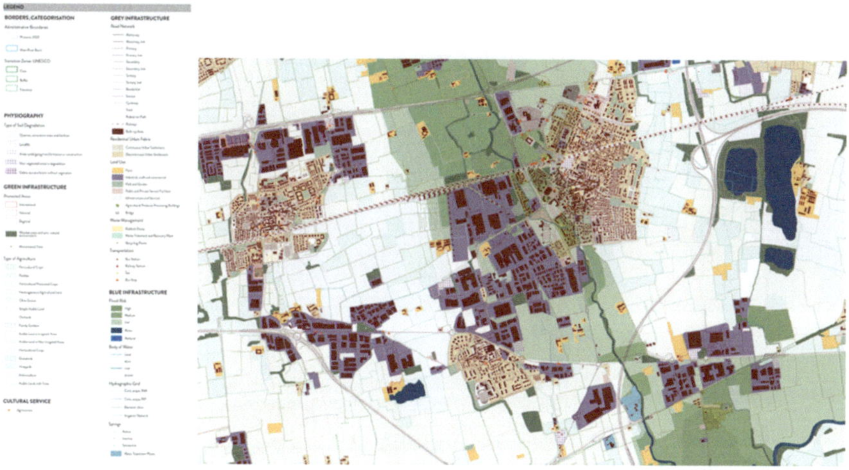

Fig. 9 Green-grey protocol map of Melzo—S scale. *Source* S. Keskin

Verona, and Venice but also accommodates the S5 and S6 commuter lines, further integrating Melzo into the broader Milanese metropolitan area. Nationally and across Europe, Melzo strategic location at the crossroads of East, West, North, and South has elevated it as an Italian hub for maritime and continental traffic. The synergy of road, rail, and the addition of TEEM has undeniably catalysed Melzo industrial growth. As these transportation networks expanded and modernized, they brought with them a surge in industrial activities, trade opportunities, and economic prospects (see Fig. 10). The town proximity to Milan, coupled with its enhanced connectivity, has made it an attractive locale for businesses and industries, further cementing its role in the regional industrial narrative. In essence, Melzo industrialization is intrinsically tied to its infrastructural evolution, showcasing how transportation networks can shape and drive a town economic destiny.

As Melzo navigates the complexities of industrial expansion and urban development, the town stands at a crossroads, balancing economic growth with environmental sustainability. The encroachment of industrial activities on green spaces and agricultural lands has raised concerns, highlighting the need for innovative solutions to preserve the town agricultural heritage while fostering industrial prosperity. In this context, the potential for partnership and collaboration between the traditional farming community and the industrial sector becomes a beacon of hope for sustainable urban development. By forging synergies, these two seemingly disparate entities can create a mutually beneficial relationship, ensuring the coexistence and flourishing of both sectors. The adoption of regenerative agricultural techniques plays a pivotal role in this partnership, offering a pathway for traditional farms to enhance their productivity and sustainability. These practices, rooted in ecological stewardship, aim to rejuvenate the soil, optimize water usage, and promote biodiversity, creating a resilient agricultural landscape capable of withstanding the pressures of urbanization (see Fig. 11).

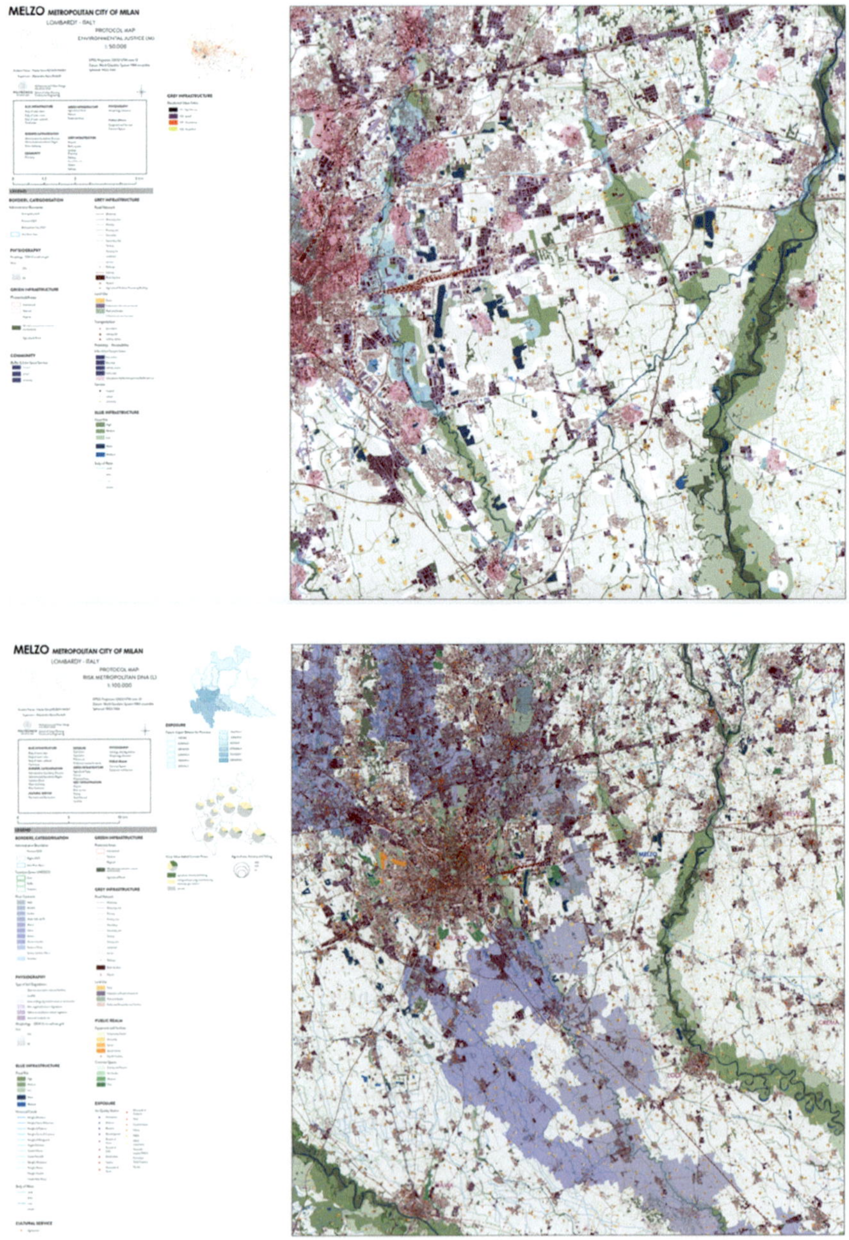

Fig. 10 Environmental justice and risk metropolitan DNA protocol maps of Melzo—L scale. *Source* S. Keskin

Fig. 11 Satellite views of Melzo. *Source* Google Earth

The town intricate water and irrigation networks, including the Torrente Molgora and the minor hydrographic networks of roggia and fontanile, are integral to this collaborative effort. By modernizing these networks and integrating sustainable water management practices, Melzo can ensure that both agriculture and industry have access to the vital resources they need to thrive. This collaborative approach, fostering a partnership between traditional farming and industrial activities, serves as a model for sustainable urban development. It underscores the importance of collective action and shared responsibility in building a future where economic growth and environmental preservation go hand in hand. As Melzo leverages its strategic location and rich agricultural heritage, the town is poised to become a beacon of sustainability, demonstrating that through collaboration and innovation, it is possible to create a resilient and prosperous urban landscape.

Against this backdrop, the site of Cascina Grande becomes especially important. Understanding the broader framework of the Parco Agricolo Sud Milano is essential to appreciate the strategic and environmental significance of Cascina Grande. This protected agricultural belt plays a crucial role in preserving traditional farming practices, safeguarding ecological corridors, and managing the urban–rural interface within the Milan metropolitan area. Characterized by the traditional 'cascina' system, these farmsteads embody the agricultural typologies of the Po Valley [6]. Historically self-sufficient, cascine were central to large-scale farming operations, integrating residential and utility buildings designed to support agricultural productivity. Within the park, the structure and distribution of farm typologies differ between the northern and southern sectors, reflecting variations in environmental conditions and historical land use. In the north, smaller farm plots and a greater presence of industrial development have led to more compact, vertically designed farmsteads, optimized for solar exposure. In contrast, the southern sector, benefiting from more favourable agricultural conditions and larger land holdings, hosts expansive farmsteads that often remain under single ownership, integrating both traditional and modern agricultural practices.

The origins of Cascina Grande can be traced back to the seventeenth century, with records suggesting its construction around 1650 (www.lombardiabeniculturali.it/architetture/schede/MI100-03883/). Historically, farmsteads like Cascina Grande were integral parts of the socio-economic and agricultural fabric of the Lombardy region, operating as self-sufficient agricultural units. The cascina underwent transformations in the eighteenth century, as reflected in its depiction in the Teresian cadastre.

Today, Cascina Grande retains its dual function as a residence and an agricultural production site. Its architectural integrity including the roof, walls, and structural elements has been well preserved [7], reflecting broader restoration trends in the Liscate area, where historical rural assets are increasingly maintained. Studies by Pergami and Antonacci (www.lombardiabeniculturali.it/architetture/schede/MI100-03883/) further document the farm evolution over time. The layout follows the traditional 'Corte Monoaziendale' model, with residential buildings, stables, and barns arranged around a central courtyard to optimize farming operations.

Strategically located between SP13 and SP14, and surrounded by growing industrial settlements, Cascina Grande sits at a critical intersection of agricultural heritage and metropolitan expansion. Despite pressures from urbanization and logistics development, it continues to function within an agricultural matrix, preserving the historical landscape structure. Nearby, Cascinello del Dosso exemplifies the diversification of modern agriculture, hosting operations such as cattle breeding and equestrian activities (https://www.coobiz.it/azienda/melzo-allevamento-bovini/%20co1034565#descrizione), highlighting how traditional and specialized farming practices coexist in this peri-urban environment.

The hydrological context also plays an important role. Located approximately 500 m from the Molgora Torrent (https://it.wikipedia.org/wiki/Molgora), Cascina Grande benefits from irrigation opportunities but faces potential flood risks, as identified by regional hydrological studies [8]. Traditional irrigation systems, including Fontanile Mora and Fontanile Santo Stefano, although partially degraded, continue to influence local agricultural viability and ecological dynamics. Managing these water resources is essential for ensuring the long-term sustainability of both agricultural practices and environmental health in the area.

Considering these dynamics, a new model aims to integrate green and digital transitions through a closed-loop system, positioning Cascina Grande as a prototype for fostering resilient, connected, and sustainable peri-urban landscapes, a model relevant not only for the Milan Metropolitan Region but also for future smart metropolitan systems globally.

The evolving landscape surrounding Cascina Grande illustrates the growing tension between traditional agricultural spaces and expanding urban-industrial settlements. Industrial clusters and residential areas are steadily encroaching upon agricultural lands, threatening not only the physical space available for farming but also the socio-ecological fabric that sustains rural heritage. Cascina Grande, positioned between SP13 and SP14 within the Parco Agricolo Sud Milano, sits at the heart of this dynamic, facing mounting pressures from metropolitan growth while retaining its agricultural identity.

In response to these challenges, the project "Integrating Twin Transition in Traditional Agriculture: Digital and Green Evolution of Cascina Grande in the Peri-Urban Industrial Context" envisions Cascina Grande as the nucleus of a sustainable, digitally integrated agricultural model under the AgriHub Cascina Grande concept. The project introduces a closed-loop system that spans the entire agricultural cycle, from regenerative production to processing, distribution, and resource recycling, aiming to foster resilience within the peri-urban landscape.

At the foundation of this vision is the implementation of the Twin Transition, where green and digital innovations are integrated into a unified strategy. The green transition is embodied in ecological farming techniques such as agroforestry, organic farming, cover cropping, no-till practices, composting, and rotational grazing. These approaches enhance soil health, biodiversity, and productivity while reinforcing environmental stewardship. Concurrently, the digital transition supports the project through precision farming, environmental monitoring, and the development of collaborative digital platforms that connect farmers, communities, and local industries. This integration fosters collaborations, partnerships, and community engagement, enabling a shared commitment to sustainability and circular economy principles.

The closed-loop system at Cascina Grande (see Fig. 12) begins with regenerative agricultural production, underpinned by sustainable water management strategies linked to nearby systems like Fontanile Santo Stefano. Harvesting is optimized through the introduction of automated and renewable-energy-powered machinery, improving efficiency and reducing environmental impact (see Fig. 13). Harvested products are processed and stored on-site in modular facilities, designed for flexibility and expansion based on demand, following spatial principles inspired by the Centuriazione Romana.

A critical element of the AgriHub vision is its logistical integration. Through the establishment of a direct connection to Provincial Road 14, Cascina Grande is transformed from a traditional farmstead into a logistically strategic node, facilitating the efficient distribution of agricultural goods to regional markets and supply chains.

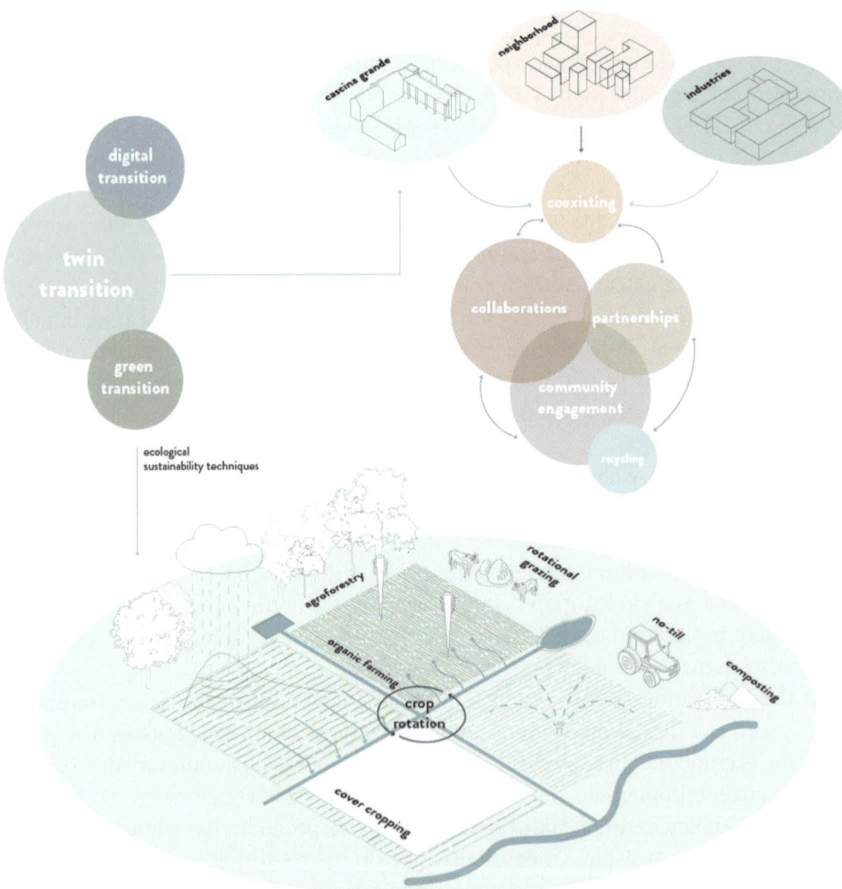

Fig. 12 Mapping the twin transitions with a focus on ecological practices and community coexistence. *Source* S. Keskin

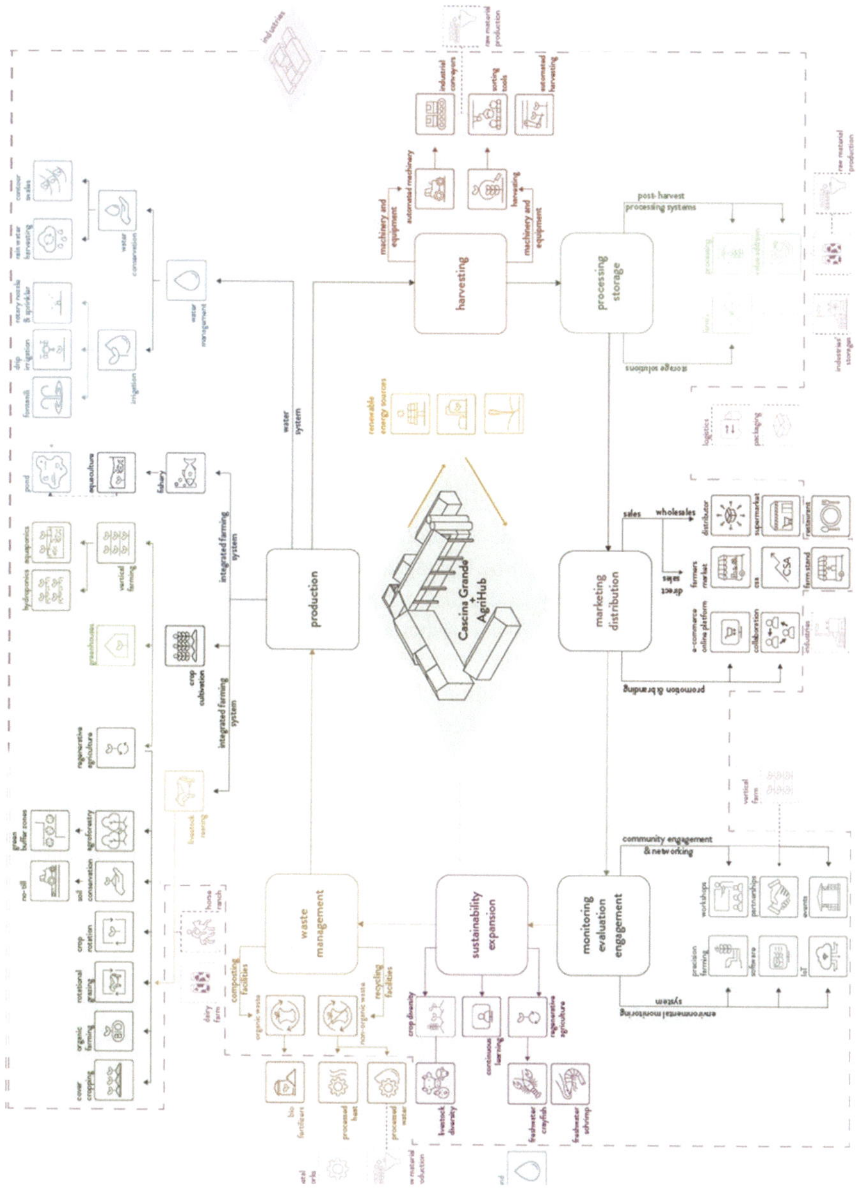

Fig. 13 Closed-loop farming system scheme. *Source* S. Keskin

This infrastructural link strengthens Cascina Grande role within a broader smart metropolitan system, ensuring that rural production remains economically viable and resilient.

Resource management is closed through comprehensive waste recycling processes: organic waste is composted and reintegrated into farming activities as natural fertilizer, while non-organic waste is processed through dedicated recycling facilities. This cycle minimizes external dependencies, reduces environmental impact, and reinforces the sustainability of the local agro ecosystem.

Beyond the agricultural operations, the AgriHub project prioritizes community engagement and social sustainability. The proposal includes the development of public green spaces and cycle paths that connect Cascina Grande with neighbouring municipalities such as Melzo and Liscate. These infrastructures not only enhance mobility and recreation opportunities but also strengthen the cultural and ecological integration of the farm within the metropolitan community.

Through this integrated model blending regenerative agriculture, smart technologies, strategic logistics, and community-driven initiatives, Cascina Grande stands as a prototype for future peri-urban development. The AgriHub project demonstrates how agricultural heritage sites can be reimagined as dynamic, resilient, and sustainable hubs, contributing actively to the broader goals of metropolitan smart systems while preserving the vital link between urban society and its rural roots.

This study, grounded in the principles of metropolitan discipline, has explored the integration of twin transitions within the context of traditional agriculture, using Cascina Grande as a focal point. The application of metropolitan cartography and spatial analysis has been instrumental in visualizing and strategizing the transformation of this peri-urban agricultural landscape into a resilient and adaptive node within the Milan metropolitan system.

Through the proposed AgriHub Cascina Grande model, this work demonstrates how modular design, regenerative practices, and digital innovation can be combined to redefine agricultural processes in the face of urban expansion and technological evolution. By embedding these strategies into a closed-loop system supported by smart logistical connectivity, Cascina Grande not only safeguards its local agricultural heritage but also contributes to broader metropolitan resilience. This transformation echoes the principles underpinning the E-Global Gateways vision, where digitally connected, ecologically sustainable, and economically vibrant nodes collectively reinforce smart metropolitan systems.

Looking ahead, several avenues for further research and exploration emerge. The scalability and adaptability of the AgriHub model across diverse peri-urban contexts present promising opportunities: how might regenerative agriculture and modular strategies be customized to different ecological, cultural, and socio-economic environments? Moreover, the long-term impacts of integrating digital monitoring, precision farming, and circular resource management into traditional agricultural systems warrant continuous observation. Understanding the evolving dynamics of farmer-industry-community collaboration will also be critical to ensuring the sustainability, social acceptance, and replication of such models. Ultimately, this study lays the groundwork for a transformative approach to peri-urban agriculture, one that balances tradition and innovation, rural continuity and metropolitan integration. By aligning local action with global strategies, projects like AgriHub offer valuable contributions to rethinking the future of agriculture within the framework of resilient, digitally connected, and sustainable global urban systems.

2. **Piacenza, Italy**

Piacenza is a medium-sized city in the Emilia-Romagna region, north of Italy. It is strategically located for its geographical, transport, and economic position.

The city is between the Po River to the north and metropolitan city like Milan and Bologna. It has historically served as a gateway city, a critical node for railways and road transportation, and has an important role in the logistics and food production sectors. The city economic development is directly dependent on accessibility and infrastructure strength, especially due to its proximity to the Po River and the country motorway network.

However, like many post-industrial cities, Piacenza has faced contemporary challenges like urban sprawl, underutilized industrial lands, and fragmented green spaces, which have limited sustainable growth. The historical relationship between Po Rive and Piacenza has become weak. Reducing the role of water and the river in economic, urban life, and ecology. Simultaneously, climate change introduces new risks: frequent floods, increasing heat waves, and rising levels of soil pollution.

Today, Piacenza must face environmental issues, and the broader digital and economic transformation reshaping cities across Europe. The transition toward a knowledge-based economy, the rise of digital infrastructure, and the need for climate adaptation all converge in this context.

This project, "Climate change and water diplomacy: giving a new role of the Po River in Piacenza" positions Piacenza as a case study where ecological resilience, digital transition, and landscape-led regeneration can converge. The proximity to the Po River becomes a catalyst for rethinking infrastructure, creating new public spaces, and integrating ecological and technological systems. In this vision, Piacenza is no longer a peripheral city but a pilot for climate-responsive and digitally enabled urbanism (see Fig. 14).

The project proposes a vision that redefines the urban-river relationship between Piacenza and the Po River. Based on the pressing need to respond to climate change, it introduces a holistic landscape strategy using the Sustainable Urban Drainage Systems (SUDS) approach. The project identifies four strategic zones—each corresponding to different levels of proximity to the Po River—and integrates them into a connected ecological corridor (see Fig. 15).

The landscape is designed not only to purify water and soil but also to become a digital and economic catalyst. Rainwater is collected, stored, and filtered naturally, while areas are adapted to support public uses like walking, resting, or temporary digital working spots powered by renewable energy. In the station zone, smart benches, digital kiosks, and energy-efficient micro-hubs support a digital economy around mobility and waiting zones.

Each zone builds upon a core ecological function—buffering noise, collecting runoff, reducing heat island effect, and preventing floods. These natural services are

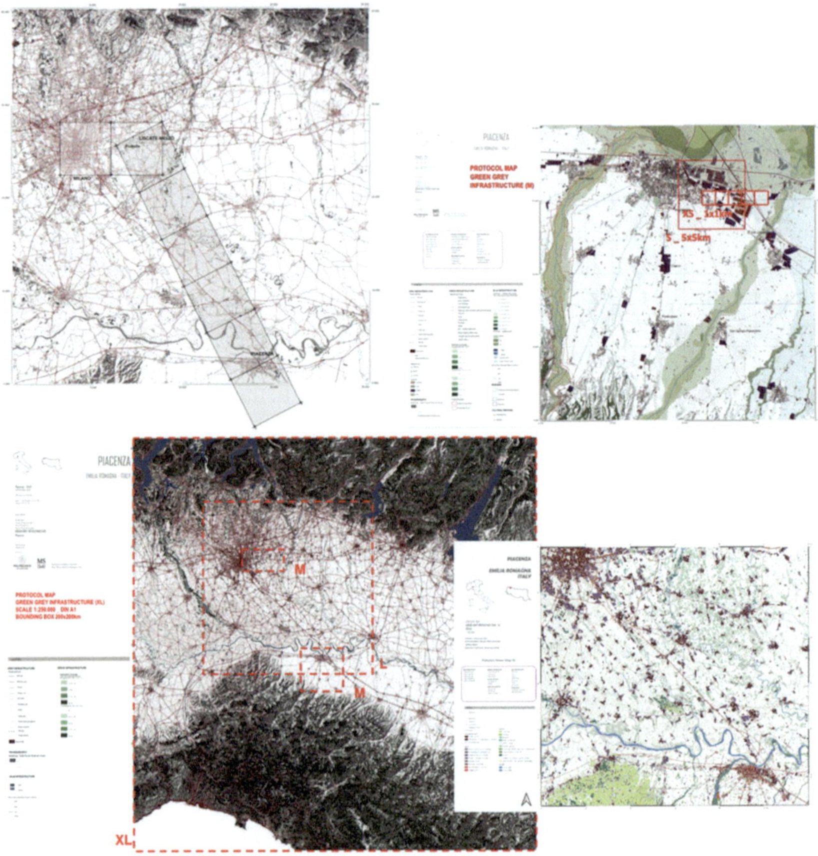

Fig. 14 Protocol maps of Piacenza. *Source* MsLab

integrated with digital interventions to promote real-time environmental monitoring and public accessibility. Through this syn-project, landscape becomes not only a climate adaptation tool but a spatial platform for new forms of economy and digital life (see Fig. 16).

The general master plan of this project builds a coherent vision across four different zones, linking the city of Piacenza to the Po River through a gradual ecological and urban transition. Each zone contributes to the Sustainable Urban Drainage Systems (SUDS) strategy while responding to different spatial, ecological, and social functions. Rather than designing fragmented spaces, the master plan offers continuity through green corridors, water management systems, and multi-functional landscapes.

At the heart of the master plan is the transformation of abandoned and underutilized spaces—railways, industrial zones, and buffer lands—into productive urban green infrastructure. These interventions not only address environmental risks like

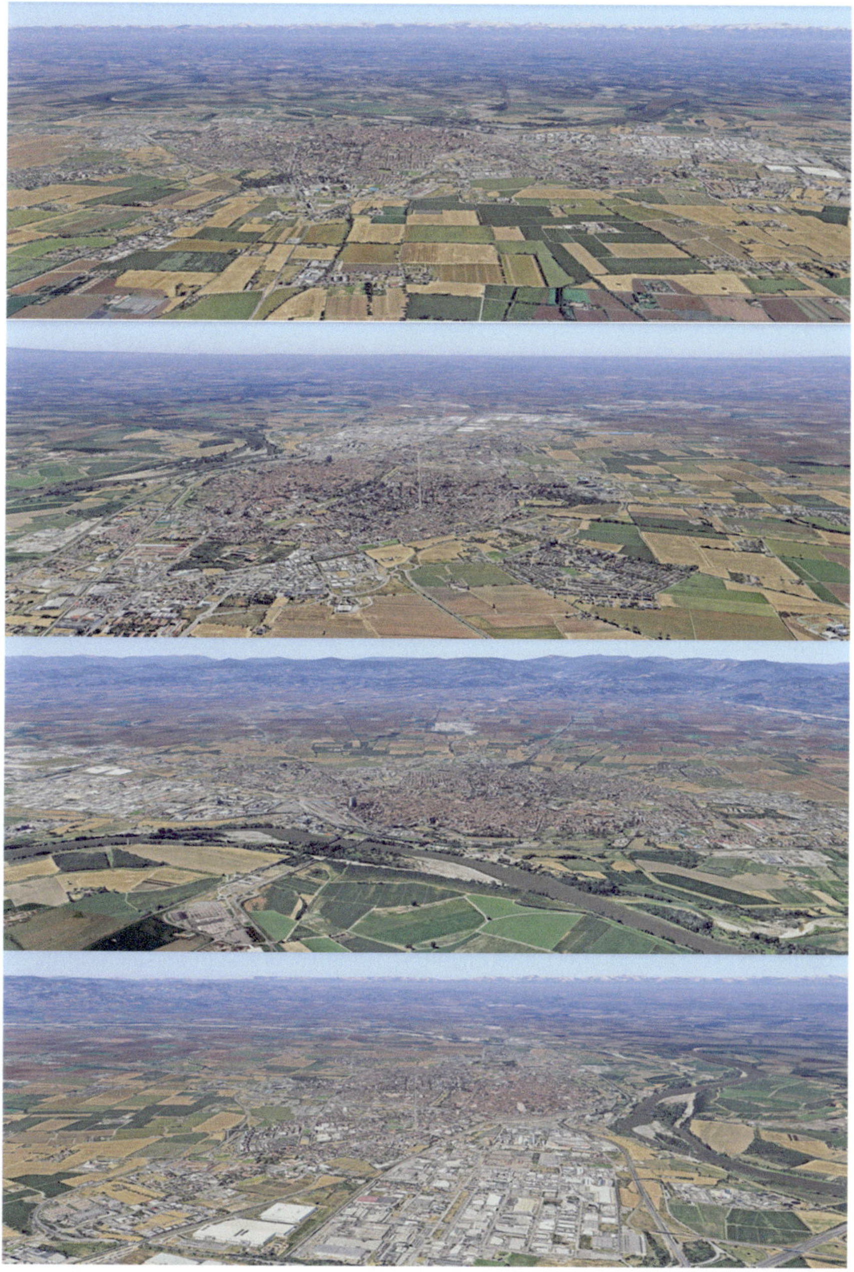

Fig. 15 Satellite views of Piacenza. *Source* Google Earth

Fig. 16 Masterplan of the station area in Piacenza. *Source* J. Kianisadr

flooding and soil pollution but also encourage biodiversity, support microclimates, and introduce new digital amenities (see Fig. 17).

From the compact green entrance near the station to the wide flood-retaining landscapes near the river, the plan defines a gradient of use: urban proximity, ecological purification, water storage, and natural regeneration. The connectivity between zones is maintained through walking paths, permeable surfaces, and vegetation patterns.

Digitally integrated elements such as sensor-activated lighting, solar benches, and digital information boards connect people to data and to place. In this way, the general master plan is a syn-project of landscape, water diplomacy, climate action, and digital innovation.

The station hub is a Digital and Ecological Gateway. The station area in Piacenza has the potential to function as a multifunctional hub, integrating ecological design with digital innovation and transforming it into a catalyst for local economic development. The station area can connect the city and the Po River, providing a blend of mobility, sustainability, and digital services.

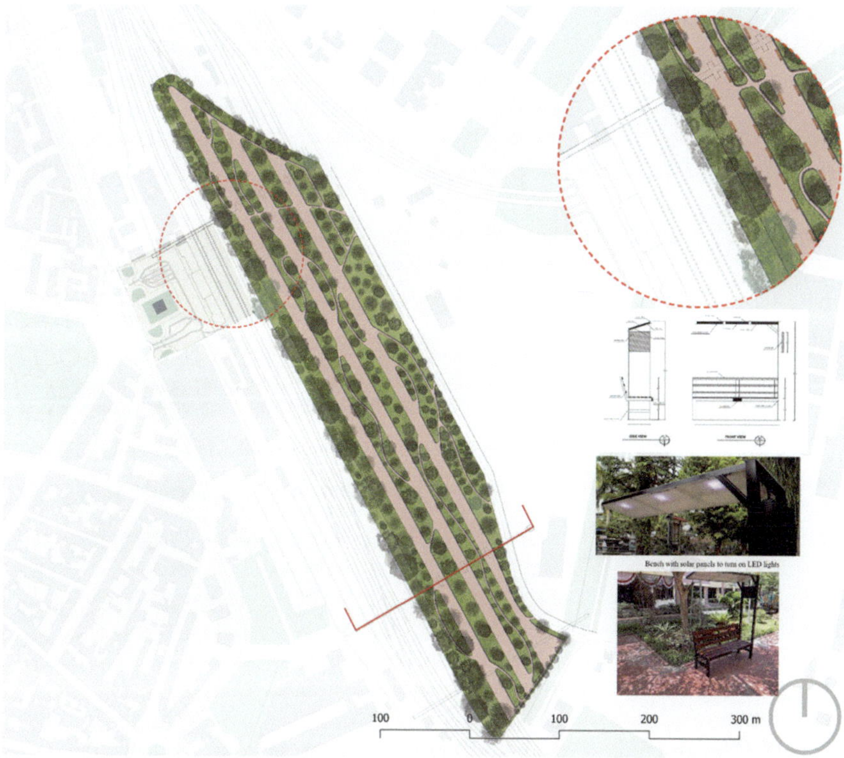

Fig. 17 Community gardens and vertical agriculture in Piacenza. *Source* J. Kianisadr

Its key features are:

- A smart waiting park equipped with solar-powered benches (see Fig. 18) offering Wi-Fi connectivity and wireless charging, allowing passengers to stay connected while waiting for their trains. Additionally, for those who wish to connect with nature, a resting area is available where they can relax or take a walk in a place designed for rest.
- Digital agricultural market app (see Fig. 19): a mobile application that connects local farmers with consumers and allows passengers to order fresh products for pickup at designated lockers within the station area. This supports local agriculture and ease users who likely lack time for grocery shopping after their working day.
- Urban Farming Spaces: Areas that include community gardens and vertical agriculture to promote local food production while offering educational opportunities about sustainable farming.
- Interactive Information Kiosks: Touchscreen kiosks providing real-time information about trains, their stops, local issues, events, and environmental data, as well as information on local market products, enhancing the experience for people and passengers beyond just being in a station.

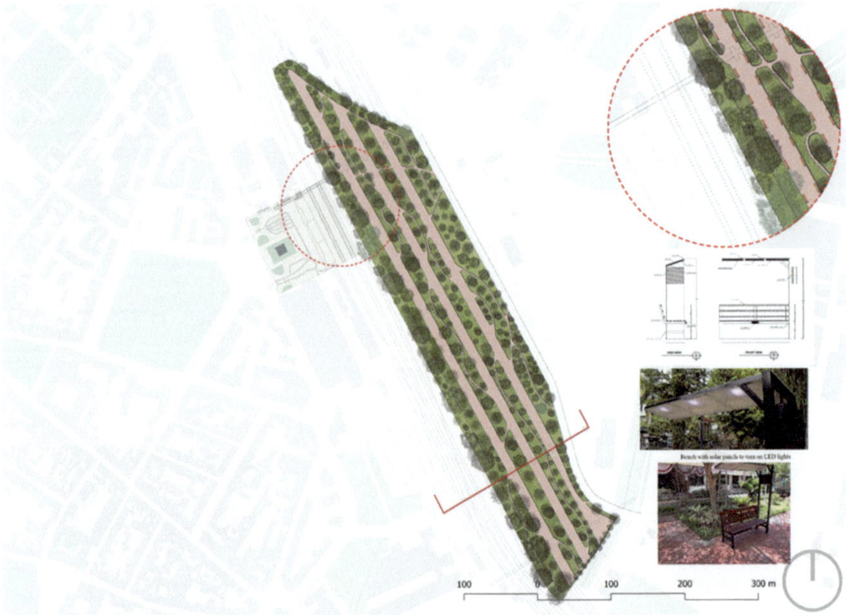

Fig. 18 Solar power benches in Piacenza. *Source* J. Kianisadr

- Green infrastructure: Provides shade during hot seasons, collects rainwater, reduces runoff, manages stormwater, and improves local climate and microclimate around the station.

By integrating these elements, the station area becomes more than just a transit point—it evolves into a vibrant community space that fosters economic activity, environmental stewardship, and digital connectivity while encouraging community engagement around the Po River. Due to the landscape and urban planning, Piacenza needs to reconnect its relationship with the Po River. This project shows how, by addressing the challenges of climate change, urban fragmentation, and economic transition, Piacenza can improve a resilient, connected, and vibrant urban environment. Considering ecological resilience, implementing sustainable urban draining system (SUDS) in the four strategic zones where it can improve urban connection with the Po River and give to Piacenza, mitigate flood risks, enhance biodiversity, and improve microclimates. This solution makes the ecological restoration function and reconnects the city with its natural surroundings.

In terms of digital innovation, bringing digital platforms, such as mobile applications, touchscreen kiosks, open Wi-Fi networks, and environmental monitoring, can help the area and finally the city to improve the connection between local agriculture and its civilians, improve the local economy, and develop smart infrastructure that transforms public spaces into interactive and responsive environment. For the community engagement, this project serves ecological, social, and economic

Fig. 19 Digital agriculture e-market app in Piacenza. *Source* J. Kianisadr

purposes by making multifunctional spaces. Improving a sense of civilians and participation in the city.

Piacenza strategic location and historical significance position it as a model for mid-sized cities facing similar challenges. This project illustrates that through the convergence of ecological design, digital technology, and community engagement, cities can navigate the complexities of climate adaptation and economic transformation. The renewed connection between Piacenza and the Po River serves as a testament to the power of integrated, forward-thinking urban planning.

3. **Chennai, India**

This project explores the intricate relationship between climate dynamics and landscape architecture in the context of Chennai, a rapidly urbanizing coastal city in Southern India. With a focus on mitigating the impacts of climate change and fostering sustainable urban development, the research investigates the challenges and opportunities presented by Chennai unique climate and topography (see Fig. 20).

The study begins by analysing the prevailing climatic conditions in Chennai, including temperature fluctuations, rainfall patterns, and the increasing frequency of extreme weather events. Drawing from historical data and future climate projections, the research aims to elucidate the evolving climatic scenario and its implications on the city landscape. Subsequently, the thesis delves into the existing landscape architecture practices in Chennai, examining how urban planning and design have responded to climatic challenges thus far. Emphasis is placed on identifying successful strategies and acknowledging shortcomings, with the goal of informing future interventions.

The core of the research involves proposing innovative climate- responsive landscape design principles tailored to Chennai specific ecological and cultural context. Through a multidisciplinary approach, incorporating environmental science, urban planning, and cultural studies, the thesis aims to develop adaptable design strategies that enhance the city resilience to climate change.

Case studies of successful international and regional examples serve as benchmarks, providing insights into best practices that can be contextualized for Chennai unique landscape. The proposed design interventions encompass green infrastructure, water management, and urban green spaces, aiming to create a harmonious balance between human activities and the natural environment.

Water holds profound cultural significance in Tamil Nadu, a state in southern India with a rich cultural heritage. The region culture is deeply intertwined with its rivers, ponds, and coastal areas. The Tamils have historically revered water sources, considering them sacred and integral to daily life. Temples and shrines are often situated along riverbanks, emphasizing the spiritual connection with water. Festivals like 'Thai Pongal' celebrate the harvest season, and part of the ritual involves boiling the newly harvested rice in fresh milk, symbolizing the importance of water in agriculture. The classical Tamil literature, Sangam poetry, vividly portrays the significance of rivers, lakes, and even rain in the lives of the people. Traditional water conservation methods, such as '*oorani*' (community tanks), have been employed for centuries, showcasing the community understanding of the value of water. As Tamil Nadu faces water challenges in contemporary times, preserving and respecting the

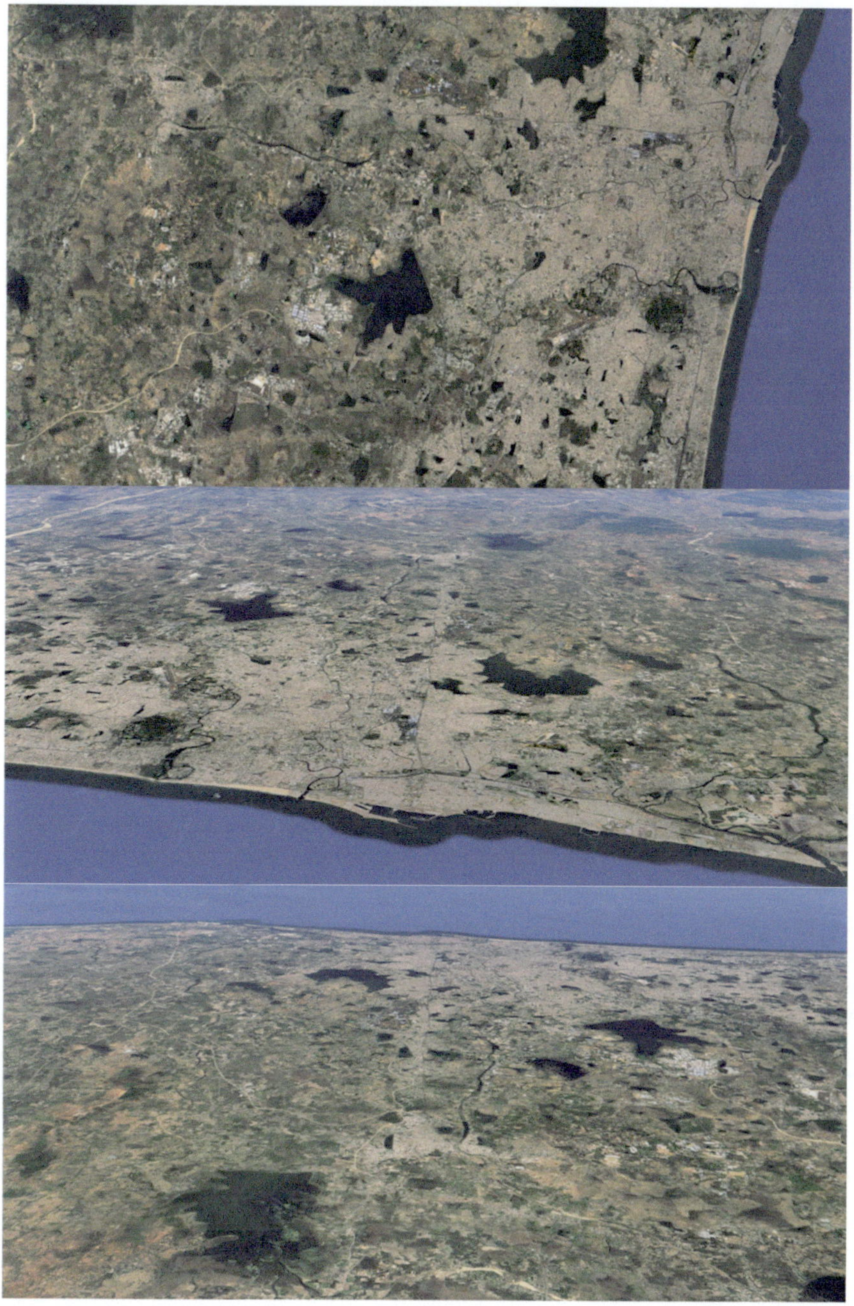

Fig. 20 Satellite views of Chennai. *Source* Google Earth

cultural significance of water can play a vital role in promoting sustainable water management practices for the future.

Chennai, the capital city of Tamil Nadu, is a major urban centre with a diverse economic landscape that encompasses both agriculture and the information technology (IT) sector. Historically rooted in agriculture, Tamil Nadu capital has transformed into a significant economic hub. This transformation has seen Chennai contribute substantially to Indian Gross Domestic Product (GDP), owing to its robust IT and manufacturing sectors.

The city hosts a thriving IT industry, with numerous technology parks and software companies playing a crucial role in Indian digital transformation. At the same time, Tamil Nadu retains its agricultural importance, with fertile lands supporting the cultivation of rice, sugarcane, and other crops. This unique combination of a burgeoning IT sector and a strong agricultural base makes Chennai and Tamil Nadu distinctive within Indian economic landscape. This diversity not only boosts the nation GDP but also highlights the resilience and adaptability of the state economy, making it a vital player in Indian growth story.

Tamil Nadu has no perennial rivers; they are all fed by the monsoons. These rivers must cross state boundaries before they irrigate the fields downstream. Several hundred years ago, a system was devised to utilize the water flowing in the rivers to the fullest before it reached the sea. It was a simple act of engineering that diverted river water into tanks through dug-out earthen channels, which in turn took care of the irrigation needs of neighbouring villages.

While it was easy to divert water to nearby villages, connecting those farther away from the source was more challenging. Engineers devised a simple solution—a series of cascading tanks. The outflow from one tank would serve as the inflow for the next in the series since the tanks were designed to allow excess water to flow out after reaching capacity. The thought and effort put into designing these massive chains of receptacles and overflow channels hundreds of years ago is awe-inspiring. No concrete, no high-tech machinery; all mud and manpower.

Eris are of two types: system and non-system *eris*. System *eris* are fed by streams of rivers through channels, while non-system *eris* are stand-alone tanks fed by rain. Most of the tanks in Tamil Nadu are system *eris*. Most *eris* in Chennai were built by rulers, either the king of the local chieftain, and looked after by the villagers. The *eri* system in the Chennai Metropolitan Area is an ancient water management network of interconnected tanks and canals. Primary canals divert monsoon water from rivers like Kosasthalaiyar, Cooum, and Adyar into system *eris*, which then distribute water to secondary tanks through overflow channels. This cascading setup ensures efficient storage and controlled release, aiding irrigation, flood mitigation, and groundwater recharge. Historically maintained by local communities, the system effectiveness has waned due to urbanization. Revitalizing these *eris* and canals is crucial for enhancing water security, supporting agriculture, and bolstering resilience against climate change.

The methodology for the analysis made in this project follows the process of protocol maps for understanding the regions of Chennai Metropolitan. These protocol maps serve as a visual tool to delineate various zones, infrastructure, and demographic patterns within the metropolitan area. By applying David Governor principles of actors, protectors, and stewards, the analysis systematically identifies key stakeholders, their interactions, and the protective measures in place for sustainable development. This comprehensive approach ensures a detailed understanding of Chennai urban landscape, facilitating in- formed decision-making and strategic planning for future growth and development in the region.

The traditional Eri system in Tamil Nadu, including Chennai, offers significant benefits by conserving water, recharging groundwater, controlling floods, restoring ecosystems, and preserving cultural heritage (see Fig. 21). Integrating *Eri* principles into Metropolitan water management can enhance the sustainability, resilience, and liveability of Chennai. Employing David Govere theory of actors, protectors, and stewards can further support these efforts, involving community participation, regulatory enforcement, and sustainable planning. This holistic approach balances ecological preservation with economic growth, ensuring long-term sustainable development. It will help support ecological and economic restoration. Actors can include community members and local businesses who actively participate in conservation efforts. Protectors, such as government bodies and environmental organizations, ensure that laws and regulations are enforced to safeguard these ecosystems. Stewards, including scientists and policy makers, guide long-term planning and sustainable practices.

Reinforcing the traditional *Eri* water system and creating an interrelated ecological network in the Chennai metropolitan region can help revive lost cultural systems and strike a balance amid economic growth and ecological sustainability. This comprehensive approach ensures that both the environmental and economic needs of the region are met, fostering sustainable development for the future.

Considering urban strategy (see Fig. 22), the concept involves, for Phase 1, reconstructing the lost *Eri* system by reestablishing connections between water bodies through water channels, thereby reintroducing water to the marshland and revitalizing the ecology in the area. Phase 2 proposes to develop the road and rail network, which divide the region into an irregular grid system. Informal settlements primarily inhabit the urban areas surrounding the marshlands and adjacent regions, resulting in an irregular configuration of Receptor system. For the phase 3, the project is proposing transformative spaces such as Urban parks, community gardens and Urban Farmlands not only sustains the diversity of productive and income-generating activities but also strategically placing them within the protective belt help limit further encroachments.

Considering water, there are 3 different kinds of water strategies proposed (see Fig. 21):

- Water at rest: water plays a pivotal role in shaping the regional, social, and economic identity of a region. The regeneration of lost wetlands facilitates the reconnection of lost relationships, as it becomes a hub for various social and economic activities.

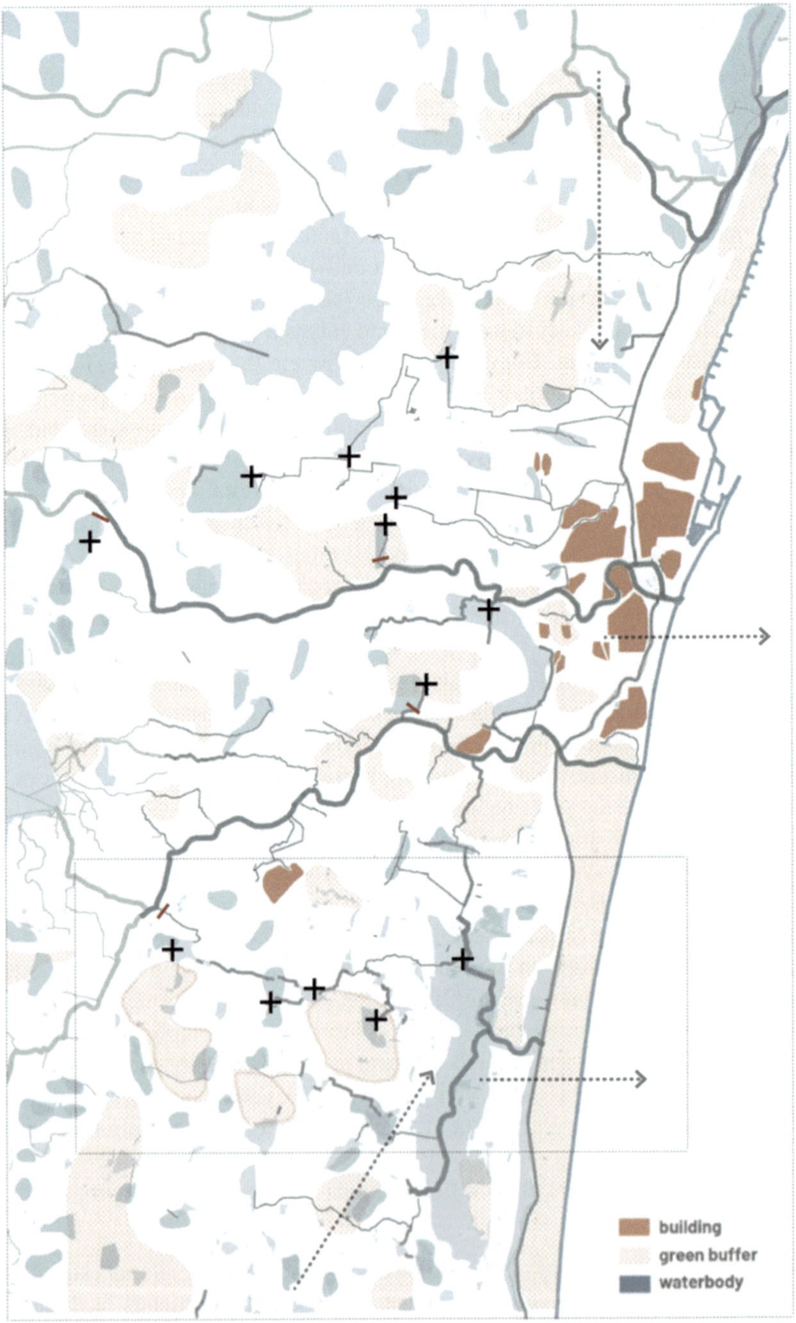

Fig. 21 Water ecosystem in Chennai. *Source* A. Saleem, P. Ulaganathan

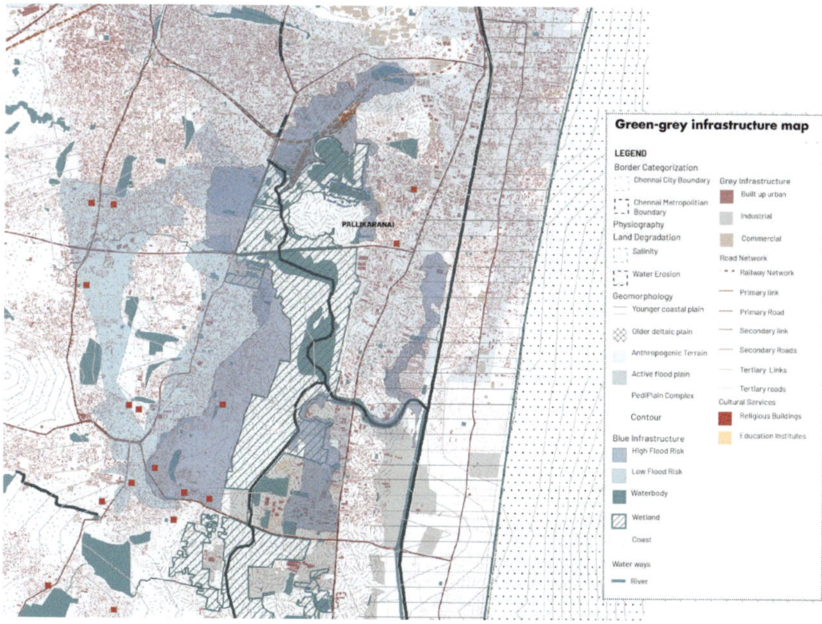

Fig. 22 Green-Grey protocol map in Chennai. *Source* A. Saleem, P. Ulaganathan

- Water in action: water from the marshland will contribute to irrigating the croplands, thereby boosting productivity and directly impacting the income of households reliant on agriculture. Additionally, the phyto-remediation process involves cleaning the waterbody and removing any impurities.
- Water in Motion: additionally, designated floodable zones encircle the marshland, serving as catchment areas during rainfall and transforming into public interactive spaces during drier periods.

Similarly for vegetation, there are 3 different kinds of strategies proposed (see Fig. 23):

- Vegetation at rest: the regeneration of lost wetlands facilitates the reconnection of lost relationships, through vegetation.
- Vegetation in action: croplands and phyto-remediation will have a central role.
- Vegetation in motion: marshland will serve as catchment areas during rainfall and transforming into green public interactive spaces during drier periods.

The project proposes leveraging socio-cultural networks for environmental transformation and restoration, recognizing social and cultural capital as the critical link between economic goals and environmental capacities (see Fig. 24). The initiative aims to restore the Pallikaranai marshland as a flood sink by creating water retention regions, enhancing its ability to retain water and form a protective green edge against

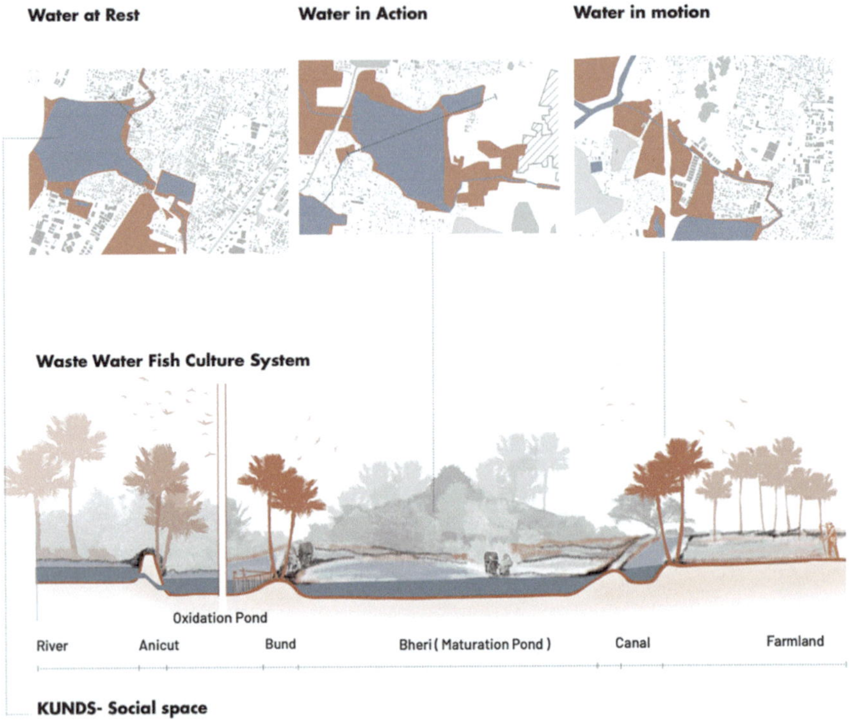

Fig. 23 Water strategy in Chennai. *Source* A. Saleem, P. Ulaganathan

floods. This approach will also establish Pallikaranai as a carbon sink, utilizing ecological and economic principles to ensure long-term sustainability.

The Pallikaranai marshland holds immense significance in Chennai as one of the city last remaining natural wetlands. Its preservation is vital for several reasons. Firstly, it serves as a haven for a diverse range of plant and animal species, including endangered birds, contributing to regional biodiversity conservation efforts. Secondly, the marshland acts as a natural buffer against flooding, absorbing excess rainwater and mitigating flood risks for nearby communities. Additionally, it plays a crucial role in recharging groundwater aquifers, ensuring sustainable water availability in the region. To prevent this marshland from further degradation multiple strategies have been put into place to ensure revitalization and requalification of this important ecological hotspot. The social link, the pathways weave together transformative and steward spaces, enriching the essence of the local functionality. Revitalizing the marshland will revive the exquisite fauna of the region, fostering a renewed connection between users and with the nature.

Okkiam Maduvu, also known as Okkiyam Maduvu or Oggiyam Maduvu, is a 2.8-km-long water channel situated in the Kanchipuram district, in the southern suburbs of Chennai, India. Originating from the Pallikaranai Wetland, it ultimately flows into the Buckingham Canal. During the 2015 floods, Okkiam Maduvu played

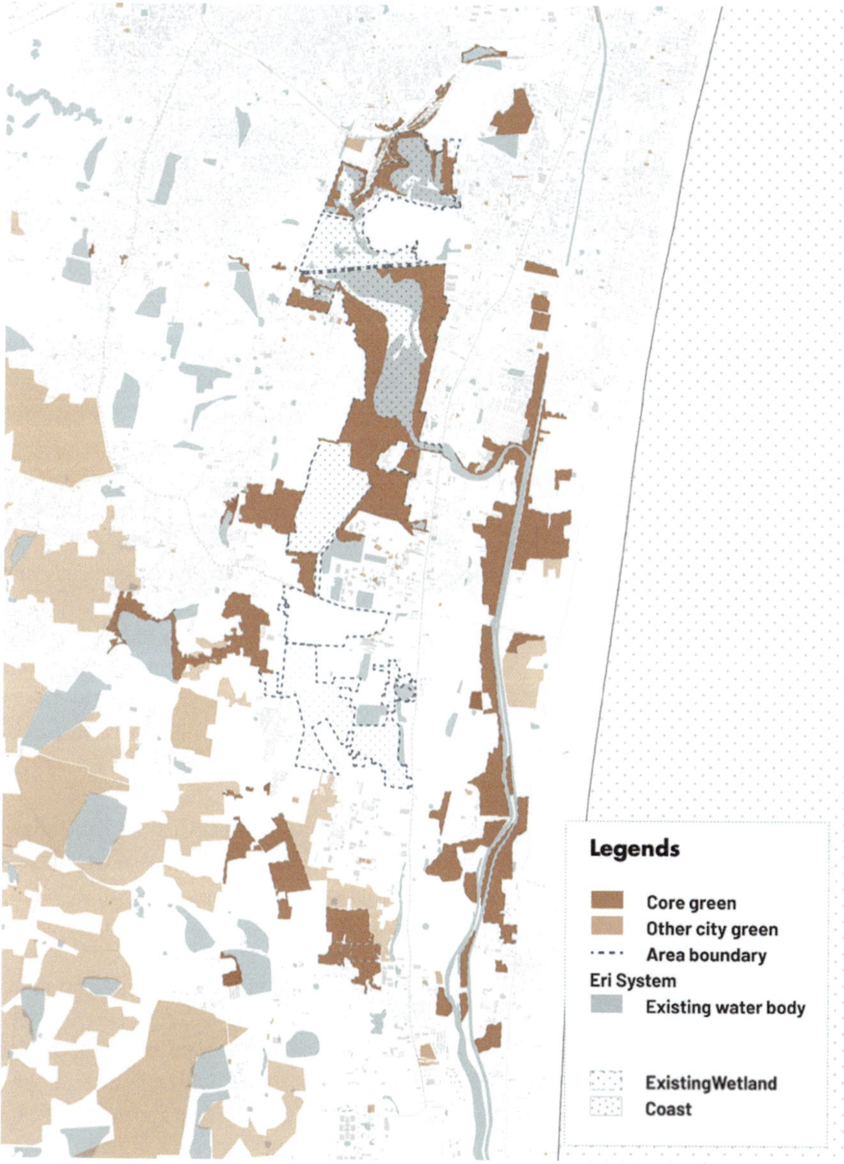

Fig. 24 Existing ecological dynamics in Chennai. *Source* A. Saleem, P. Ulaganathan

a crucial role in flood prevention due to its capacity to absorb excess rainwater and its strategic location for draining water away from vulnerable areas. This demonstrates its resilience, making it an ideal area to implement our water conservation and management strategies.

However, the site faces several challenges, including disconnection from the local community, unprotected edges, and neglected zones that need to be reconnected with the surrounding ecology and hydrology. Our proposal aims to create an integrated matrix involving actors, protectors, and stewards to establish robust social, economic, and cultural connections. By doing so, we seek to restore the lost linkage between hydrology and ecology through active cultural engagement.

The region is characterized by two strong vertical axes that remain disconnected. The urban matrix consists of institutions, residential areas, commercial blocks, and green wasteland pockets that are currently fragmented. This fragmentation, along with unprotected edges, contributes to the area susceptibility to flooding and drought. By addressing these disconnections and integrating the roles of various stakeholders, our project seeks to create a cohesive and resilient environment.

The project will focus on reconnecting these fragmented urban elements, enhancing the protective infrastructure along the channel, and revitalizing neglected zones. This will involve community participation and collaboration with NGOs and government bodies to develop and implement policies that support sustainable water management and conservation practices. By fostering a sense of ownership and responsibility among residents, we aim to ensure the long-term success and sustainability of these initiatives.

The area surrounding Okkiam Maduvu faces several significant challenges that contribute to frequent flooding and drought problems. Residences and institutions near the water channel frequently experience severe flood issues, primarily due to inadequate maintenance and lack of protective infrastructure (see Fig. 25). The unprotected edges of Okkiam Maduvu, combined with neglected zones, exacerbate the regional vulnerability to water overflow during heavy rains.

Additionally, urban encroachments and the proliferation of invasive species have further disrupted the natural flow and absorption capacity of the waterway. Unplanned urban development has intruded upon natural waterways, reducing the available area for water retention and increasing the risk of flooding. These urban encroachments hinder the channel ability to manage excess rainwater effectively.

Climate change and rapid urbanization have intensified these issues, leading to more frequent and severe flooding in the region.

The increased rainfall variability and extreme weather events associated with climate change have overwhelmed the existing infrastructure, which is ill-equipped to handle such volumes of water. During the dry season, the area faces significant drought problems. The lack of designated retention regions for capturing and storing rainwater exacerbates water scarcity. Without sufficient measures to retain rainwater, the region struggles to maintain adequate water levels, impacting both the local ecosystem and water supply for the community.

The Ecological connection aims to unify the two disconnected vertical axes by establishing horizontal sponge axes that link all green waste areas currently turning into swamps. These sponge axes will include green pockets that act as retention points and open green spaces that serve as flood sink zones during calamities. Additionally, these areas will function as collection points during dry seasons, ensuring continuous ecological flow. By absorbing excess rainwater, these green pockets and

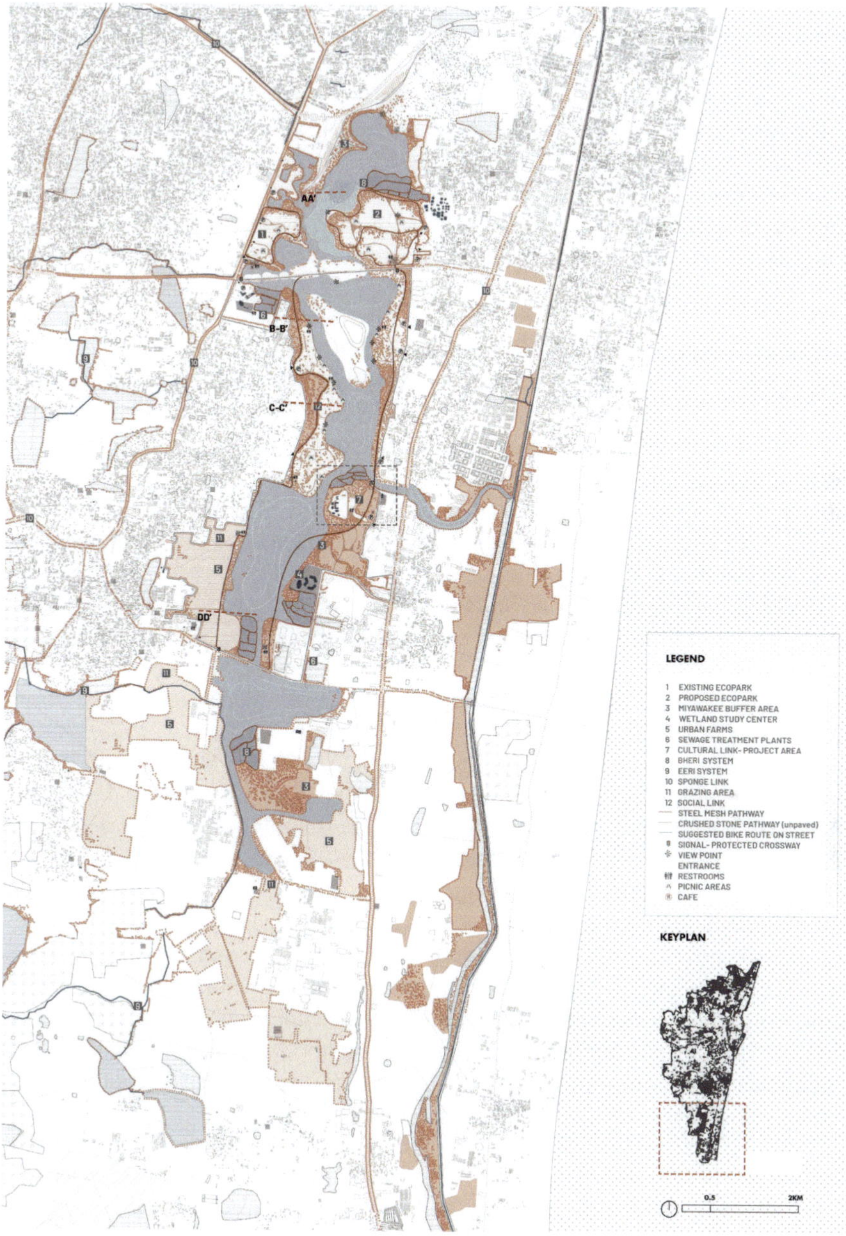

Fig. 25 Masterplan in Chennai. *Source* A. Saleem, P. Ulaganathan

open spaces will mitigate flooding and ensure water retention for drought periods. Strategically integrating these elements will enhance the natural water management system, promoting resilience and sustainability in the region.

The *bheri* are an integral part of the hydrological system, working as a series of interconnected ponds. They play a crucial role in cleaning greywater and regulating water flow toward the wetland. Positioned along the drainage flow direction, these ponds act as retention points during floods, helping to control and mitigate flooding. This strategic placement ensures that excess water is managed effectively, reducing the risk of flood-related damage and promoting a balanced hydrological cycle.

The *bheris* also provide significant economic benefits to the local community, particularly to farmers and fishers who lease these areas. Excess water hyacinths harvested from the ponds are used to create marketable products, contributing to local commerce. Additionally, the *bheris* are stocked with fish that are harvested and sold, providing a vital source of income. Market blocks around the *behris* serve as points of community celebration and trade, highlighting the cultural and economic importance of these areas. This dual role of the *bheri* in water management and economic activity underscores their importance in maintaining both ecological and economic stability in the region.

The kunds, which collect water during the rainy season, transform into social hubs during the dry season. Alongside these, the *bheris* that merge into the social maturation ponds become popular community meeting centres. Socio-cultural centres and open plazas serve as vibrant marketplaces and celebration spaces where people gather for various events.

Educational centres play a crucial role as learning hubs, providing insights into the *bheri* system and educating the community on flood zone protection and the maintenance of flood sink regions. Additionally, architectural blocks such as workshops and recreation centres attract tourists, further enhancing the area appeal.

The architecture facilitates a seamless connection between the landscape and the urban fabric, integrating these diverse spaces into a cohesive whole. This integration of kunds, socio-cultural centres, educational hubs, and recreational zones establishes a balanced and harmonious environment. By attracting people back to the region, it enhances social and cultural connectivity, strengthens community bonds, and promotes a vibrant and resilient community. A circular economy emerges with a balanced provision of all activities—provisioning, regulating, supporting, and cultural—paving the way for a sustainable future.

Waste collected at the collection points is treated, and the remnants flow to the next step, which involves the *bheris*. This treated waste and grey water are directed into the *bheri* ponds, where local fishermen and farmers breed and stock fish, integrating them into the wastewater treatment system. Consequently, the *bheris* act as ecological filters, purifying the water before it is released into the wetland. In these ponds, water hyacinths thrive, further assisting in the treatment of greywater by blocking the flow and absorbing impurities (see Fig. 26). The sludge produced at the bottom of the ponds is utilized as compost for urban rice and paddy farms. These agricultural products are then sold in the socio-cultural market, thus creating a circular economy. Additionally, socio-cultural activities designed around the *bheri* waters enhance community engagement and interaction, allowing people to experience these multifunctional spaces while contributing to a sustainable and self-sufficient ecosystem.

The success of the Bheris system in soil and water conservation hinges on a multi-faceted approach that involves foresight, community engagement, and a robust enabling stream comprising NGOs, government bodies, and policy-making institutions. Foresight and vision allow for the anticipation of future challenges, while active community participation ensures culturally appropriate and sustainable practices. NGOs play a critical role in organizing user groups and facilitating capacity building, while government bodies provide the necessary regulatory frameworks and supportive policies. Effective organizational structures are essential for systematically planning, executing, and managing conservation initiatives. This collaborative effort ensures that water conservation and development initiatives are implemented effectively and sustainably in the long term, creating a resilient system capable of adapting to future challenges and ensuring the well-being of local communities.

4. **Xiangtan, China**

The value of landscape is often reflected in how it serves people, not only by ensuring a better quality of life and sensory experience to them, but also creating more significant values for the society, including economical ones. Due to climate change, natural disasters and flooding events are becoming more frequent and intense, and only by constantly reinventing and innovating can communities reduce crises and bring stability, health, wealth, and economic development to people. Landscape architects have come to realize that only built infrastructures will not solve the immediate crisis of communities, but also innovative soft skills and electronic tools facilitating global multifunctional integration (also in economic terms). Walls against floods and concrete levees will only solve the need for temporary flood control, but they will not work for the creation of a digital sharing base for community knowledge.

However, resilient landscape design approaches present opportunities and benefits for communities. Taking Xiangtan section of Xiangjiang River in Hunan Province, China, as an example, this chapter describes how to use nature-based solutions, digitally integrated devices and related artificial equipment to reduce the impact of unstable water levels on the ecology of the site and create a multi-dimensional landscape corridor as a 'resilient' strategy against floods in such a lack of safety and vitality. The strategy of living in harmony with nature and building multiple layers of defence will bring more efficient benefits to the community.

The Xiangjiang River in southern China Hunan Province, bounded by Dongting Lake to the north and surrounded by mountains on three sides, resembles a giant horseshoe. It is one of the main tributaries of the Yangtze River. The project site is in Xiangtan City in the middle reaches of Xiangjiang River, which is one of the important industrial and economic cities in Hunan.

The Xiangjiang River is the largest river in Hunan province and the largest tributary of the Dongting Lake system. Xiangjiang River flows from south to north through

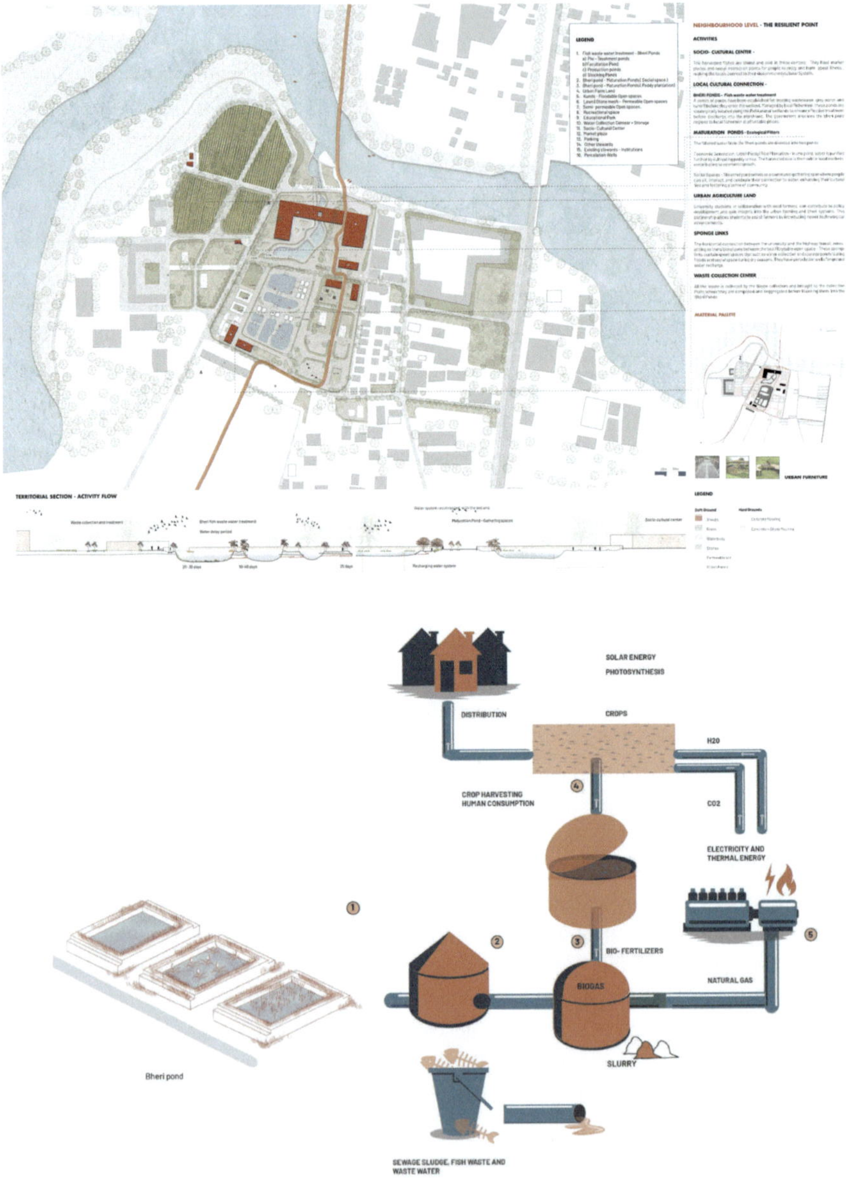

Fig. 26 Masterplan in Okkiam Maduvu. *Source* A. Saleem, P. Ulaganathan

Yongzhou, Hengyang, Xiangtan, Zhuzhou, Changsha, Xiangyin, Haohekou in two branches into Dongting Lake (see Fig. 28). It has a total length of 948 km, a watershed area of 94,721 square kilometres and an average annual runoff of 79.16 billion cubic meters. The average annual passenger volume of Xiangtan section of Xiangjiang River is 54.949 billion cubic meters, and the water surface width is 460–750 m. Xiangtan section of Xiangjiang River is in the 'S' shaped channel (see Fig. 27), the river flow rate is slow, easy to silt. As a result, there are many sandbanks, such as Arbutus Island and Gusang Island.

Xiangtan is a subtropical monsoon humid climate area, which is dry in summer and autumn, and susceptible to cold waves and strong winds in winter and spring. Light energy resources are relatively rich, the average sunshine hours over the years 1640–1700 h. Abundant heat resources, the average temperature 16.7–17.4 °C. Precipitation is abundant, but the seasonal distribution is uneven, the annual change is large, the annual precipitation is 1200–1500 mm.

The Xiangjiang River flows from south to north through Xiangtan City, with an average width of 500–700 m, a depth of 10–20 m, and a maximum depth of 30 m. The surface water system in the area is developed, with Juanshui and Lianshui rivers as the main tributaries. In the past 20 years, affected by sediment deposition, the area of Yangmei Island in the channel is increasing day by day, and the highest water level is also increasing. Riverbanks have long been underdeveloped or semi-developed, especially in rural areas, and are generally in a natural state, so their resistance to flooding is very low.

As the Xiangtan section of the Xiangjiang River is in the 'S' shaped channel, the flow rate of the river slows down, it is easy to accumulate sediment, resulting in the continuous rise of the riverbed, so this area is extremely prone to the danger of breaking the bank. As can be seen from the software simulation chart below, in the rainy season, the western and southern parts of the river are the worst-hit areas.

The Xiangtan River passes through Xiangtan City from south to north, and the terrain of Xiangtan section is generally low in the north and high in the south. Affected by the subtropical monsoon, there are distinct dry and rainy seasons, which are often disturbed by floods. The average water level in the dry season is 26.3 m, and that in the rainy season is 35.6. The river channel of Xiangtan section of Xiangjiang River shows an 'S' shape, with relatively flat terrain and slow flow rate. Therefore, the river channel near Yangmei Island is easy to deposit sediment, and the highest water level is getting higher and higher, so the nearby area is in the frequent flood area (see Fig. 29). It is worth noting that most of the riverbanks in this region are in an undeveloped or semi-developed state, especially in rural areas, most of the riverbanks are in a natural state, no matter from the perspective of landscape or safety, the development potential of the riverbanks is huge.

Xiangtan has developed land transportation, and there are convenient ways to get here from surrounding cities, but the condition of the road along the river is not good, so it needs to be constructed and separated from vehicles. Crowd gathering places are mainly concentrated in the north and south of Xiangjiang River, so landscape centres can be arranged in these areas as much as possible. Along the Xiangjiang River are connected buildings of different ages, including old towns, new towns,

Fig. 27 Satellite views of Xiangtan. *Source* Google Earth

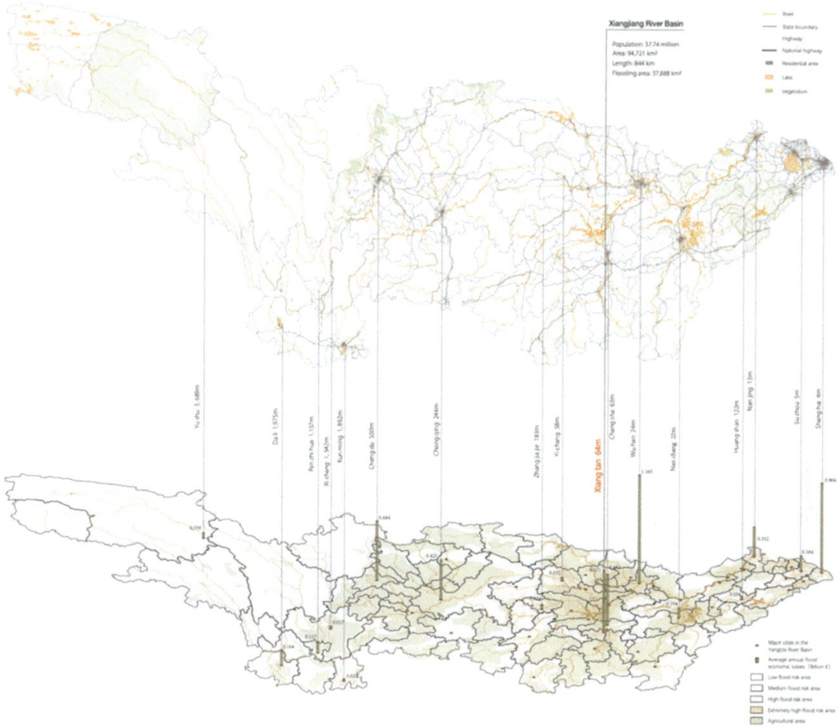

Fig. 28 DongTing Lake water system. *Source* T. Yan

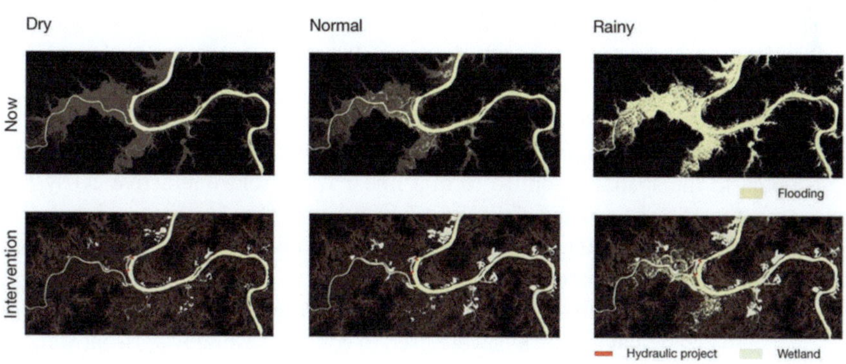

Fig. 29 Software simulates flooded areas. *Source* T. Yan

heavy industry areas and rural areas. Therefore, the design should be tailored to local conditions, protect the architectural ruins of the old city, and carry out renovation and upgrading without destroying the original landscape. In terms of greening, the natural form along the Xiangjiang River has been maintained for a long time, and the artificial traces are not obvious. The native aquatic plants can be used. There are many wetlands near the riverbank. Fishponds, artificial lakes, and channels can be preserved and utilized. At the same time, there are many docks on the bank, the design content needs to consider the impact on ship berthing and cargo.

As events become more frequent and intense due to climate change, communities must adapt and redevelop to reduce risk and improve ecological and human health. Now is also the time to stop placing communities and infrastructure in high-risk areas. We need to reduce sprawl, which further exacerbates risks. Resilient landscape planning and design offers a way forward for communities. We can now use multi-layered protection systems with diverse, scalable elements, any of which can fail safely in the event of a disaster.

For such a changeable site, multi-dimensional landscape design is necessary. Highly resistant plants and associated artificial devices help to reduce the impact of unstable water levels on the ecology of the site. The new elements of the plan include two small dams, a total of about 30 km of trails at the top of the dam, 30 km of trails in the middle of the dam, 15 km of swamp trails, 2 landscape buildings, 12 landscape squares, and landscape bridges 2, 12 artificial wetlands, 3 wetland parks, 3 beaches, 19 Diaoyutai, 2 wind farms, 224.2 hectares of new flood control green infrastructure, etc. The project site is located on the west bank of the Xiangjiang River in Xiangtan. It is about 8 km long from north to south and 5 km wide from east to west. The green corridor along the river is about 15 km long and it is shaped like a 'C'. The total construction area is about 224 hectares.

In this project, the landscape will be upgraded without destroying the original ecological appearance as much as possible. Since the Xiangjiang Riverbank is a narrow strip with a height difference, it is difficult for conventional landscape ecological design to meet the design objectives. Therefore, the form of green corridor should be adopted in the spatial layout to form a linear waterfront interactive ecosystem extending for 15 km, and the diversity of landscapes should be considered. The diversity of landscape is the result of the cooperation of biological and geological diversity and human interactions.

Local plant species with strong stress resistance were selected and planted along the river, and ecological greenways were used to connect several waterfront wetland parks to become habitats for fish and birds, which contributed to the stability and species richness of the ecosystem. In addition, relevant artificial devices are placed on the green space, three roads of different elevations provide people with activities and viewing requirements of different water levels, swamp walk, fishing platform, wooden plank road, observation platform, and beach further deepen the connection between people and the water. When the rainy season comes, excess flood water will be discharged to the nearby constructed wetland through the newly built ecological channel on both sides of the riverbank and stored, which will relieve the pressure on the riverbanks to a certain extent. Meanwhile, the small dam designed by Yangmei

Island will also play an important role at this time, which will use the difference between the internal and external water levels to generate electricity until the end of the rainy season, and the power generation time will last for three months.

Furthermore, water stored in constructed wetlands and dams will provide water during the dry season, boosting the local economy, that can be monitored through a digital platform where resources are bought and exchanged by the community. In addition, to cope with the once-in-a-century flood, different types of landscape squares are distributed along this corridor, providing temporary relief camps for emergency situations, providing tents and storing disaster relief materials. In the upper reaches of the river, a new wind power device, a bladeless turbine, is proposed to provide clean energy, and at the same time, it is equipped with a water level alarm, that can flash red light and connect to a digital platform in case of danger, to remind people to pay attention to safety. These interventions will maximize the value of urban open space, creating potential ecological value and economic benefits.

The goal is also to improve the attractiveness of waterfront space (see Fig. 30). The project plans to build a style street on the riverbank to attract investment, carry out appropriate economic activities, and drive the development of local tourism; actively promote the spread of park culture, vigorously develop cultural peripheral industries, do a good job in logistics management of the park, and improve the experience of tourists visiting the park. Through technological innovation, the project will also maximize the value of urban open space and create potential ecological value and economic benefits.

During the period from 2010 to 2020 alone, four large-scale floods occurred in the Xiangjiang River, which caused serious damage to the lives and property of residents near the riverbank (see Fig. 31). It is a great threat, so reasonable early warning devices and adequate safety zone settings are necessary. It is planned to design and construct landscape plazas of different sizes and specifications along the Xiangjiang River according to the density of people. These plazas are distributed along the river and arranged reasonably according to the population density. Generally, one is built at about 300–500 m, mainly to provide leisure and entertainment places for residents. During the flood season it can also provide emergency shelters for residents along the coast.

The main purpose of the dam is to coordinate the water balance between the dry and rain seasons and use the water level to generate electricity. Xiangtan from July to September is the water abandonment period (Rainy season), October to June is the non-abandonment period (Dry season), the annual average power generation is expected to be 3.5 billion KWH. Another important use of the dam is to provide landscape access for visitors, it can connect the riverbank with the Yangmei Island, forming a landscape closed loop. The lotus sculpture references iconic sculptures in the city centre, which can leave a deep impression for tourists visit here, at the same time, it gives residents a sense of belonging and cultural identity (Figs. 32 and 33).

The proposal considers, therefore, the importance of combining multiple disciplines. Not only do you need to use knowledge related to energy engineering, mechanical manufacturing, and electrical engineering, but you also need to consider material design, aesthetics, and ecological principles. For example, the size and distribution of

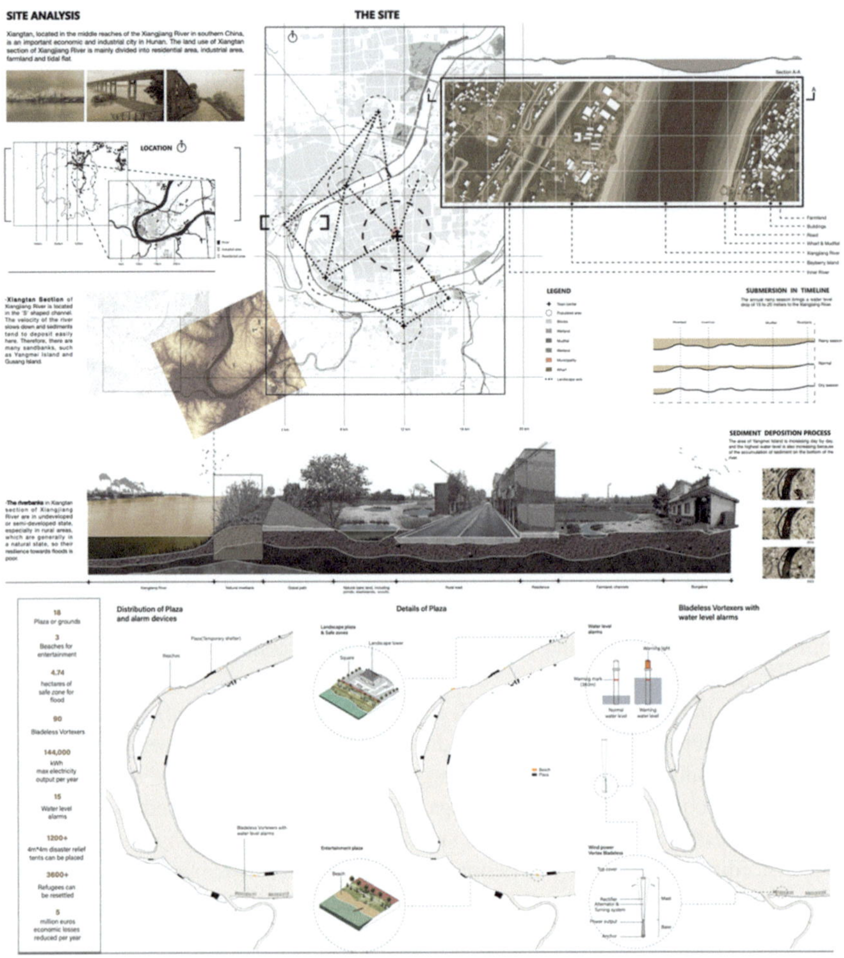

Fig. 30 Riverbank project site. *Source* T. Yan

bladeless turbines need to consider wind direction, wind speed and wind utilization efficiency, and power generation materials need to be wear-resistant and environmentally friendly (Fig. 34). At the same time, the selection and matching of installation colours also affects the harmony of the landscape. The choice of blue is to blend in with the river and sky, minimizing the impact on the surrounding ecology. Therefore, this is a multi-disciplinary technology application result.

The aerial view at the top of the page shows the 'U'-shaped bend of the Xiangjiang River, showing the Yangmei Island, an independent site located in the middle of the project: it is not only a 'performance' platform for the two hydroelectric power stations, but also serves as an island, bringing a different tour experience to tourists, that could be booked and organized through the same digital platform that will

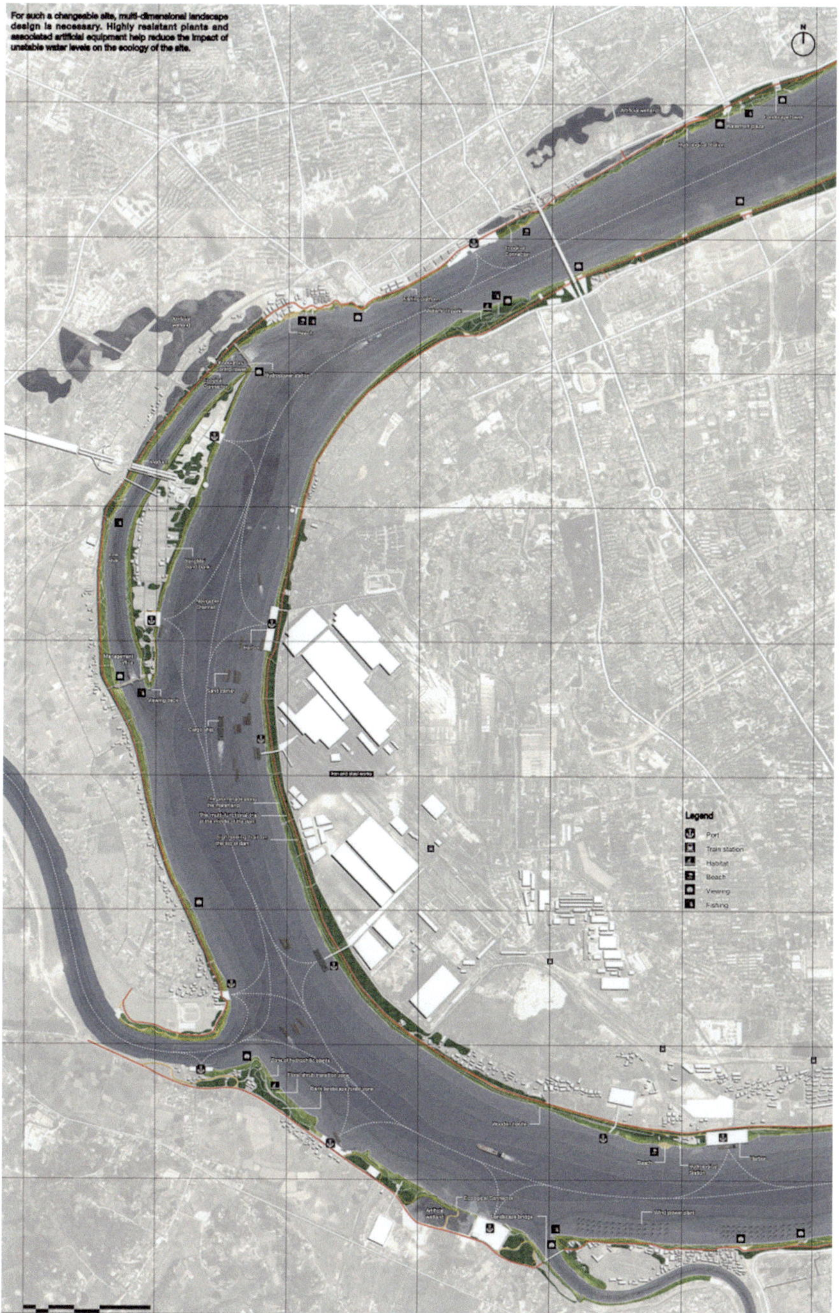

Fig. 31 Masterplan for the riverbank in Xiangtan. *Source* T. Yan

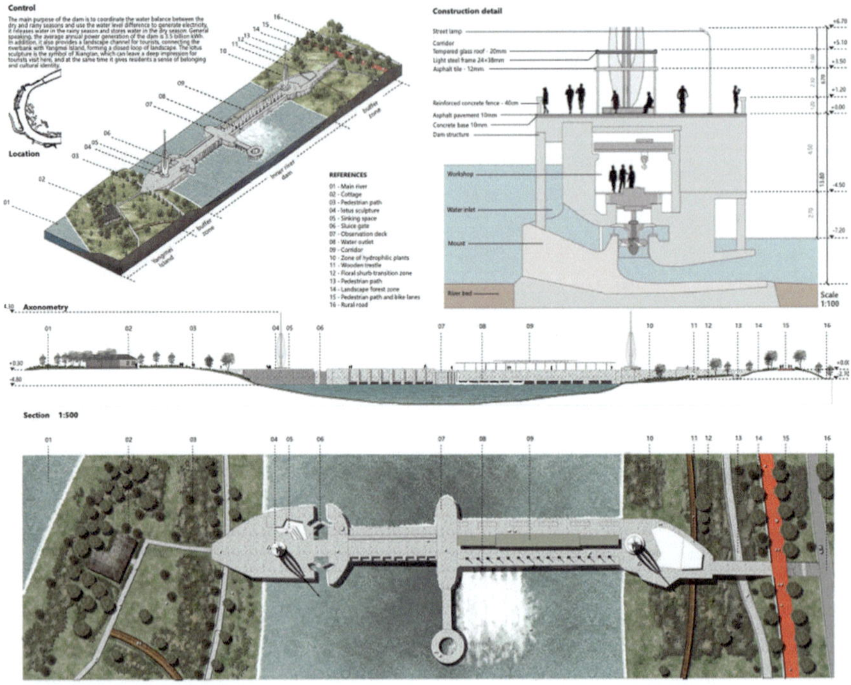

Fig. 32 Technology for a new digital community marketplace. *Source* T. Yan

provide alarm messages, a digital marketplace for the exchange of the produced resources (electric energy and water), but also leisure opportunities for the local community and visitors. Piers, fishing platforms, and trails on dams will also bring more opportunities for people. That platform could also provide information about the local cultural identity, the relevant heritage sites, and the events in the community, serving as a potential job marketplace, too.

The landscape square and wind power generation device emphasize the sustainable development, comprehensiveness, and innovation of the project, organically integrating the cultural landscape and natural landscape power generation device. It is proposed to build 18 landscape squares, 3 recreational beaches, provide a 4.74 ha flood control safety area, 2 floating power plants, and a total of 90 bladeless turbines. If the power generation efficiency of the bladeless turbine is calculated as 70% of the efficiency of the traditional rotary alternator,[4] then the power of each bladeless turbine is about 420–1000 kWh. Based on 10 h of working time per day, it is expected that the Xiangjiang Wind Farm It can generate a minimum of 137,970,000 kWh of electricity per year. In addition, the landscape square can provide space for more than 1200 tents, which can accommodate more than 3600 refugees. These interventions

[4] See http://www.polycae.com/materials/strcae/nonblades.pdf.

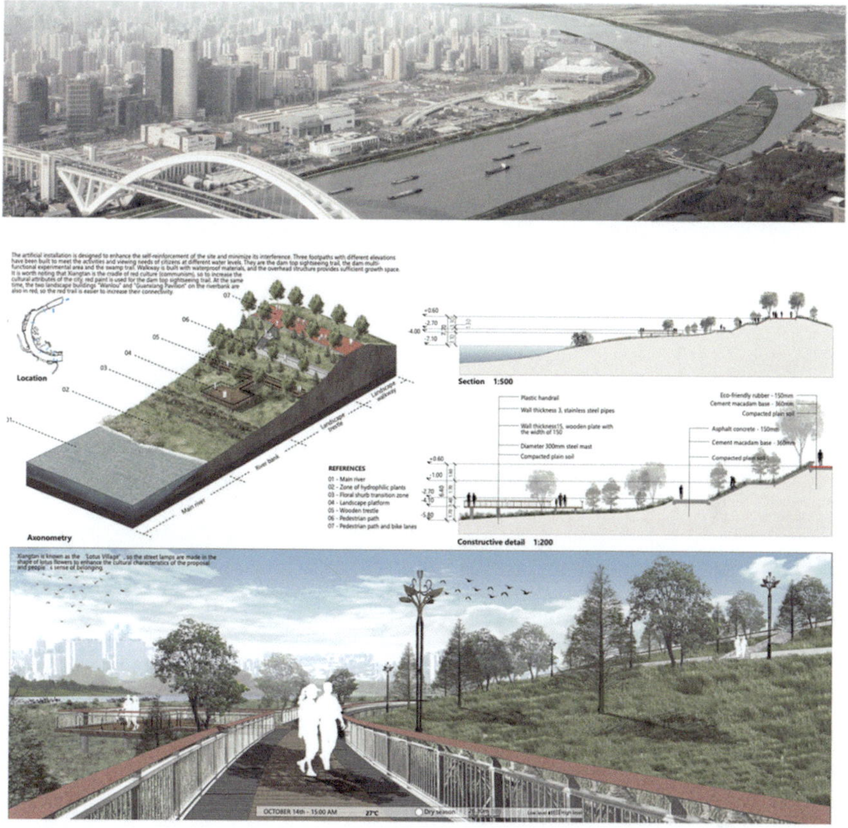

Fig. 33 Digital multifunctional technology for flood control. *Source* T. Yan

are expected to reduce economic losses caused by flooding by 5 million euros per year.

The proposal will reduce the damage caused by floods to the ecological environment, allowing more animals and plants to survive here, so the number and quality of species are greatly improved. The walking trail will make transportation convenient, where people will be able to walk, ride bicycles, and explore leisurely in the park. The marsh trail will attract visitors to stroll on the carefully planted river beach, and they can also enjoy the scenery from the observation deck.

The dam on Yangmei Island and the new wind farm will become new landmark buildings in Xiangtan (see Fig. 35). Visitors in the park will become potential consumers of the two tourist attractions, bringing some economic benefits to the city. The landscape in the park has been further enhanced and it provides visitors with a healthier way to get closer to nature, allowing the area to be visited by at least 100,000 people every day. The project will also play an important role in cultural dissemination. The lotus sculptures on dam and the lotus shapes of urban furniture

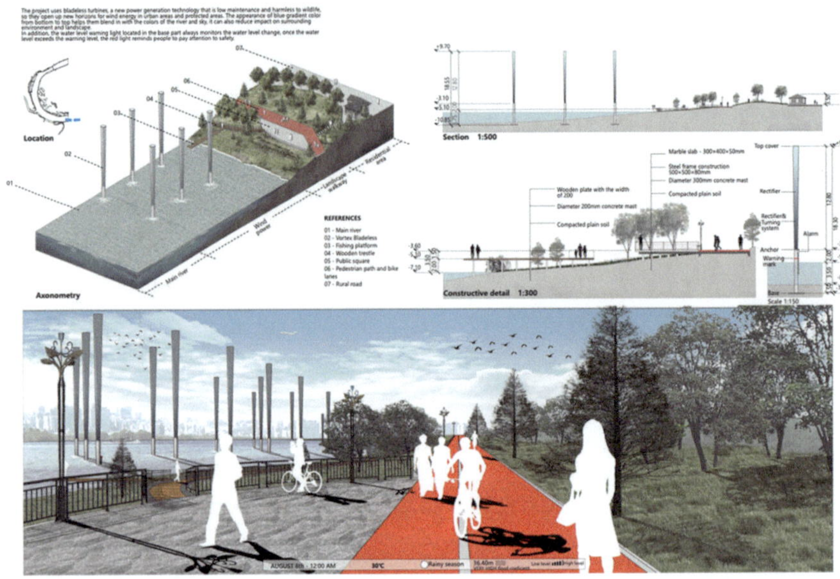

Fig. 34 Energy marketplace and alarm system. *Source* T. Yan

should leave a deep impression on tourists here, and at the same time give residents a sense of belonging and cultural identity.

The project was also analysed through a Cost–benefit analysis, showing that the income situation of roads, landscape bridges and urban furniture is lossmaking in the 5-year period, although other incomes are positive, especially hydropower stations, landscape vegetation, landscape plazas, artificial wetlands, and wind farms, with a positive the total Net present value for the proposal. Therefore, the Xiangjiang Riverbank Resilient Improvement Project is feasible within a 5-year period.

5. **Manila, Philippines**

Metro Manila, officially known as the National Capital Region, has a population of over 12 million people, making Metro Manila the most populous metropolitan area in the country and one of the most populous urban areas in the world (see Fig. 36). It is a region of 16 cities and a municipality with a land area of 619.57 square kilometres. The region comprises the seat of the country political power, economic movement and cultural life, and acts as the symbol of the Philippines rapid modernization and the temerity of socio-environmental challenges it raises.

Already home to more than 13 million people and still growing, the region struggles under intense pressures on land, water and infrastructure. Over the past decades, rapid urbanization has caused an increase in informal settlements, particularly those living on riverbanks, canals, and coasts. Many are living precariously on reclaimed land or in flood-prone areas, facing climate risks such as typhoons, storm surges, coastal flooding, and sea level rise.

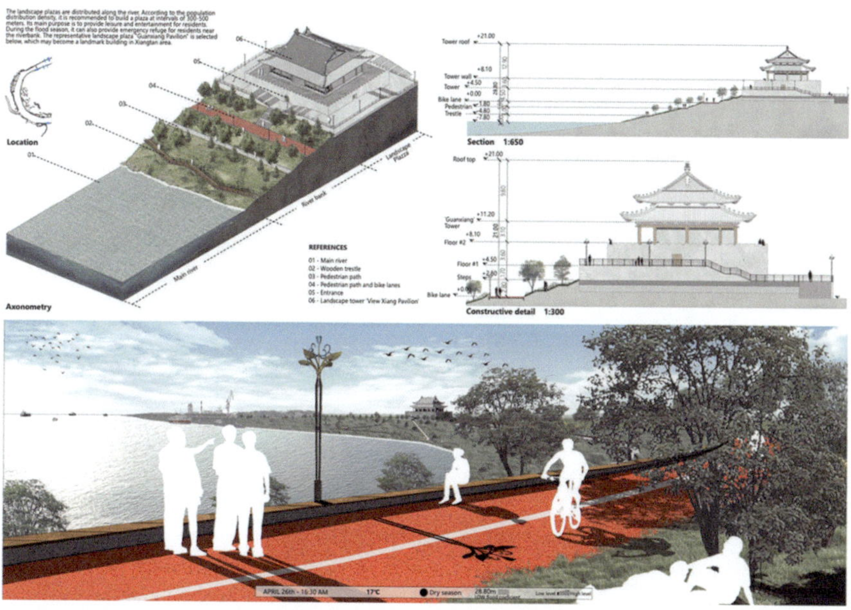

Fig. 35 Emergency site and interactive landscape. *Source* T. Yan

Historically, Manila thrived because of its strategic coastal position along Manila Bay, with access to rich fisheries and maritime trade routes for its growth. But its development has been marred by ecological challenges. Massive land reclamation, loss of natural mangrove buffers, and fouling of waterways have eroded the region defences against natural calamities.

Metro Manila has a dense urban environment with stark contrasts: towering office buildings and high-end housing developments coexist with congested and underserviced slums (see Fig. 37). These inequalities are especially felt more keenly in coastal areas such as Navotas, where economic life is deeply tied to the sea. Here, fisherfolk families face the volatility of fish stocks while having to deal with new threats of environmental disasters, rising waters, and the uncertainty caused by climate change.

Efforts to manage the metropolitan vulnerabilities have frequently relied on engineered, large-scale infrastructure projects such as sea walls, flood gates and land reclamation. Despite providing temporary relief, they often fail to address the ecological and social elements of resiliency, leading to fragmented landscapes and further marginalization of vulnerable communities. In this context, there is an increasing awareness of the need for holistic, responsive and regenerative approaches to urban development—approaches that do not imply resist nature but work in harmony with it.

This project is aimed at Redefining resiliency for coastal communities through aqua silviculture in Navotas, Philippines, using an E-gateway approach that can develop a unified platform to manage emergency, climate change effects, and economic development for the local community.

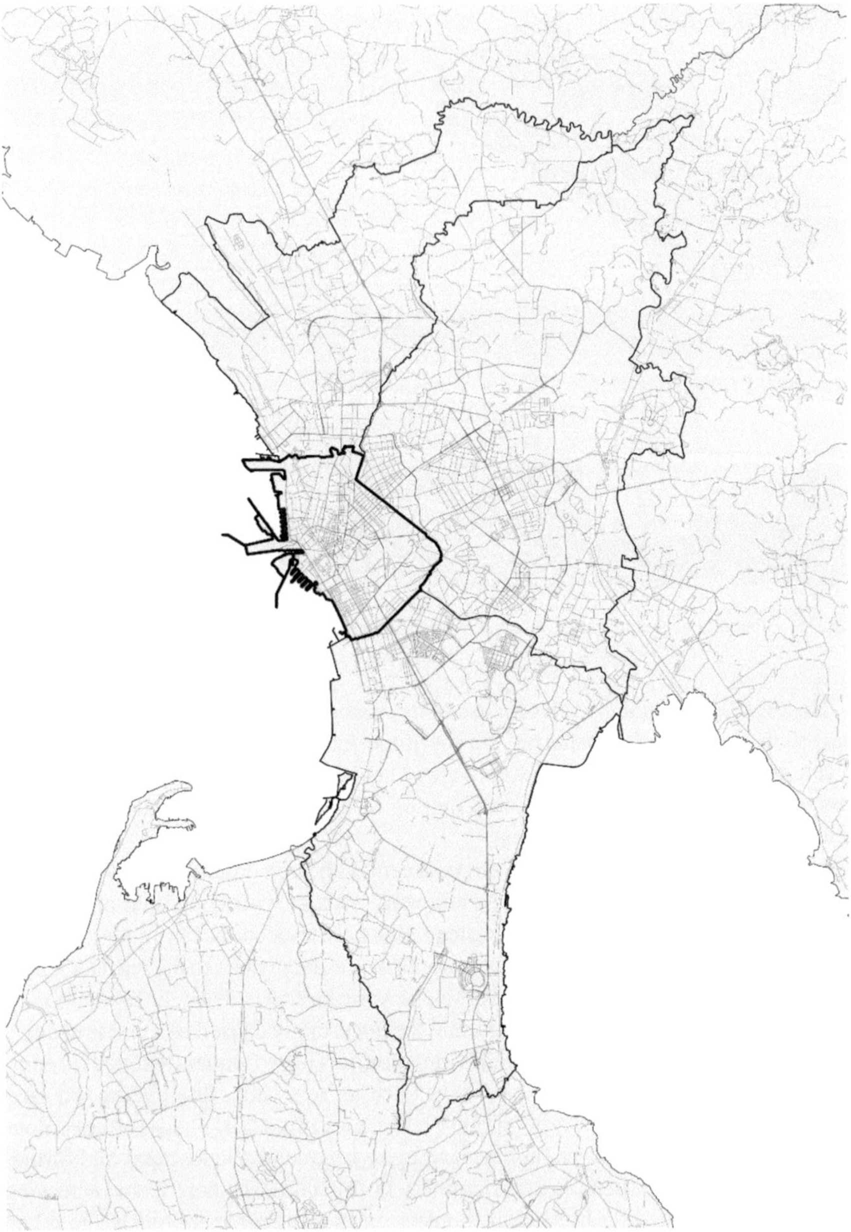

Fig. 36 Metro Manila. *Source* G. C. Canlas

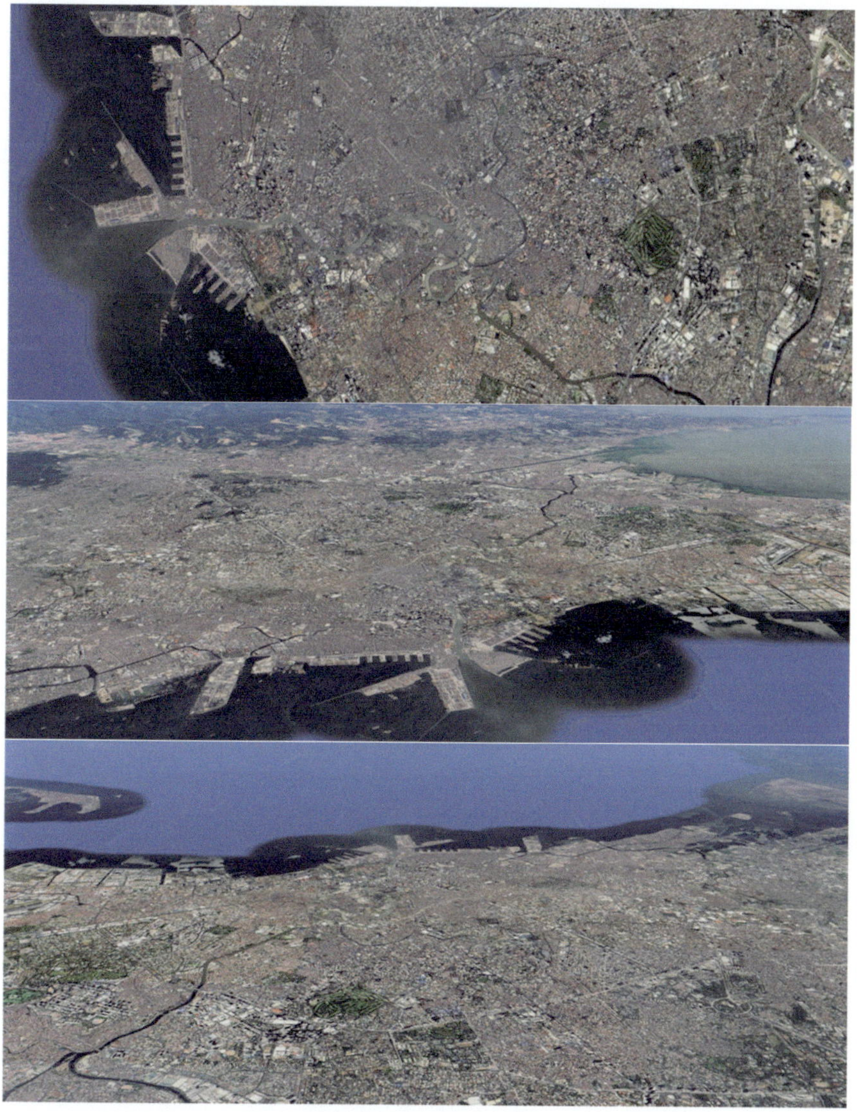

Fig. 37 Satellite views of Manila. *Source* Google Earth

"May Nilad" is a response to the multiple vulnerabilities of coastal communities. The name references the indigenous Nilad mangrove, which the city of Manila is named after, representing the project aim to reintegrate urban and coastal life with regenerative ecosystems (see Fig. 38).

The project redefines resiliency for the fisherfolk families by reinforcing them with an environment that supports their ability to adapt to changing environmental

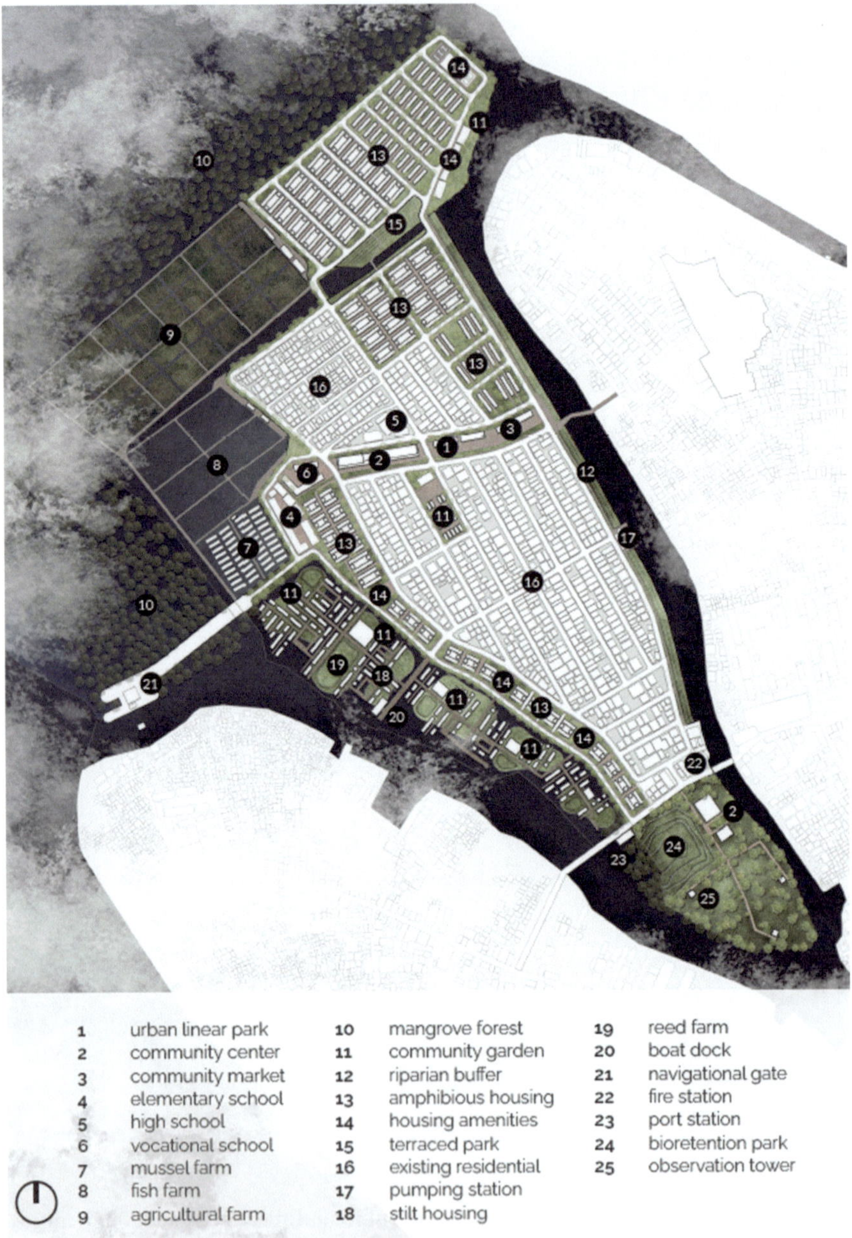

Fig. 38 General masterplan of "May Nilad". *Source* G. C. Canlas

conditions and their pursuit for sustainable livelihood. It focuses on these landscape design strategies:

- Transformation of Coastal Environment: Development of ecologically sound and adaptive landscapes that respond to the changing environmental conditions.
- Restoration of Social Infrastructure: Re-imagination of public spaces as critical spaces for community engagement, social cohesion, and cultural continuity.
- Strengthening Sustainable Aquaculture Systems: Creating resilient ecosystems that foster sustainable and productive aquaculture practices that sustain fisherfolk families.

At the urban scale, "May Nilad" aims to intertwine the community socio-economic fabric with its natural environment. The strategy focuses on urban interconnectivity—reinforcing spatial, ecological, and social connections through a network of green corridors that seamlessly connect Navotas to the adjacent neighbourhoods. The project promotes the reforestation of denuded mangrove areas to stabilize the coastal environment and raise awareness about their need for conservation. Lastly, the project introduces a green-grey infrastructure approach that restores polluted waterways. This hybrid approach combines engineered solutions (e.g., constructed wetlands, stormwater management systems) with natural systems (e.g., riparian buffers, vegetated floodplains) to enhance water quality, reduce flooding, and transform canals into vibrant community spaces. The renewed waterways are conceived not as barriers, but connectors, between the urban, community, and landscape.

At a neighbourhood level, "May Nilad" aims to enhance urban mobility, accessibility, and environmental quality. The strategy proposes wide, tree-lined, and pedestrian-friendly streets that connect the different zones, including open recreational areas and the city network of rivers. These green corridors create an uninterrupted urban landscape for both the locals and wildlife, enhancing biodiversity and supporting urban cooling. The project also ensures the proper management of the wetlands in support of the diverse wetland ecosystems—natural bird sanctuaries that provide refuge to migratory and endemic bird species and strengthen the identity of Navotas as an ecologically preserved area.

"May Nilad" presents a holistic vision for Navotas as a more attractive, integrating ecological restoration, community development, and sustainable livelihoods into a cohesive landscape framework (see Fig. 38). The masterplan is organized around key functional and programmatic zones:

- Green and Public Spaces Network: At the heart of the masterplan is the urban linear park (1) that weaves through the neighbourhood, linking residents to essential services and ecologically sensitive landscapes. Community centres (2) and community markets (3) are interspersed within the neighbourhood as key social nodes: spaces for daily life, culture, commerce to manifest, that could be managed and organized through a digital set of tools. Terraced parks (15) are strategically located along these edges and are designed to manage tide levels as they rise and fall, offering high vantage points for recreation and refuge.

- Productive and Resilient Economies: The masterplan integrates different productive landscapes that contribute to economic resilience:
  - Mussel farms (7) and fish farms (8) reinforce regional aquaculture systems that offer food and income that are sustainable.
  - Agricultural farms (9) diversify food production and thereby reduce their reliance on external supply systems.
  - Community gardens (19) contribute to food security, community engagement, and urban space activation.
  - Reed farms (21) are added for ecological and economical services, providing reeds that may be harvested as construction materials and bio-filters.
- Ecological Restoration and Protection Zones: These focus on the restoration and proposed natural ecosystems:
  - Mangrove forests (10) and riparian buffers (11) are restored to act as storm buffers, cleanse water, and provide habitats for wildlife.
  - The bioretention park (24) combines stormwater management with public green spaces. Integrated in this park are observation towers (25) that enable environmental monitoring and ecotourism.
  - Housing Typologies: Recognizing the community vulnerability to rising tides and floods, the masterplan introduces resilient housing models such as the amphibious (12) and stilt houses (18) which allow homes to float up and down according to water levels, creating a safe, versatile living space.

Through the integration of ecological restoration, productive landscapes, adaptable dwelling typology and resilient infrastructures as one cohesive system, the masterplan for "May Nilad" develops a vibrant urban ecology. Every component is designed to be in synergy with each other, forming that everything-holistic regenerative loop, where human, economic, ecological, and natural systems are reinforcing one another—the essence of a circular economy approach for coastal urban resilience.

Considering circular economy and digital communities, "May Nilad" is envisioned as an active and effective planning approach for Navotas and as a replicable example of true resilience for future development of coastal communities (see Fig. 39). Central to this vision is the principle of circular economy—an economic system that focuses on regenerative systems, resource productivity and a production/consumption system.

Considering Navotas, the project applies circular economy through:

- Restoration of Coastal Ecosystems: Rehabilitation of denuded mangrove forests and tidal wetlands acting as natural protection against storm surges and nurseries for fish and crustaceans.
- Regenerative Aqua silviculture: Introduction of the productive and sustainable aquaculture practice of integrating fish farming with mangrove cultivation, creating an environment that enhances productivity without straining natural resources. This could be managed through a specific platform that can help the

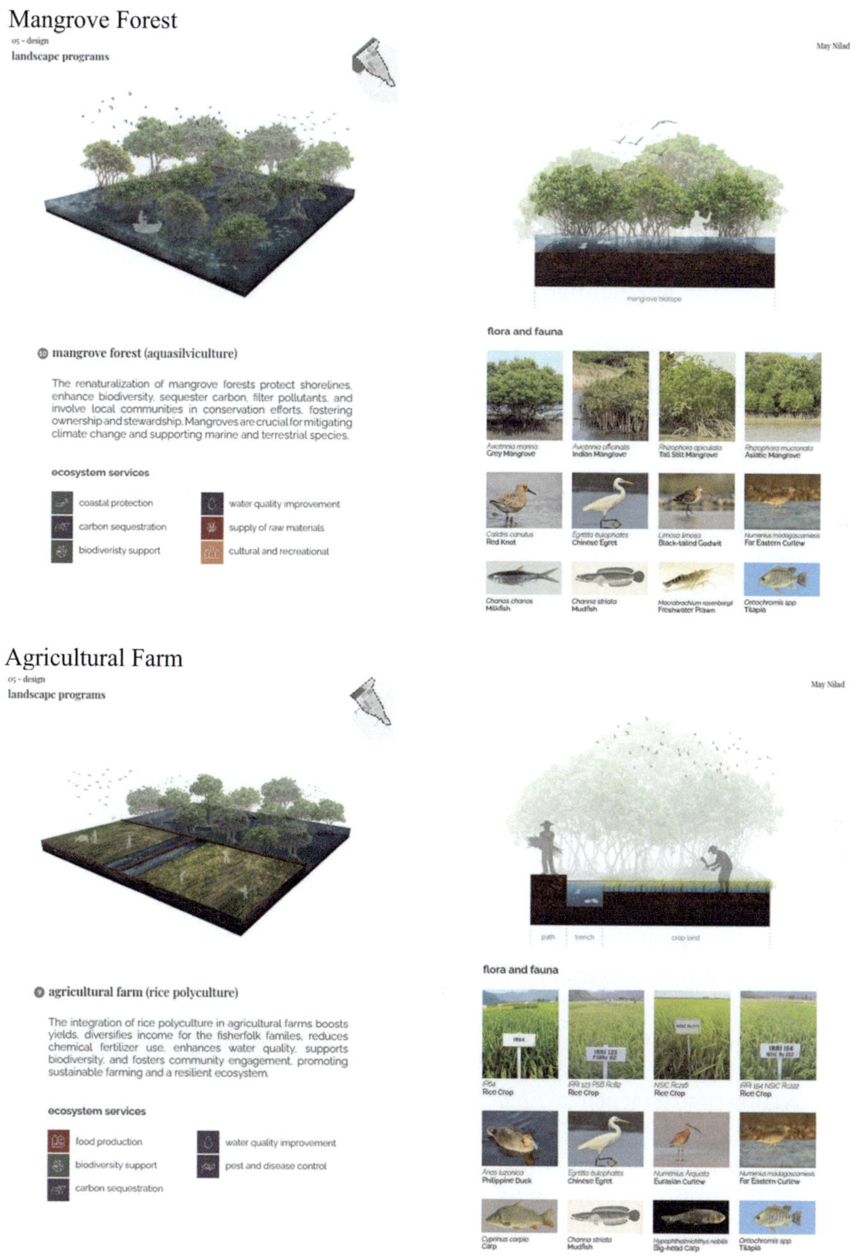

Fig. 39 Landscape planning for mangrove forests and agricultural farm. *Source* G. C. Canlas

local community managing farming with integrated data on weather, climate, ecosystems, and economic requests (demand).

- Waste-to-Resource Innovations: Development of systems that transform organic waste and aquaculture waste into value-added products (e.g., fertilizer, feed, energy), reducing pollution and adding value to by-products. Similarly to aqua silviculture, also this part of the project can be managed digitally, integrating information in one single management platform.
- Community-Led Local Resource Management: Enhancement of the local community members through training and capacity-building programs on sustainable fishing, aquaculture, and ecosystem management, ensuring that the economic benefits circulate within the community.
- Resilient Public Spaces: Development of open multi-use public spaces that are resilient to environmental change (e.g., rising tides) and are equipped to host education, social, and economic activities.

From landscape design down to livelihood systems, the project offers a paradigm shift from the conventional cycle of environmental degradation and vulnerability to one of regeneration, self-sufficiency, and resilience (see Fig. 40).

It transforms vulnerability into strength through the proposed system where resources are recycled and renewed, and livelihoods are equipped with sustainable practices, and the community is empowered to be the steward of their environment.

"May Nilad" proves that resilience lies not just in withstanding disasters but about thriving and continuously regenerating and growing with one another; it lays out a blueprint for future construction of coastal communities.

"May Nilad" applies the circular economy is a living framework, an alternative development model subverting the traditional approach to development, which is fuelled by linear extraction and consumption of resources, to become a "closed loop" approach where natural, social and economic resources are continually restored, reused, and regenerated (see Fig. 41).

At the ecological level, the project restores critical natural systems, such as mangrove forests, wetlands, and riparian buffers. These landscape systems are not preserved as isolated conservation areas but are integrated into the fabric of daily community life—providing food, shelter, refuge from natural disasters and essential ecosystem services to its people. The restoration of these habitats ensures the regeneration of natural capital, increase of biodiversity, and resilient local economy.

At the community level, livelihoods are centred on sustainable, placed-based industries in harmony with environmental stewardship. Mussel farms, fish farms, agricultural farms, reed harvesting, and eco-tourism initiatives develop a diversified local economy. Waste from one becomes input to another system, like organic farm waste or aquaculture by-products, which reduces impact and boosts productivity.

At the urban design level, the spatial structure of Navotas is redesigned to optimize resource cycling and social resilience. Proposed green corridors, terraced parks, and community gardens act as multi-functional infrastructures that manage stormwater, provide better micro-climate, stimulate food production, and enhance social cohesion. Amphibious and stilt housing are both examples of resilient housing typologies,

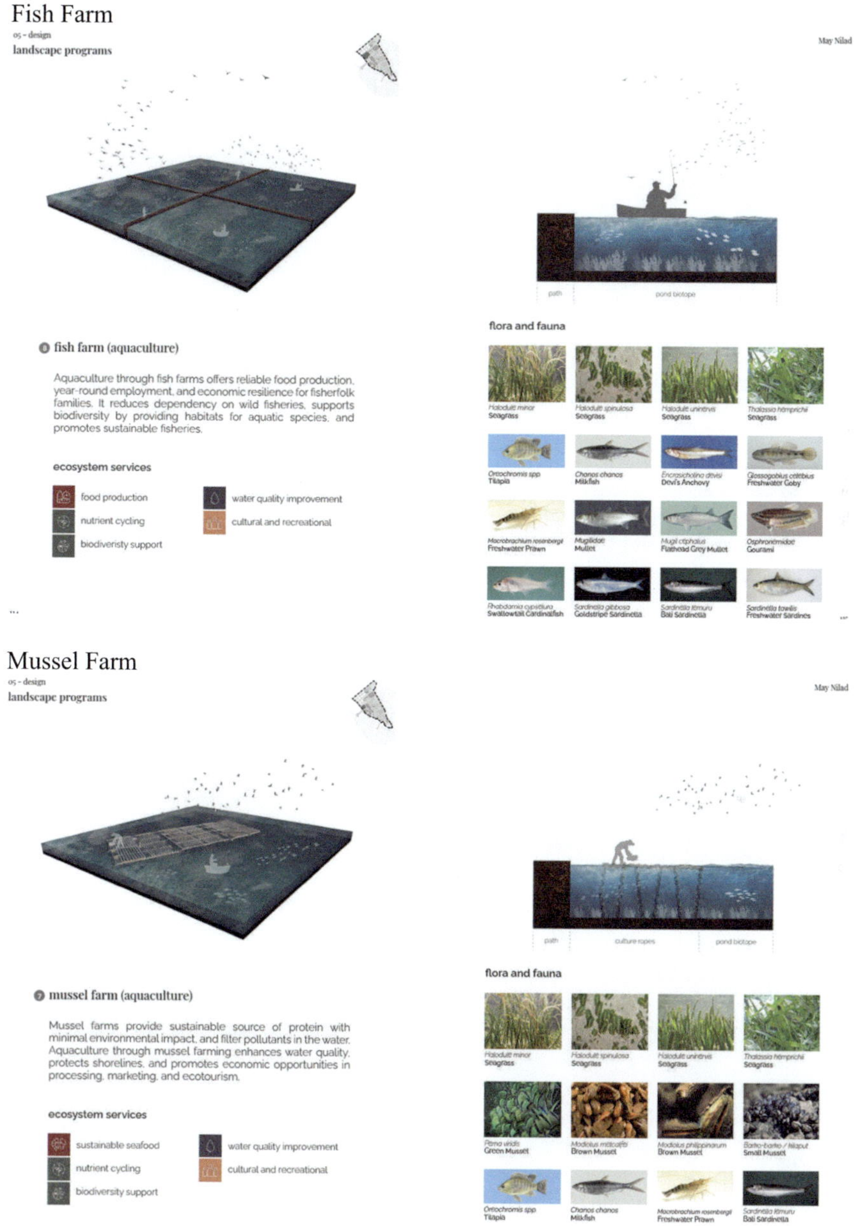

Fig. 40 Landscape planning for fish and mussel farm. *Source* G. C. Canlas

Fig. 41 Landscape planning for riparian buffers and community gardens. *Source* G. C. Canlas

designed to cope with environmental changes, minimizing the need for expensive disaster recovery and rebuilding.

Knowledge, skills, and cultural identity are also treated as renewable resources that can be managed digitally by the community. Educational facilities, community centres, and community markets are planned to promote the circulation, not only of the economic, but also of the restorative, extending circularity from materiality into that of social innovation.

Through this circular and regenerative approach, "May Nilad" demonstrates that true resiliency in coastal communities is achieved only if circular economy principles are applied at every scale: ecological, economic, and social.

"May Nilad" is envisioned as a replicable model for future development of coastal communities, an example of how circular design can enable communities not just to withstand change, but to flourish in harmony with their environment.

6 Case Studies in Other Continents

Several countries in the Americas have developed their own national Indo-Pacific policies, starting from Canada and USA, which aim at increasing their international collaborative projects with the most dynamic economic region in the next century. South America is now developing their own policies, though their commitment may seem more recent. The Indo-Pacific region is becoming a global hub of economic dynamism and strategic challenge, impacting Canadian national security, economic prosperity, and human rights. By 2040, the region will account for over half of the global economy, with China, India, and Japan being major economies.

The Indo-Pacific Strategy of Canada, launched in November 2022, aims to deepen relationships with the region and promote peace, resilience, and security. The strategy focuses on 24 initiatives across 17 departments and agencies across the federal government, organized under five interconnected strategic objectives: promoting peace, resilience, expanding trade, investment, supply chain resilience, investing in and connecting people, and building a sustainable and green future. The Strategy outlines five interconnected strategic objectives:

1. Promote peace, resilience and security,
2. Expand trade, investment and supply chain resilience,
3. Invest in and connect people,
4. Build a sustainable and green future,
5. Canada as an active and engaged partner to the Indo-Pacific.

Regional security is crucial for global stability, and Canada is engaged in bilateral and multilateral security consultations and frameworks across the Indo-Pacific. The strategy recognizes that Canada must engage as a regional security partner to protect its national interests. Canada will contribute to regional stability through an enhanced military presence, expand existing military capacity building initiatives, and launch new training programs that advance joint priorities and interoperability with regional partners.

The Indo-Pacific region makes up more than one-third of all global economic activity, with three of the world largest economies being in this part of the world. By 2040, the region will account for more than half of the global economy and two-thirds of the global middle class. Canada has prioritized key trade initiatives, including the Comprehensive and Progressive Agreement for Trans-Pacific Partnership, Canada-Japan Economic Partnership Agreement, Canada-India Comprehensive Economic Partnership Agreement negotiations, Canada-Korea Free Trade Agreement, Canada-ASEAN Free Trade Agreement, Canada-Thailand exploratory free trade discussions, and Pacific Alliance free trade agreement negotiations.

Canadian development assistance to the Indo-Pacific region provides targeted assistance to support and respond to the needs of Indo-Pacific countries, aligned with its Feminist International Assistance Policy. Canada will continue to deliver feminist international assistance and defend human rights through partnerships with local governments, regional groups, civil society, academia, and other stakeholders in the region and beyond.

Canadian strength lies in sectors like education, health services, food, agriculture, fisheries, natural resources, energy, financial services, advanced manufacturing, and green infrastructure. People-to-people ties with the Indo-Pacific are vital for this opportunity and enrich the Canadian social and economic fabric: Canada welcomes more international students from India than any other country. However, strategic challenges such as inter-state tensions, China assertive pursuit of economic and security interests, and China coercive treatment of other countries and economies are reshaping the strategic outlook of the region. Canada is engaging in the Indo-Pacific in coordination with its allies, recognizing its unique contribution to the region. The region ongoing development challenges, poverty, and inequality are exacerbated by climate change and biodiversity loss.

Latin America faces the challenge of maintaining diplomatic balance between China and the US, which is the region number one trading partner. As pressures to choose sides increase, the region would benefit from envisioning its own version of the Indo-Pacific. Only two of the eleven countries in Latin America have an official policy document that mentions the term 'Indo-Pacific'. Colombian Ministry of Foreign Relations Action Plan 2021–2022 aims to include Latin America in the Indo-Pacific visions of Australia and New Zealand, emphasizing the importance of fostering relations with India to strengthen political and economic ties with the Indo-Pacific. Chilean Defence Policy in 2020 classifies the Indo-Pacific as 'vital' for Chilean development and prosperity, justifying this emphasis by framing the region as a security concern due to its potential for conflict.

Brazil has omitted the term 'Indo-Pacific' in official policy documents, even though it enjoys status as a regional power with its extensive Atlantic coast and strategic ties to India and Lusophone Africa. A Latin American vision for the Indo-Pacific would prompt states to rethink regional schemes of integration and cooperation, aligning these institutions with respect to their external orientations.

Argentina has relaunched the National Committee for Asia and the Pacific (CONAPAC) to strengthen its relationship with the Asia–Pacific region. The committee, which includes government, academic, and business representatives, aims to achieve concrete foreign policy recommendations to strengthen bonds with the region. Argentina has also participated in the Pacific Alliance, APEC, and ASEAN summits, and has held meetings with heads of state from China, Japan, Australia, New Zealand, India, and Indonesia. In 2016, Argentina began the accession process to the Association of Southeast Asian Nations (ASEAN), a successful regional integration and coordination organization. Argentina has diplomatic relations with the Solomon Islands and Nauru, and the Asia–Pacific region is home to 60% of the world population.

1. **Toronto, Canada**

Toronto, the largest and one of the most diverse cities in Canada, is known for its vibrant communities, strong economy, and global reputation for inclusion. But alongside its many strengths, the city faces ongoing challenges, especially when it comes to homelessness and waste management. High real estate prices, limited affordable housing, and growing inequality have made it increasingly difficult for many residents to find stable and safe places to live. Thousands of people in Toronto experience homelessness each year, often dealing with other challenges like unemployment, mental health struggles, or lack of access to support services. At the same time, Toronto rapid growth has increased the amount of waste being produced, adding pressure to the city efforts to manage it sustainably. Since 2000, Canada has been working toward national waste reduction goals, and Toronto has introduced recycling, composting, and education programs to reduce environmental impact. However, more innovation is needed to make a real difference.

In this context, digital platforms offer new opportunities to support both social and environmental goals—particularly in addressing the growing issue of homelessness in Toronto. With thousands of individuals facing housing insecurity due to rising costs, limited shelter space, and systemic barriers, technology can play a key role in bridging the gap between vulnerable populations and the services they need. Digital tools can streamline access to housing applications, mental health support, employment resources, and community outreach programs, helping people navigate complex systems more easily. For those experiencing homelessness, apps and online platforms can offer vital connections to emergency shelters, food banks, medical care, and long-term support networks. At the same time, these technologies can support environmental initiatives by coordinating waste collection, promoting recycling habits, and tracking sustainability data. Community-based digital platforms can further engage residents by allowing them to report issues, share resources, and participate in local problem-solving. By improving communication, increasing efficiency, and ensuring that everyone—especially those most at risk—is included in the conversation, digital technology can help Toronto move toward a more inclusive, sustainable, and resilient future.

'Hope' is a promising and feasible project that combines a micro-circular economy approach with the power of digital platforms to address homelessness and promote sustainability in Toronto. Positioned within the E-Global Gateway of the Digital Economic Community, 'Hope' exemplifies how digital innovation can integrate sustainability and social inclusion to create resilient, equitable, and connected urban communities.

In the initial phase of investigating the issue of homelessness, we identified that its root causes are primarily social in nature, closely linked to architectural and urban conditions. Rather than viewing homelessness solely as a lack of accommodation, we approached it as a complex social issue. Our first step was to establish the relationship between these social challenges and broader urban strategies. Building on this foundation, we then explored architectural responses by proposing innovative approaches to both temporary and permanent forms of accommodation.

The project addresses the interconnected challenges of homelessness and waste management in Toronto through the innovative reuse of materials sourced from urban waste streams, possibly relying on a digital community approach (see Fig. 42). We propose the development of affordable housing that can be built by individuals experiencing homelessness using these recycled resources. By directly involving them in the construction process, the initiative fosters both employment opportunities and practical skills development. This participatory approach not only empowers individuals to generate income but also enables them to take an active role in shaping their own living environments. This initiative offers a sustainable, community-driven alternative to traditional housing programs by transforming waste materials into affordable housing. By engaging individuals experiencing homelessness in the construction process, we reduce environmental impact while promoting job creation, skills development, and financial independence. The project provides a holistic response to homelessness, fostering empowerment, inclusion, and a sense of belonging within the community, while contributing to the environmental and social sustainability of Toronto.

Homelessness in Toronto remains a critical and complex issue, with over 12,000 individuals experiencing homelessness on any given night. The crisis is driven by a combination of factors, including a severe lack of affordable housing, high rental costs, and a low vacancy rate. In 2024, the city reported 135 deaths among the homeless population in the first half of the year, with drug toxicity being the leading cause. The median age of death is significantly lower than the general population, reflecting the harsh realities of life without stable housing. Indigenous peoples are disproportionately affected, underscoring systemic inequities. While the City of Toronto continues to invest in shelter services, outreach, and public health interventions, the scale of the crisis highlights the urgent need for comprehensive, sustainable answers that tackle the social and structural extents of homelessness.

This project investigates homelessness in Toronto, highlighting mental health issues and the loss of social identity as key challenges, with lack of shelter as a secondary concern (see Fig. 43). We propose a three-phase strategy: first, involving individuals experiencing homelessness in a government-backed waste management initiative to provide employment and foster community interaction; second, allocating land for the construction of self-built temporary shelters to promote autonomy; and third, developing permanent social housing integrated with public spaces

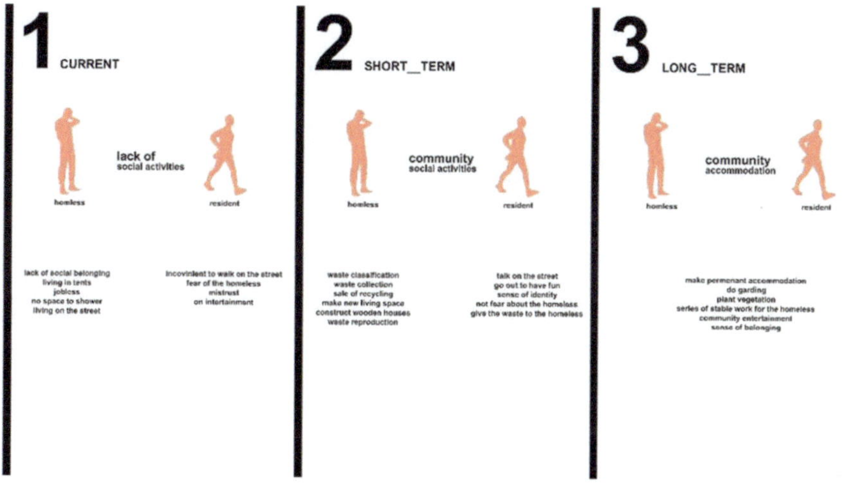

Fig. 42 A homelessness strategy. *Source* M. Khajehahsani, N. Haghpanah

designed to encourage tourism and social inclusion. This approach aims to address homelessness through a combination of empowerment, community engagement, and sustainable housing solutions.

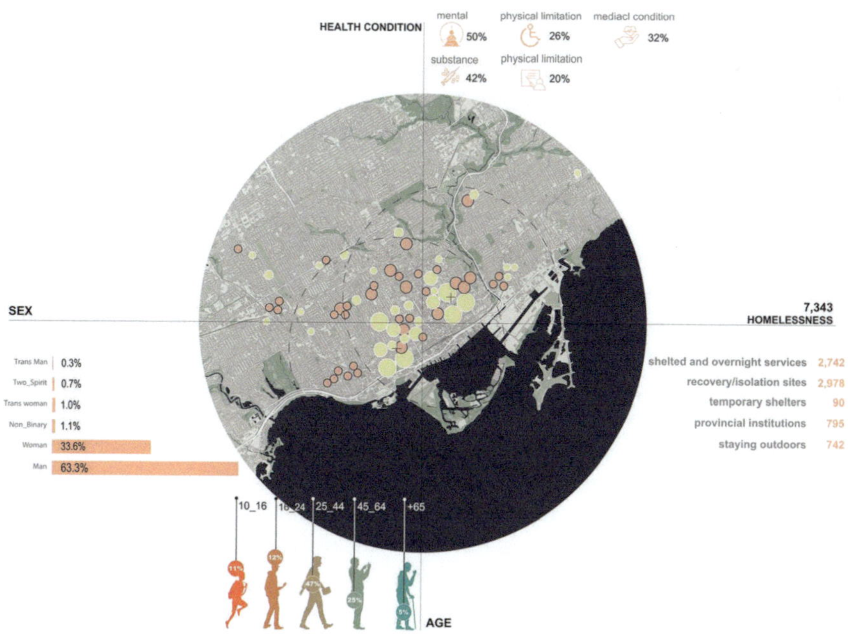

Fig. 43 Homelessness in Toronto: key factors. *Source* M. Khajehahsani, N. Haghpanah

For the temporary accommodation, repurposing parking lots using a modular design approach offers a flexible and scalable solution to homelessness (see Fig. 44). This strategy transforms underused spaces into temporary housing and community areas, while integrating units between existing buildings improves land-use efficiency. It promotes inclusivity and social integration within the urban fabric.

For long-term accommodation, we developed social housing that incorporates a semi-private environment within the urban core to enhance interaction and connectivity with the surrounding city (see Fig. 45). This design strategy fosters a sense of community while promoting integration with the broader urban context.

The graph map outlines the designated zone and chosen parking sites, with light and dark squares indicating areas of low and high connectivity. Key landmarks are marked to provide spatial context, while parking locations are clearly identified. This visualization emphasizes the strategic positioning of parking areas to enhance accessibility, connectivity, and opportunities for future infrastructure development.

In the second phase of the project, which focuses on temporary accommodation, we selected parking lots as intervention sites due to their potential for fostering connectivity with local communities. This phase involves a flexible, scalable strategy that can be expanded or limited based on spatial needs. For the third phase, dedicated to permanent accommodation, we targeted one of the most touristic areas of Toronto, aiming to integrate social housing with public spaces that also serve as cultural or touristic attractions. This approach seeks to promote social inclusion while enhancing the visibility and acceptance of social housing within the urban landscape. As demonstrated in the previous section, we designed the platform solely to organize the placement of tiny houses constructed by individuals experiencing homelessness during the second phase of the project. We believe that, as architects, our role extends beyond identifying isolated sites for temporary shelters; instead, we must create opportunities that restore social identity and foster community integration. By enhancing parking lots with pergolas and green spaces, this approach not only improves the quality and usability of urban infrastructure but also supports the reintegration of marginalized individuals into the social fabric of the city.

The strategic framework (see Fig. 46) of the project encompasses three integrated approaches, each designed to address key aspects of homelessness while promoting urban sustainability, community integration, digital approaches and environmental enhancement.

The first approach focuses on improving waste management, building upon the city long-established system initiated in 2000, and boosted through a specific digital management platform approach. By involving individuals experiencing homelessness in the collection, sorting, and reuse of materials, this strategy provides both income-generating opportunities and skill development, while encouraging the creation of functional items such as furniture from recycled resources. This not only supports environmental goals but also fosters social reintegration through collaboration with local communities. The second approach emphasizes improving the city appearance by introducing flower boxes and utilizing repurposed materials to beautify public spaces. In parallel, it addresses informal tent settlements by offering dignified and affordable housing solutions, thereby improving both urban aesthetics

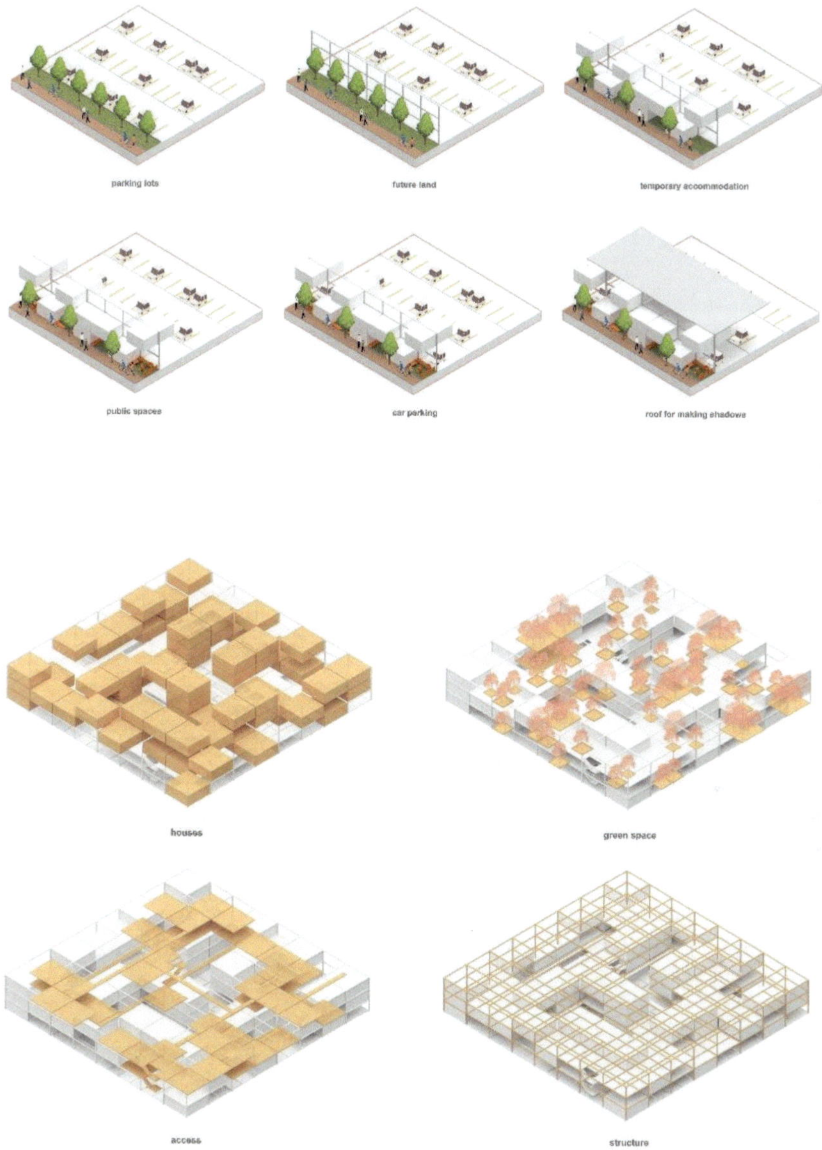

Fig. 44 Temporary and permanent accommodation solutions. *Source* M. Khajehahsani, N. Haghpanah

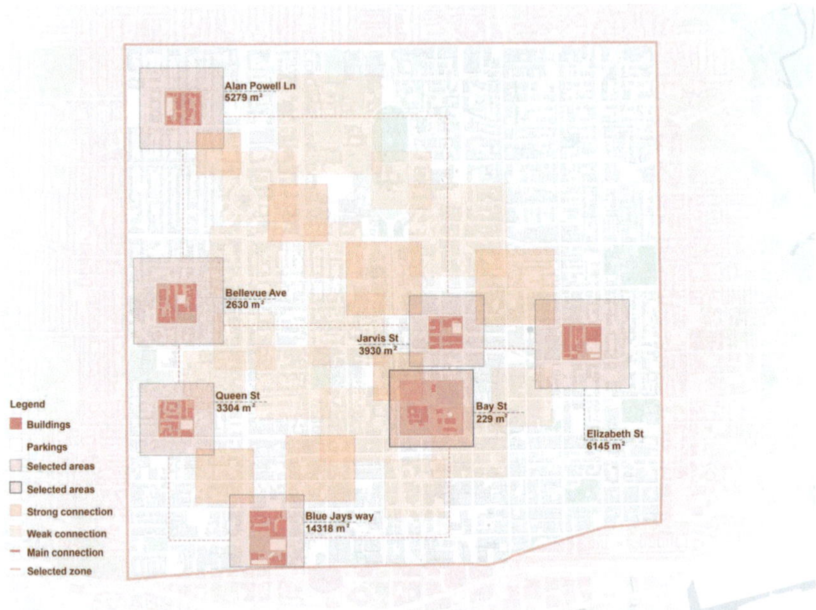

Fig. 45 Masterplan of Toronto homelessness project. *Source* M. Khajehahsani, N. Haghpanah

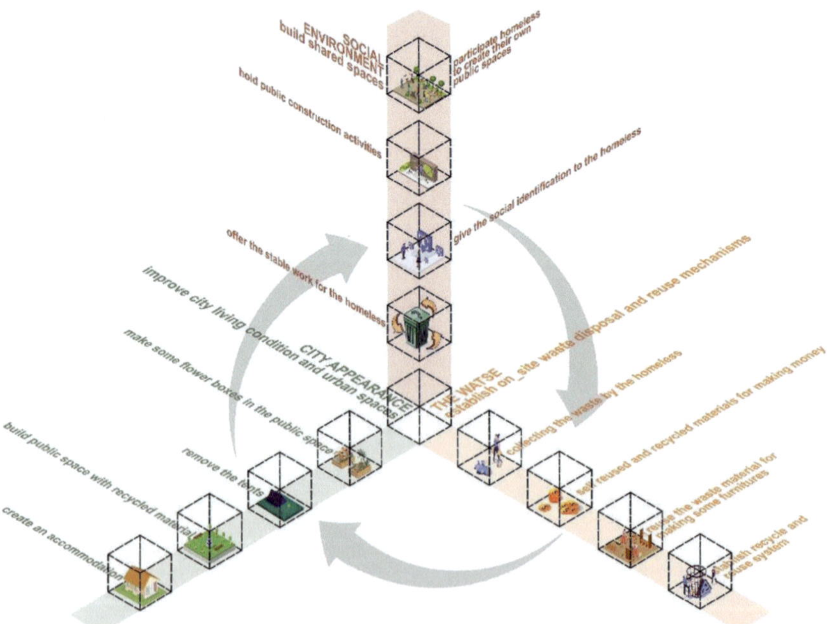

Fig. 46 Strategic framework of Toronto homelessness project. *Source* M. Khajehahsani, N. Haghpanah

and living conditions. The third approach promotes to strengthen the social fabric through the creation of inclusive public spaces and stable employment pathways, boosted by the digital community approach. By engaging the homeless population in public infrastructure projects and urban development initiatives, also through a dedicated platform, the strategy empowers them to contribute meaningfully to the city growth. Collectively, these interconnected approaches promote social inclusion, environmental stewardship, and urban resilience, advancing a holistic response to homelessness.

Architecturally, the tiny houses (see Fig. 47) are designed for simplicity and feasibility to ensure they can be built by individuals experiencing homelessness. Using accessible materials and easy assembly methods, the design promotes practical construction while fostering autonomy and skill development.

We designed two types of flower boxes, chairs, fences, benches, and tables to support both the housing units and surrounding public spaces (see Fig. 48). Made from repurposed materials, these elements foster social interaction, promote sustainability, and create a sense of community and belonging for individuals experiencing homelessness.

The height of the building is constrained to a maximum of 9 m, ensuring it remains in scale with the surrounding urban environment while adhering to height regulations (see Fig. 49). The length of the structure, however, is flexible and dependent on the dimensions of the existing parking lots, allowing the design to be adaptable to

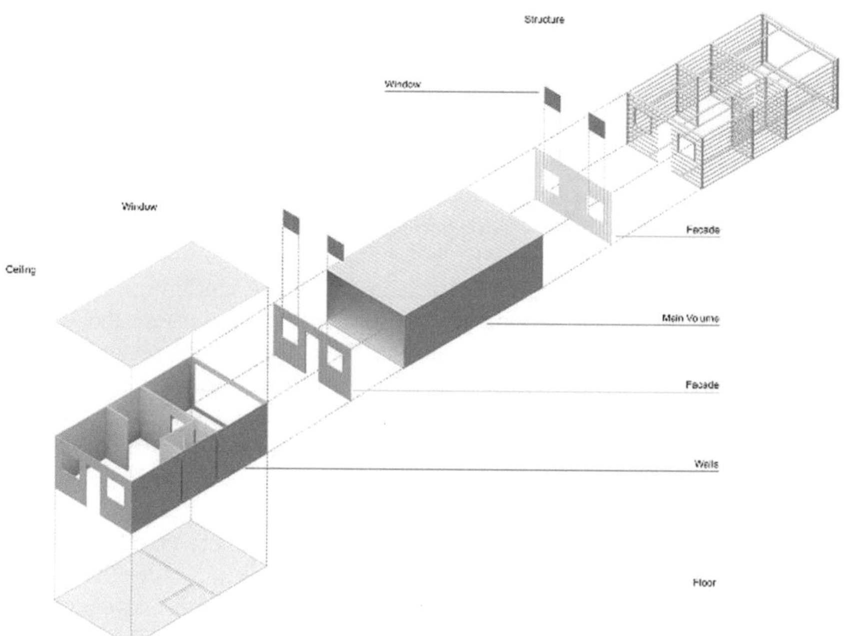

Fig. 47 Composition of tiny houses. *Source* M. Khajehahsani, N. Haghpanah

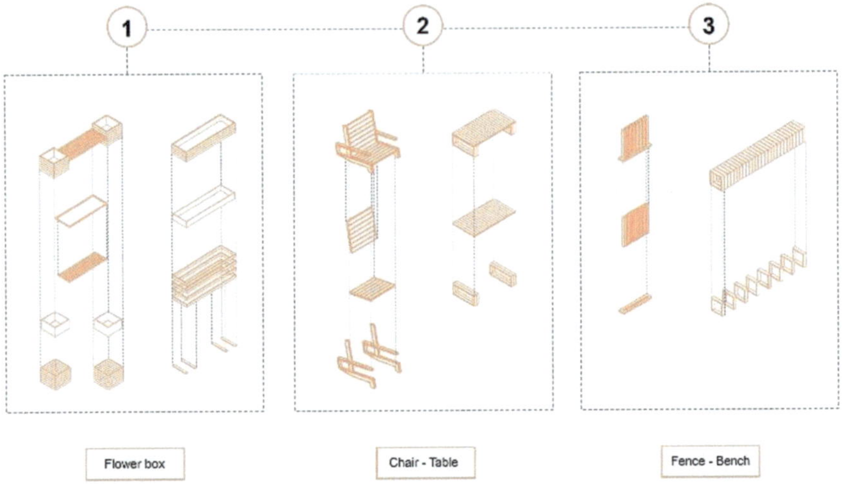

Fig. 48 Furniture design. *Source* M. Khajehahsani, N. Haghpanah

different site configurations. At the ground level, public furniture, such as benches, tables, and planters is strategically placed between the housing units to foster social interaction and create shared community spaces. These spaces not only enhance the sense of community among residents but also serve as areas for engagement.

The project proposes a digital hub or platform, where most of the information about tiny houses and furniture schemes can be shared. Similarly, the platform could also serve as a collector and a marketplace, where the sourcing of the materials is. provided (after collection activities promoted by the garbage strategy and actions implemented through the city programs). Being conceived as a marketplace, the platform can even become a job marketplace or a timesharing opportunity, where people offer and seek for help, advice and tutoring about the different stages of the project.

The project concluded with the construction of pergolas, offering shaded areas and completing the overall design. Prior to that, green spaces were thoughtfully integrated between the housing structures to enhance liveability. Ground-floor areas were activated using design elements from earlier projects, adding familiarity and functionality. The layout fostered interconnections between housing units, intentionally creating voids that introduced openness into the development. The housing was introduced in three distinct phases, each respecting the maximum height limit of nine stories suitable for the site.

'Hope' is, then, a comprehensive urban strategy that addresses the interconnected challenges of homelessness and waste management in Toronto, while aligning with the broader goals of the E-Global Gateway of the Digital Economic Community. The project begins with short-term actions, transforming underutilized urban spaces—such as vacant parking lots—into modular temporary housing units constructed from recycled materials. These units provide safe, immediate shelter for homeless families

Fig. 49 Tiny houses layout. *Source* M. Khajehahsani, N. Haghpanah

while promoting sustainability through the reuse of waste materials. Ground-floor spaces are furnished with repurposed items, and green spaces are integrated between structures to enhance liveability and environmental quality.

To create economic opportunities, the project involves homeless individuals in structured waste collection and recycling activities. These efforts are supported by digital training programs that introduce participants to basic technologies, enabling them to engage in the emerging digital economy. This inclusion supports the Digital Economic Community vision of economic empowerment through digital literacy and innovation.

A key component of the project is its digital monitoring system, which tracks housing progress, environmental impact, employment outcomes, and community well-being. This system provides real-time feedback and facilitates data-driven decisions, ensuring the project remains adaptable, transparent, and responsive to community needs. Alongside this, the project incorporates digital communication tools to enhance peer-to-peer interaction among homeless residents. These tools support emotional well-being, build community trust, and connect individuals to essential services and collaborative opportunities, strengthening the social fabric of the temporary communities. In the long term, Hope envisions the transition of these temporary settlements into permanent social housing integrated with digital infrastructure and community services. The design process is organized into three construction phases across four site zones, respecting local building regulations and emphasizing interconnected layouts that foster interaction and accessibility. Features such as shaded pergolas and communal public spaces further enhance comfort and community engagement (see Fig. 50). Importantly, Hope is designed to be scalable and replicable across different global contexts. With adaptable design principles and a strong digital foundation, the project offers a flexible model that can be implemented in other cities and countries facing similar socio-environmental challenges.

By integrating housing, employment, digital tools, and sustainability, Hope serves as a pioneering model for inclusive urban development. It contributes to global goals of digital inclusion, environmental stewardship, and social equity through the framework of the E-Global Gateway of the Digital Economic Community.

This project would perfectly fit in the Canadian strategies for the Indo-Pacific policies, as it could be used for exporting best practices in terms of E-Gateway communities for promoting peace, resilience, investing in and connecting people, and building a sustainable and green future. It could also be an opportunity to export practices that can help improving regional security, as crucial strategy for global stability, supporting also the Canadian development assistance to the Indo-Pacific region on the needs of Indo-Pacific countries, by defending human rights through partnerships with local governments, regional groups, civil society, academia, and other stakeholders in the region and beyond.

2. **Buenos Aires, Argentina**

Climate change is a very complex phenomenon that involves countless disciplines and experts. As landscape architects in the actual world, it is interesting to try to experience this problem as it will probably affect gradually the economic community ability to take advantage of more economic/financial opportunities. Even though the climate change impact is a very broad topic, that involves many technical aspects that go beyond every single field of study, the main idea of this project is to contribute by proposing a landscape vision that tries to think what kind of interventions could implement to improve the conditions of the city facing these issues, in the aim of strengthening global digital infrastructure, promoting values-based investment, and

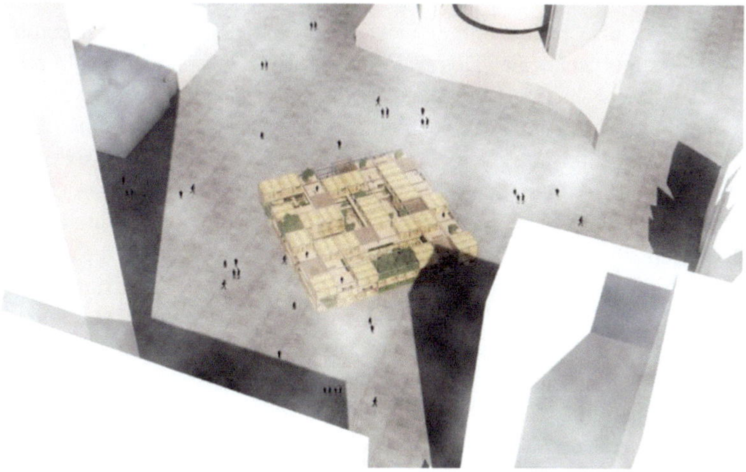

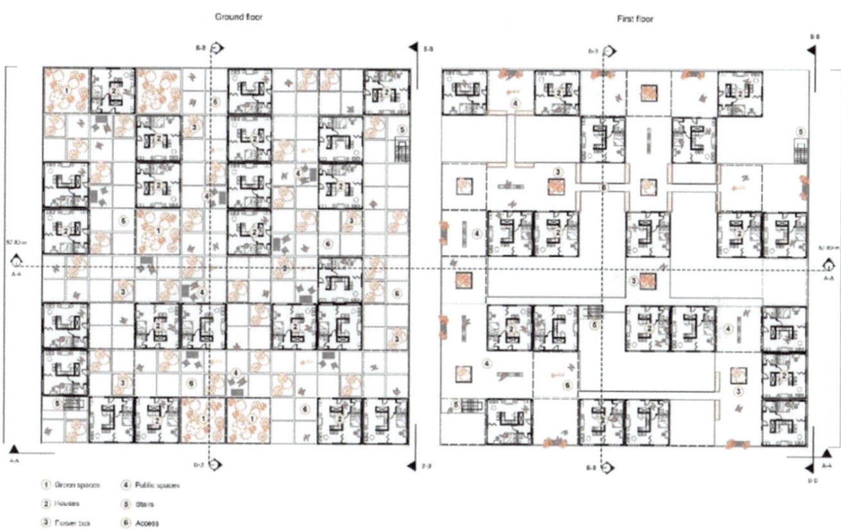

Fig. 50 General development plans. *Source* M. Khajehahsani, N. Haghpanah

advancing a human-centric approach to the digital economy adaptability to Digital Economic Communities.

The project seeks to study the main problematic factors that would affect the city and propose some interventions to improve the flooding risk in the most affected urban areas, seeking ways to support the capacity-building necessary to advance sustainable and transparent systems. Due to the characteristics of Buenos Aires, the main challenge to face in the future would be the flooding issues and urban heat island effect. This research tries to understand the broad landscape, not only of the city scale but also from a larger view. Therefore, this study is structured from a regional to an object scale. This way of organizing the project would allow having a more complete vision of the area and the subject. If landscape were a discipline includes both natural and artificial elements that shape the appearance and character of the city, the proposals developed in this project would not also assess the physical problems but also to the intangible aspects of the urban site. Traces of the invisible refers to the identity of the city that is hidden underground. By trying to bring natural interventions for minimize the impact of climate change along the areas where the nature is hidden, it would also connect the people back to the history of the city.

Buenos Aires is the largest city in Argentina and the 7th most populated city in Latin America. It is in the central-east part of the country, on the western coast of La Plata river, in the Pampas Plain. A great region with 40 municipalities surrounds the Metropolitan Area of Buenos Aires (AMBA, for its name in Spanish). These areas are in permanent contact with transportation of goods and services and constant commuting from one area to another. It has a surface area of approximately 3600 km^2 and a population of around 14 million inhabitants. Situated on the eastern shore of the Río de la Plata estuary has a temperate climate characterized by hot, humid summers and mild winters. As any city of the world, Buenos Aires faces many challenges about the impact of the climate change.

To address this topic, in 2009, the City of Buenos Aires developed a Climate Action Plan that determines that the city will become carbon–neutral, resilient, and inclusive by 2050. In this report the city states that then the effects and threats of climate change will be coastal flooding, pluvial flooding and heat waves.

A multi scalar understanding of the site is needed to tackle the aim of the project. Rivers are a complex system of water flow that cannot be understood in a small-scale view. Buenos Aires is a city located in the shore of the Rio de la Plata. The hydrographic area of influence of the Rio de la Plata is one of the biggest of the continent (see Fig. 51). It covers an area of about 3,100,000 km^2 from Argentina, Uruguay, Brazil and Paraguay. The main river of this system in the Paraná River that runs north south of Argentina. The anti-fragile design approach is organized in four different scales: regional scale, metropolitan scale, urban scale and city scale, where a case of study is developed in a small scale.

The flooding issue is a complex phenomenon that must be understood in a big picture and interventions has to be done in different scales (see Fig. 52). Río de la Plata is the last part of a complex water system. Its hydrology influences an area that goes beyond the national limits. Río Paraná is the most important river in this system, and it goes from the south of Brazil until Buenos Aires. It crosses different latitudes, climates landscapes and socio-economic systems, therefore it wouldn't be possible to design an unique solution for the flooding issues in the system. The management of the flooding risk should take in account the trace of the water since it is a raindrop until goes back to the sea.

According to the National Secretary of Environment and Sustainable Development, there is a catalogue with several strategies that could be applied in the area, to minimize the impact of flooding. Since the project aims to study nature-based solutions for flooding risk that could be implemented to create a digital ecosystem

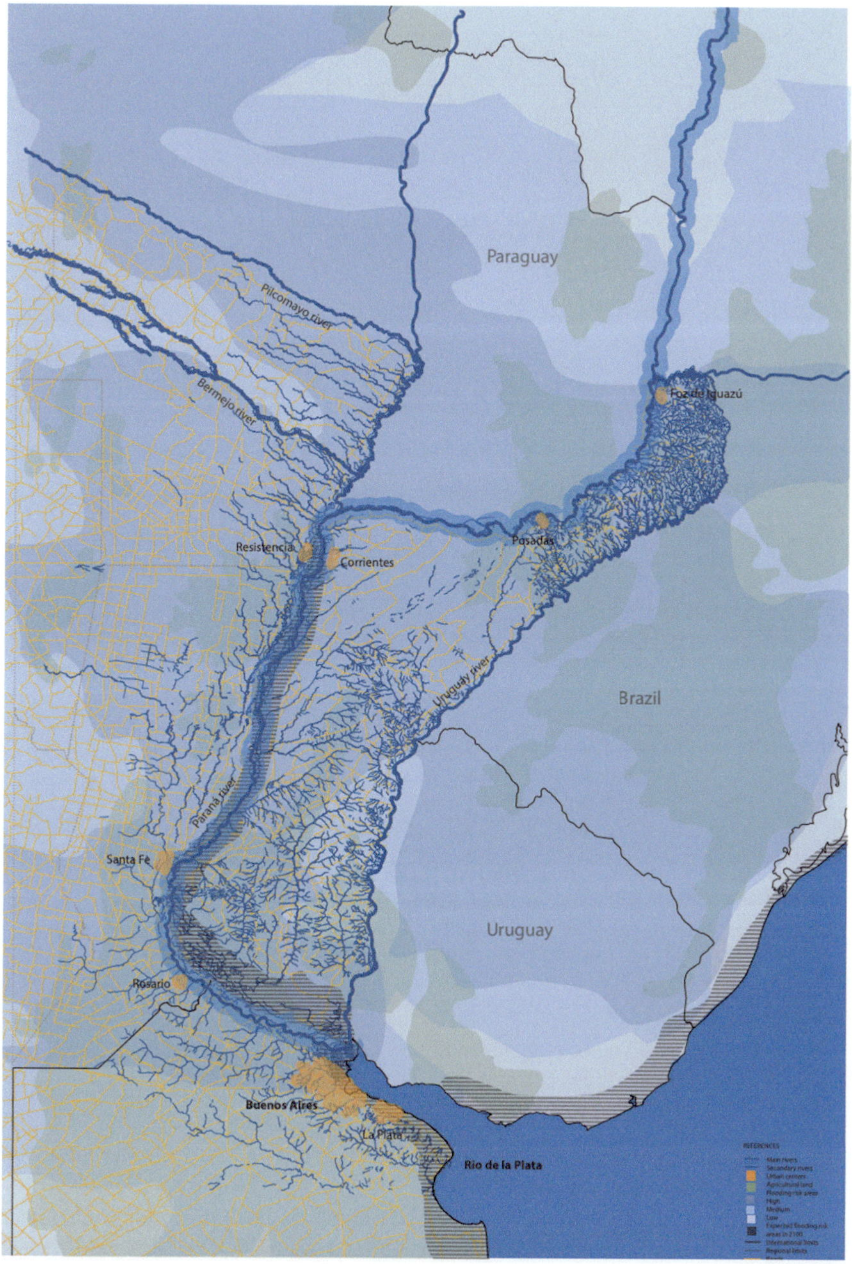

Fig. 51 The river basin of Rio de la Plata. *Source* E. Cicero

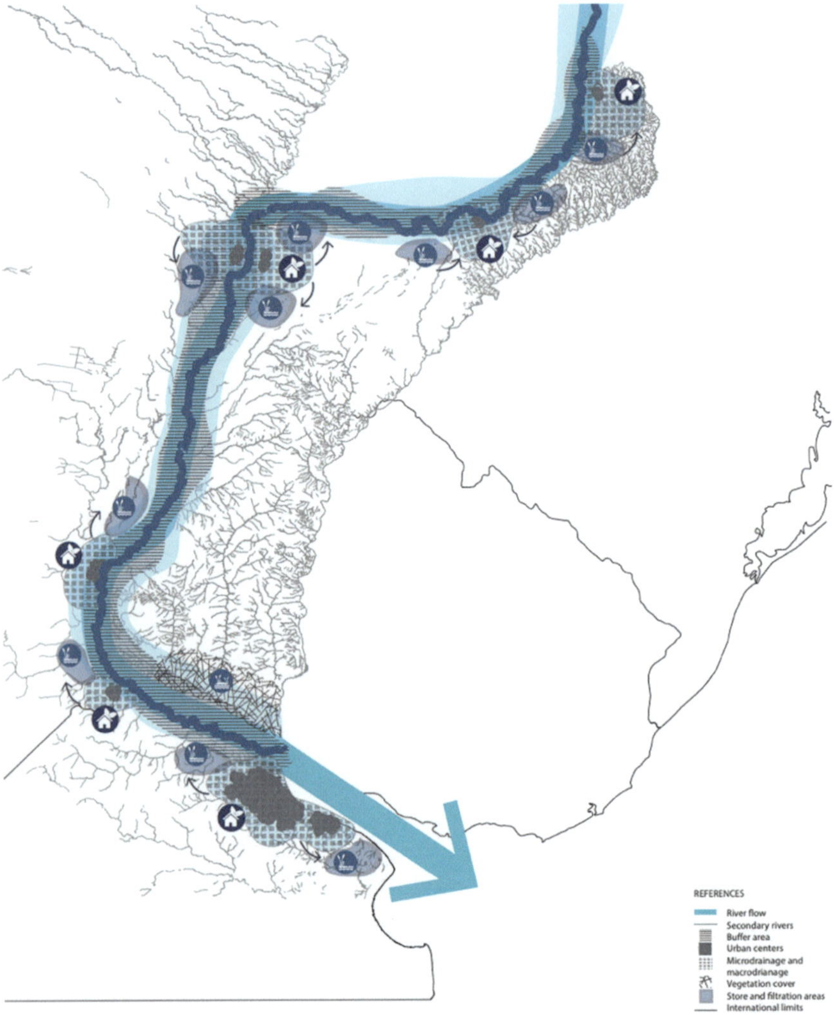

Fig. 52 International level strategy. *Source* E. Cicero

to improve the local economy, only the some of the interventions from the catalogue proposed should be taken in account.

Through the analysis of river trace, it can be found that the upper, middle and lower of the Paraná River goes by three different water landscapes, that can be very different from each other as the river goes by different topographies and climates. This would present unique challenges to propose solutions to minimize the flooding risk along the river path, mainly as it approaches the Rio de la Plata. It could also become an opportunity to create a capacity-building program and an integrated digital platform for the management of flooding risk and the economy that can be created through the proposed solutions. E.g. the proposal could benefit from the creation of an integrated digital platform where materials, like gravel and stones that are caved from the river shaping are sold, to raise the necessary funds to implement a continuous management programme, whose costs could be partially covered also by the touristic economy that could be implemented thanks to the creation of new wetlands and humid areas, propending to new leisure opportunities.

The Metropolitan region of Buenos Aires (Area Metropolitana de Buenos Aires, AMBA) was analysed as intermediate scale. The process of urbanization in fragile areas due to a continuous demographic expansion, is having a severe impact on the affected population, especially in the case of flooding caused by overflowing rivers or streams or by rainfall. A complex infrastructure network encompassing extensive roads, diverse land uses, and a significant river system characterizes AMBA. Its intricate road network, featuring major highways like the 'Autopista del Oeste' and Avenida General Paz, facilitates seamless interconnectivity between municipalities and provinces, augmented by advanced traffic management systems. In terms of land use, the AMBA exhibits a diverse mix of urban functions, from the high-rise office buildings of Buenos Aires central business district to sprawling residential neighbourhoods and industrial parks in suburban areas. However, the presence of informal settlements underscores socio-economic disparities and poses challenges for urban planning efforts. Along the eastern boundary of the AMBA lies the Rio de la Plata estuary, a vital maritime corridor influencing hydrological patterns and navigational dynamics. Although primary rivers like the Paraná and Uruguay Rivers do not directly traverse AMBA, their confluence at the estuarine delta influences the regional environment, supporting critical port facilities and providing opportunities for waterfront development and ecological restoration.

The aim is to propose a system that can help the city to face the challenges of the climate change (see Fig. 53). The idea is to implement interventions that can help manage the flood risk because of the heavy rain and to reduce the Urban Heat Island effect in the city. Natural based solutions can help these problematic situations. In one hand, more green areas help the infiltration and absorption of the rainfall and, in the other, help the natural cooling processes by evapotranspiration, possibly creating a new economic opportunity. Continuing the proposed strategy system for flooding risk management in the regional scale, the strategies should be from micro infrastructures areas to macro. Also, considering the future transportation plans, implementing a green corridor along the General Paz highway could serve dual purposes: promoting sustainable transportation and enhancing urban resilience to flooding.

Plus, a green coastal border along the Rio de la Plata is suggested to act as a buffer zone during rainy seasons, preventing river overflow and safeguarding other parts of the city from flooding hazards. These interventions collectively aim to create a more resilient urban landscape that can adapt to the challenges of climate change and extreme weather events.

At the urban scale, the cultural landscape of Buenos Aires and its metropolitan region has undergone significant transformation, resulting in the loss of much of its original natural areas (see Fig. 54). To address this, the city needs infrastructure to meet the demand for public services. Improving access to these services can enhance

Fig. 53 Metropolitan level strategy. *Source* E. Cicero

the quality of life and overall health of the inhabitants. A prominent consequence of climate change in Buenos Aires is the increased frequency of heat waves, associated with elevated minimum, average, and maximum temperatures. Heat waves, defined as periods with minimum temperatures exceeding a certain threshold for more than three consecutive days, occur with alarming regularity.

The most affected areas of the city by the future events are the ones along the piped rivers impacting in big part of neighbourhoods of the city. The strategy for the urban scale aims to help the city of Buenos Aires become more antifragile, therefore, the proposed vision for the city should incorporate interventions that can be flexible to face the future climate challenges. The implementation of the system already explained in the project, taking in account the unique characteristics of Buenos Aires. The interventions should address both the flooding risks and the urban heat island effect by creating more permeable areas in existing and new green areas.

These spaces should improve the drainage and create a new productive and economic landscapes in the city that can adapt to the seasons. The interventions should engage with the urban context and the neighbourhood. In the Climate Action plan for Buenos Aires 2050, the city government proposes limitations of the use of vehicles and the promotion of a sustainable transportation system. Therefore, the avenues and highways of the city can be reconceived as green economic corridors with extensive permeable surfaces, focusing on the economy of clean and reusable water, that will be the scarcest resource of the third millennium.

The city-scale strategy aims to implement and test the system proposed at the metropolitan level to manage rainfall water effectively and minimize the risks of

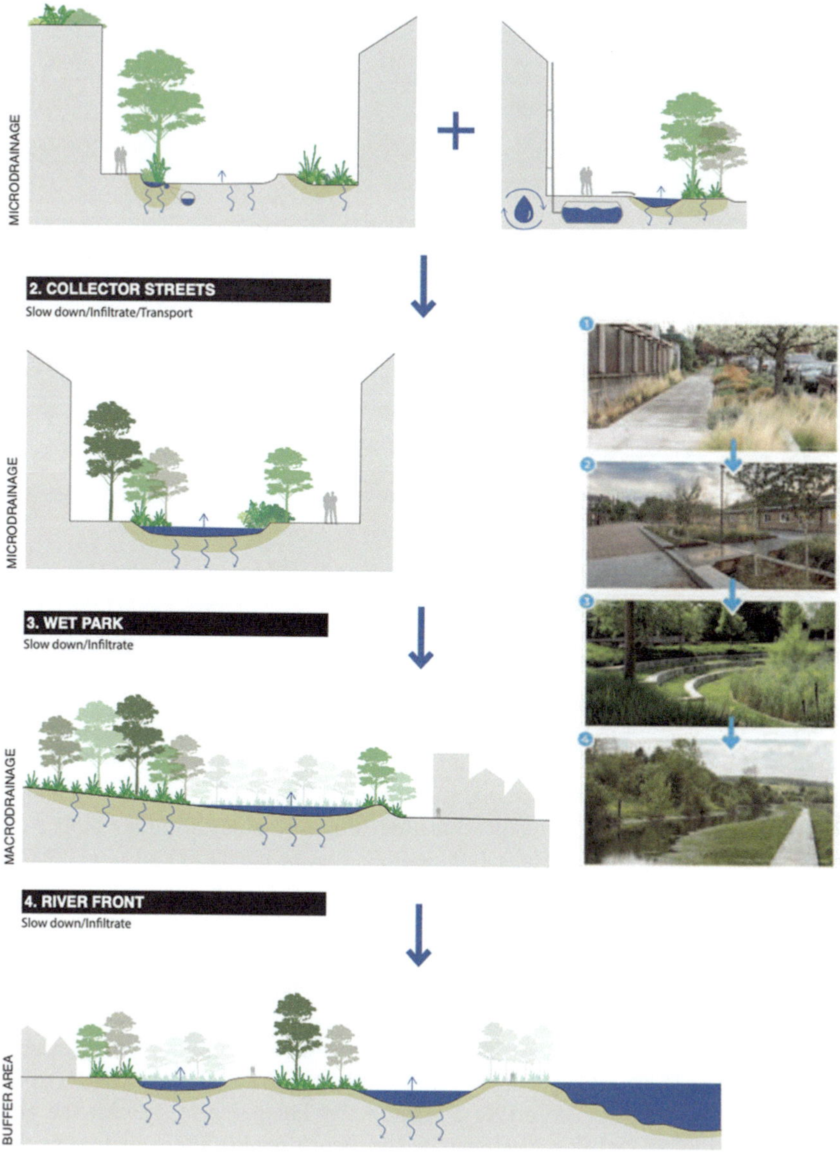

Fig. 54 The anti-fragile city actions. *Source* E. Cicero

flooding (see Fig. 55). By emphasizing natural-based strategies, the focus would be on reducing stress on the city sewage system due to extreme rainfall patterns exacerbated by climate change. This approach not only would address flooding risks but also tackles the urban heat island effect by increasing permeable areas in existing and new green spaces. Flooding risk can be also managed through a digital platform.

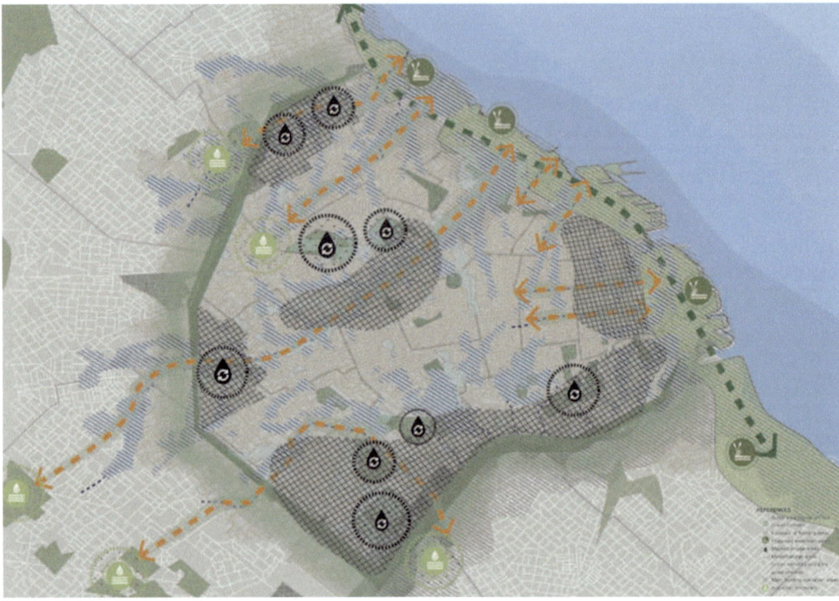

Fig. 55 City level strategy. *Source* E. Cicero

managing all the solutions in an integrated way, including alarm and emergency situations. Firstly, residential areas would be encouraged to adopt small-scale rain gardens and rainwater reuse systems, particularly in flood-prone areas. Secondly, green corridors with microdrainage areas are proposed along avenues built over stream traces to mitigate flood risks. Larger green spaces like parks and sport centres would serve as macrodrainage areas, storing and gradually releasing heavy rainfall.

Additionally, water reservoir systems in suburban areas would complement macrodrainage efforts during extreme rainfall events. The creation of a green corridor along the city perimeter highway aligns with the Climate Action plan for Buenos Aires 2050, promoting sustainable transportation and enhancing permeable surfaces to mitigate the urban heat island effect. Plus, a green corridor along the riverbank of the Rio de la Plata estuary is suggested. This area would act as a buffer zone for potential river growth and create a linear water landscape to connect the city with its waterfront, fostering engagement and resilience against future environmental challenges.

At the neighbourhood scale (see Fig. 56), the proposal approach involves creating an antifragile spine along the Medrano stream, spanning from Sarmiento Park to the Rio de la Plata riverfront. This linear system comprises three main types of interventions on a city scale: infiltration, buffering, and storage of water. The design strategically utilizes different types of green areas, with unused green spaces along the stream serving as rain gardens for infiltration, Saavedra Park and Sarmiento Sport Field as storage areas, and the waterfront as a buffering zone along the river.

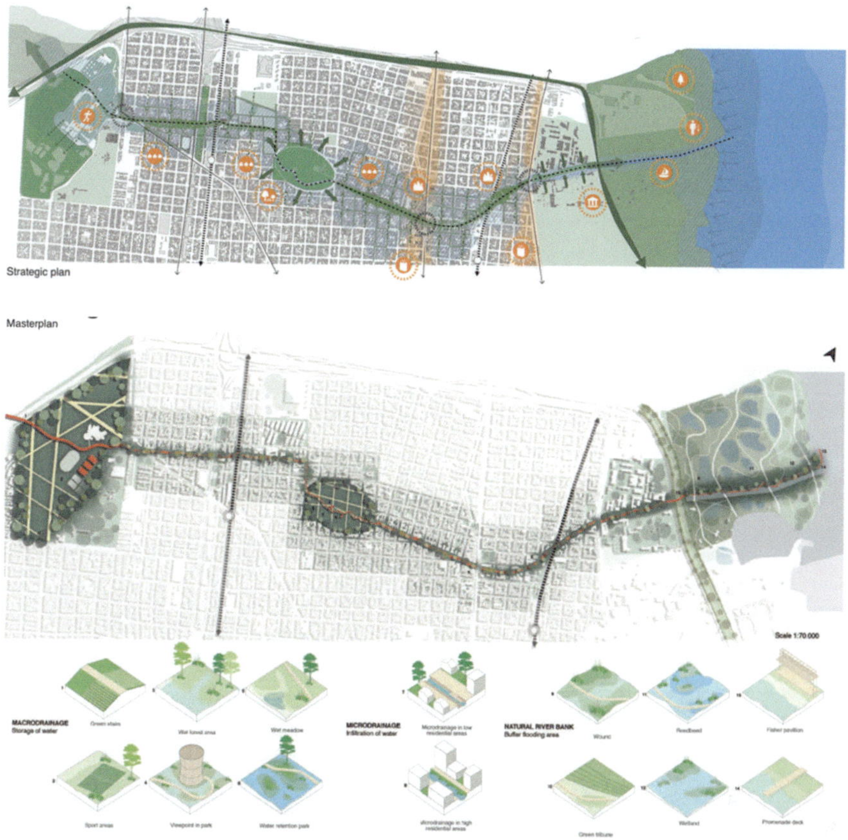

Fig. 56 Local level strategy. *Source* E. Cicero

Significantly, the strategy considers the landscape value and context of the site, ensuring that interventions are not only functional during rainfall but also enhance the urban environment and engage the community. Programmatic uses are integrated into the design to ensure a harmonious fit within the urban fabric. Through this approach, the strategy aims to propose antifragile water management systems that would help to improve both the built environment and would address the impact of the risk of urban flooding.

The riverfront in Buenos Aires serves as the final segment of the Medrano stream, representing the only section where the stream remains in the open air (see Fig. 57). Currently, there is a disconnection between the various activities taking place along the riverfront, such as sailing and fishing. To address this, there is a proposal to reimagine the landscape to better integrate these activities while mitigating the impact of flooding in the area, creating another possible opportunity for local economic activities (see Fig. 58).

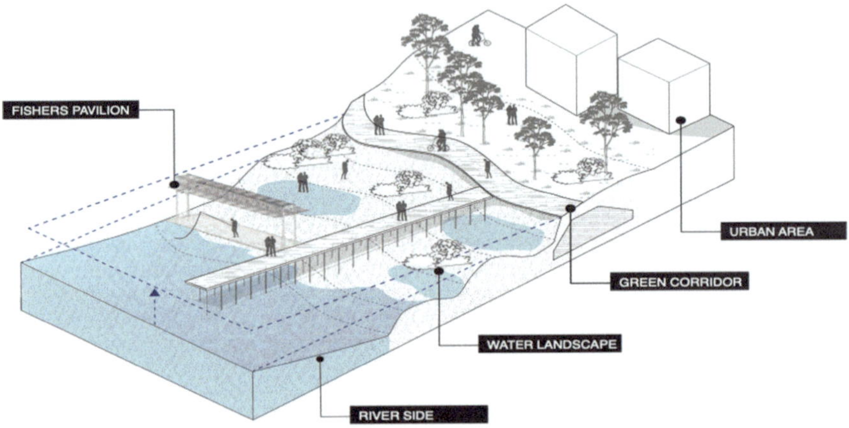

Fig. 57 The Medrano riverfront project. *Source* E. Cicero

Fig. 58 Engaging the community with the landscape. *Source* E. Cicero

This area would form part of the broader green corridor envisioned in the urban scale strategy. As outlined in the urban strategy, a green corridor would serve as a transitional zone between the urban environment and the natural landscape of the riverfront. An elevated pathway would ensure access to the area regardless of the season. The design of the landscape areas along the riverfront would be tailored to the frequency of flooding. This could include the implementation of diverse habitats such as river forests, grasslands, wetlands, reed beds, and floating communities. Local vegetation species capable of adapting to varying flood frequencies would be utilized to enhance resilience and ecological sustainability. By integrating these elements, the riverfront can evolve into a cohesive and resilient space that harmonizes with both its natural surroundings and the urban tissue of Buenos Aires.

The new landscape will also provide new engagement opportunities, creating a local touristic economy that could benefit from the creation of an integrated management platform, using flooding risk management data to plan events and reduce emergency effects. The integrated platform could also offer new job opportunities and access to community activities, which could be connected also to non-profit initiatives. These best practices could be a significant example to export potential E-Gateway experiences for the Indo-Pacific Metropolitan systems investments.

The concept of antifragile design should be increasingly relevant in urban planning, particularly in the face of rising challenges like storms, floods, fires, and heat waves brought about by climate change. Climate change is a multifaceted issue that transcends disciplines, including landscape architecture. Understanding and addressing this challenge would be crucial for the future of our profession.

While climate change is a complex issue, this project aims to propose landscape interventions to mitigate its impacts, focusing particularly on flooding in Buenos Aires. The research spans from a regional to an object scale, providing a comprehensive understanding of the land- scape and its challenges. Beyond addressing physical problems, the project also considers the intangible aspects of urban sites, aiming to reconnect people with the city hidden history through natural interventions. By uncovering the invisible traces of the city identity, we can create a more resilient and harmonious urban environment for the future. Perhaps cities shouldn't face climate challenges but benefit and grow from them.

With this approach in mind, a new idea of metropolitan community can be created. Due to access to digital infrastructure and public goods, all economic communities are progressively becoming digital economies, a trend that has been approved and supported by both international organizations and local governments. The multiscalar approach could determine the goods and services that the community can provide internationally in the subregions of the Metropolitan system and the economic corridors that extend outside the subregions based on an analysis of its economic and financial specificities, developing a plan based on the specialization of local land use systems for the digital economic community.

7 Conclusions

The concept that emerges from the case studies aims to develop an AI-powered cartographic tool model to map territorial assets, focusing on regional cohesion and sustainable development. By integrating historical, political, economic, environmental, and social data, the project provides a multidisciplinary framework for regional planning and policy-making decision-making [9]. Innovation is embedded in using Metropolitan Cartography [10], Development and E-gateway Strategies, and the Quintuple Helix Model, ensuring a place-based approach [11] that considers environmental and societal dynamics. Unlike traditional mapping, this approach could generate synthetic maps rather than thematic ones, offering a holistic view of territorial interactions. AI models will analyse territorial variables selected by different stakeholders, enhancing data-driven governance.

Contributions from political science, history, urban planning, artificial intelligence, and design ensure the multidisciplinary nature of the approach [12], fostering cross-sectoral insights [13]. Potential beneficiaries include:

- Public authorities and policymakers, gaining tools for strategic regional development.
- Urban and regional planners, accessing dynamic territorial analyses.
- Researchers and academic institutions, exploring new methodologies in spatial analysis.
- Local communities, benefiting from improved governance and sustainability planning.

This approach not only enhances territorial knowledge and decision-making, but also serves as a model for future applications in regional policies, addressing a complex, multidisciplinary challenge with high innovation potential. Such an approach could deliver innovative, multidisciplinary results that align with the objectives of EU territorial cohesion, sustainable development, and data-driven policymaking. It will also be benefitting for the whole Indo-pacific policy environment, promoting a more cohesive view of the territorial development.

In fact, this approach stems from the meeting of different skills—management engineering, civil engineering, sustainable architecture, and industrial production—with the aim of developing digital and interactive tools to enhance metropolitan areas, starting from its identifying characteristics: water, agriculture, landscape, and local entrepreneurship. Through applied research work, consolidated in collaboration with institutions and international research centres, such an approach could develop an interactive mapping platform, smart itineraries, and service models to connect local producers, citizens, and visitors. This approach could link to the topics of territorial invariable elements for a more conscious development, as it proposes a new territorial narrative based on enabling technologies, participatory platforms, and zero-kilometer e-commerce practices, creating a dynamic network between local and international actors.

The multi-scale approach [14], focused on sustainability and the enhancement of cultural and environmental capital, represents a replicable model for other metropolitan contexts, because this approach goes far beyond mapping: it is a proactive system, a digital territorial infrastructure that can integrate the physical dimension with the virtual one, restoring the metropolis identity as a city of water and agriculture.

This becomes even more important in the light of the urgence of climate change [15], which is a complex issue affecting economic communities, and landscape architects are tasked with proposing interventions to improve city conditions. The

approach aims to strengthen global digital infrastructure, promote values-based investment, and advance a human-centric approach to the digital economy. By focusing on the metropolitan dimension, this approach aims to address physical problems and reconnect people with the city hidden history through natural interventions. By uncovering the invisible traces of city identity, a more resilient and harmonious urban environment can be created. This approach could create a new metropolitan community, focusing on the provision of goods and services internationally in subregions of metropolitan systems and economic corridors. The multi-scalar approach could develop a plan based on local land use systems for the digital economic community.

As shown through the proposed case studies, Metropolitan greening through digital and ecological interventions for the well-being of people in many countries can be duplicated in many countries of the Indo-Pacific, as these innovative technologies be shared and upscaled for the population of the EU and the entire Indo-Pacific region. European Digital Public Goods, therefore, represent shared, scalable and democratically compliant infrastructures that can drive digitization in partner countries in an inclusive and sustainable way. The role of SMEs, the support of public institutions and the integration of local communities are indispensable elements to make the EU not only an exporter of technologies, but a builder of collaborative and resilient platforms. More explicit political will and more agile operational tools are now needed to capitalize on this potential and position Europe as a global technology platform for the common good.

Considering the role of cities in the European digital identity [16], in the context of an increasing digitization of public services and social life, the European Union is faced with the strategic challenge of building a single, inclusive and interoperable digital identity [17]. In this perspective, metropolitan cities emerge as central hubs to implement a federated vision of the European digital infrastructure, capable of enhancing proximity to citizens, the heterogeneity of territories and the plurality of actors involved.

The European digital identity cannot be merely the result of legislation from above [18]: to be effective, it must be rooted in places, recognize diversity and, at the same time, ensure a shared infrastructure. Metropolitan cities, as multi-scalar spaces where the local meets the global, are the ideal context for testing and implementing distributed governance of digital identity, where interoperability, civic participation and security coexist.

The proposal is to imagine a federation of European metropolitan cities as a living infrastructure for digital identity. Each city would act as a local node connected to the continental network [19], maintaining its own cultural and operational specificity but sharing common standards, protocols and values. This federated structure would make possible not only the efficient management of digital identity, but also the creation of public digital platforms that are truly scalable and adaptable to the needs of urban communities.

In fact, the metropolitan dimension [20] makes it possible to overcome the abstraction of centralized models, offering an intermediate scale between the national and local levels that is large enough to support complex technological infrastructures, but close enough to guarantee social inclusion, listening to needs and digital accessibility. Cities already play a leading role in the management of digital services for welfare, mobility, identity, taxation and civic interaction [21].

A federated metropolitan network would also promote a culture of interoperability and facilitate the secure circulation of digital identities between territories, avoiding duplication, enhancing transparency and reducing administrative costs. This approach, consistent with the European model centered on the rights and protection of the individual, strengthens the Union digital sovereignty and consolidates its role as a global player capable of proposing a democratic alternative to the large, centralized platforms.

Finally, multi-level and participatory metropolitan governance, based on the quintuple helix of innovation (which includes environment, citizens, research, industry and institutions), finds in metropolitan cities the space to be practiced. The city becomes laboratory and platform, node of a polycentric digital ecosystem, capable of integrating public and private resources, technological infrastructures and social capital, in the construction of an equitable, open and resilient European digital identity.

In this scenario, digital diplomacy, a distinctive feature of Europe, emerges as a fundamental pillar for science diplomacy. The Politecnico di Milano, for example, is actively engaged in scientific diplomacy, both in the sense of 'science for diplomacy'—science as a strategic axis in international cooperation—and of 'science in diplomacy'—the university as an engine of informed policies for sustainable development with global impact. In the period 2023–2025, the Athenaeum aims to consolidate paradigms for international cooperation in research and institutional capacity building, and to train young people in the ability to inform policies using the results of scientific research. Furthermore, it will actively participate in the main networks and think tanks to help bridge the gap between science and policymaking.

For all this, the federation of European metropolitan cities, through shared digital infrastructures and multi-level governance, can strengthen Europe digital identity, promoting an inclusive, sustainable and resilient development model in line with the democratic values and fundamental rights of the European Union.

8 Summary

The chapter proposes a Green and Digital Transition, a shift towards relational urban living that integrates communal existence, regenerative knowledge, and digital tools. The Twin Transition policy emphasizes digital technology in metropolitan construction and the growing role of intangible components like knowledge, local culture, and labor. The research aims to develop guidelines for sustainable habitats in vulnerable territories in metropolitan regions, focusing on data-driven innovation and strategic partnerships. An "E-Gateway for digitisation in the metropolitan economy" strategy is proposed to enhance economic potential in Italian urban systems, including digital infrastructure development, skills enhancement, supportive policies, and collaborative partnerships. E-Global Gateways are digital platforms that facilitate global economic integration, enhancing local digital economy participation and strengthening metropolitan systems' competitiveness in the global marketplace.

The proposed vision for Europe future focuses on integrating multi-level governance, civic innovation, sustainability, and digital sovereignty at the metropolitan scale. The E-Gateway will serve as a cognitive infrastructure, promoting economic, cultural, institutional, and anthropological exchanges. The Twin Transition will significantly impact the future, necessitating policymakers to address existing inequalities and socioeconomic effects. Strategies embracing de-risking and anti-fragility principles are being implemented to improve community well-being, conserve local identity, and preserve natural heritage. The study will investigate the intersection of the Twin Transition with disparities in the Far East, Mediterranean, and Latino-American metropolitan areas. Anti-fragility, a concept by Nassim Nicholas Taleb, refers to systems that thrive from stress and uncertainty, promoting growth and improvement. By applying anti-fragility to metropolitan areas, cities can transform potential challenges into opportunities for growth, ensuring their vibrant, resilient, and sustainable future.

The concept of developing an AI-powered cartographic tool model to map territorial assets, focusing on regional cohesion and sustainable development, is emerging from case studies. This multidisciplinary approach integrates historical, political, economic, environmental, and social data, providing a place-based approach for regional planning and policy-making decision-making. The approach uses Metropolitan Cartography, Development and E-gateway Strategies, and the Quintuple Helix Model to generate synthetic maps, offering a holistic view of territorial interactions.

The multidisciplinary nature of the approach fosters cross-sectoral insights, benefiting public authorities, urban and regional planners, researchers, academic institutions, and local communities. Potential beneficiaries include public authorities and policymakers, urban and regional planners, researchers, and local communities, who benefit from improved governance and sustainability planning.

This approach not only enhances territorial knowledge and decision-making but also serves as a model for future applications in regional policies, addressing a complex, multidisciplinary challenge with high innovation potential. It could deliver innovative, multidisciplinary results that align with the objectives of EU territorial cohesion, sustainable development, and data-driven policymaking.

The multi-scale approach, focused on sustainability and the enhancement of cultural and environmental capital, represents a replicable model for other metropolitan contexts, as it goes far beyond mapping and is a proactive system, integrating the physical dimension with the virtual one. This becomes even more important in the light of climate change, which affects economic communities and landscape architects are tasked with proposing interventions to improve city conditions.

The European Digital Public Goods represent shared, scalable, and democratically compliant infrastructures that can drive digitization in partner countries in an inclusive and sustainable way. The role of SMEs, support of public institutions, and integration of local communities are indispensable elements to make the EU not only an exporter of technologies but a builder of collaborative and resilient platforms.

The European Union faces the strategic challenge of building a single, inclusive, and interoperable digital identity, with metropolitan cities emerging as central hubs to implement a federated vision of the European digital infrastructure. A federated metropolitan network would promote a culture of interoperability and facilitate the secure circulation of digital identities between territories, avoiding duplication, enhancing transparency, and reducing administrative costs.

Multi-level and participatory metropolitan governance, based on the quintuple helix of innovation (environment, citizens, research, industry, and institutions), finds the space to be practiced in metropolitan cities, becoming a laboratory and platform for a polycentric digital ecosystem capable of integrating public and private resources, technological infrastructures, and social capital in the construction of an equitable, open, and resilient European digital identity. In this scenario, digital diplomacy, a distinctive feature of Europe, emerges as a fundamental pillar for science diplomacy.

Acknowledgements Antonella Contin is the supervisor of all the chapter contents, plus she authored Sects. 1, 2, 3, 4, and 7. Fioralba Vela is the co-author of Sect. 3. Arxhenda Lipovica contributed to Sect. 4, Dicheng Yang to Sect. 1. Alessandra Pandolfi is the co-author of Sects. 1, 2, 3, 4 and 7. Simal Keskin (Melzo), Javad Kianisadr (Piacenza), Tianduo Yan (Xiangtan), Glen Carlo Canlas (Manila), Aafreen Saleem and Prarthana Ulaganathan (Chennai) are co-authors of Sect. 5, Eugenia Cicero (Buenos Aires), Morteza Khajehahsani and Nika Haghpanah (Toronto) are co-authors of Sect. 6.

References

1. Taleb NN (2012) Antifragile: how to live in a world we don't understand
2. Taleb NN (2014) Antifragile: things that gain from disorder. Global Penguin Random House
3. Okano-Heijmans M (2023) The EU connectivity strategy 2.0: global gateway in the Indo-Pacific. In: Hosoya Y, Kundnani H (eds) The transformation of the liberal international order. SpringerBriefs in International Relations. Springer, Singapore. https://doi.org/10.1007/978-981-99-4729-4_4
4. Carayannis EG, Barth TD, Campbell DFJ (2012) The Quintuple Helix innovation model: global warming as a challenge and driver for innovation, in the Journal of Innovation and Entrepreneurship. Springer
5. Gates N (2025) Europe's digital public goods moment: new use cases for public sector adoption of DPGs in Europe. Open-Source Observatory (OSOR). Published 27 May 2025
6. Agostini Stella, A. (2000). Classificazione delle cascine del parco agricolo sud Milano. Franco Angeli
7. The Blossom Avenue (2017) Variante al piano di governo del territorio DP_1.2 Patrimonio edilizio rapporto di copertura. Scale 1:8,500. Comune di Liscate
8. Leoni DGC (Dott. Geol. Carlo Leoni) (2010) Determinazione del Reticolo Idrico Minore. Amministrazione Comunale di Liscate
9. Contin A (2021) The metropolitan structure for a set of metropolitan landscapes. In: Contin A (eds) Metropolitan landscapes. Towards a shared construction of the resilient city of the future. Springer, Berlin

10. Contin A, Galiulo V (2023) Methodological guidelines for metropolitan cartography projects mapping the accumulation of contested territories in southern Latitude metropolitan cities. ARDETH 9:117–133
11. Contin A, Galiulo V, Pandolfi A, Koike R, Yang D, de Chavez K (2023) Metropolitan architecture and sustainable habitats in the Indo-Pacific region to reinforce the megacity system through urban-rural patterns. In: Vinod Kumar TM (eds) Indo-Pacific smart megacity system. Springer, Singapore. https://doi.org/10.1007/978-981-99-6218-1
12. Contin A, Giordano P (2019) Training for knowledge to action: tools for the metropolitan architecture discipline. In: Lester E (eds) Education, design and practice—understanding skills in a complex world. Stevens Institute of Technology, AMPS, PARADE, Architecture_MPS
13. Contin A, Giordano P, Nacke M (2021) Training for education, learning and leadership towards a new metropolitan discipline. Inaugural book, Buenos Aires, CIPPEC
14. Heritopolis (2023) UN-habitat project UNI MetroHUB consortium. http://heritopolis.org/. Accessed 25 Oct 2023
15. UNESCO (2022) Rethink policies for creativity. Pose culture as a global public good. United Nations Educational, Scientific and Cultural Organization (UNESCO), Paris. ISBN 978-92-3-300175-6. https://unesdoc.unesco.org/ark:/48223/pf0000380479_eng. Accessed October 30, 2023
16. Contin A, Ortiz PB (2015) Qualitative metropolis. A new culturally encoded concept with qualitative regeneration of formal and informal settlements. The rule and the model in TECHNE. J. Technol. Architect. Environ. 10:120–129. https://doi.org/10.13128/Techne-17508
17. Contin A, Ortiz PB, Galiulo V, Siahkali RS, Pandolfi A, Campi P, Lutukurthi S, Sathiwada R, Kumar K, Girgaonkar P (2021) Territorial intelligence project: governance for megalopolis urban-rural linkage pattern. Springer, Singapore. https://doi.org/10.1007/978-981-16-2023-2_3
18. Contin A, Pandolfi A (2024) Metropolitan cartography: a novel approach for assessing how new morpho-types solutions impact the circular city agenda. The Ouagadougou case study. In: Okyere SA, Abunyewah M, Erdiaw-Kwasie MO, Boateng FG (eds) Urban slums and circular economy synergies in the global south. Springer Nature, Berlin. https://doi.org/10.1007/978-981-99-9025-2
19. Lanfranchi G, Contin A (2017) Cities for all: a guide for inclusive urban development in Latin America. In: Gómez-Álvarez D, Rajack R, López-Moreno E, Lanfranchi G (eds) In steering the metropolis: metropolitan governance for sustainable urban development. Inter-American Development Bank, Washington, DC
20. Geddes P (1909) City development: a study of parks, gardens, and culture-institutes. Edinburgh, Geddes and Colleagues
21. Gouverneur D (2014) Planning and design for future informal settlements: shaping the self-constructed city. Routledge, London

Websites

22. https://www.art-er.it/
23. https://www.eeas.europa.eu/eeas/eu-indo-pacific-flagship-projects_en
24. https://climate.ec.europa.eu/eu-action/european-climate-law_en
25. https://commission.europa.eu/strategy-and-policy/priorities-2019-2024/stronger-europe-world/global-gateway_en
26. https://commission.europa.eu/strategy-and-policy/priorities-2019-2024/european-green-deal_en (EU Green Deal)
27. https://commission.europa.eu/strategy-and-policy/priorities-2019-2024/economy-works-people/european-fund-strategic-investments_en
28. https://cordis.europa.eu/article/id/87501-the-smart-approach-to-territorial-intelligence/it (FP6)

29. https://ec.europa.eu/commission/presscorner/detail/en/IP_22_4004
30. https://heritopolis.org/
31. https://satoyama-initiative.org/
32. https://sdgs.un.org/goals
33. Heritage—Galbani. https://www.galbani.com/heritage.php
34. Ferrovia Milano-Venezia. https://it.wikipedia.org/wiki/Ferrovia_Milano-Venezia
35. Cascina Grande—complesso, Liscate MI100–03883, Lombardia Beni Culturali. https://www.lombardiabeniculturali.it/architetture/schede/MI100-03883/
36. Retrieved from https://www.comune.liscate.mi.it/amministrazionetrasparente/
37. Melzo Allevamento Bovini. https://www.coobiz.it/azienda/melzo-allevamento-bovini/%20co1034565#descrizione
38. Molgora. https://it.wikipedia.org/wiki/Molgora

Antonella Contin Arch. PhD is a Research Associate at DAStU/Department of Architecture and Urban Studies, School of Architecture Urban Planning Construction Engineering at the University of Politecnico di Milano. She is the Coordinator of UNHabitat MetroHub Milano and MSLab—Measure and Scale of the Contemporary city. She is a scientific and fully organisational responsible for international Workshops and Seminars. She is the Group Coordinator of some EU co-funded research projects and gives lectures at many international Universities. Her research deals with the Metropolitan Architecture Project, a new sustainable model of urban growth, smart, sensitive cities, new technologies and urban design mapping. She is scientific and full organizational responsible of international Workshop (Milan, Madrid, Seville, London, Istanbul, Teheran, NY, Cairo, Bari, Cadice, Guadalajara, Rio de Janeiro, Arua-Uganda) and International Seminars.

Fioralba Vela is Associate Professor (from 2016) is a lecturer at the Faculty of Economy, University of Vlora "Ismail Qemali" since 2000 in the field of business administration, public administration and tourism management of subjects: Quality Management, Organizational Behavior, Management of Public Institutions, Change and Innovation Management. Her experience is mainly focused on higher education and business consulting and analytics. She got the title PhD of Economics by the Faculty of Economy, University of Tirana, and in April 2004 she got her Master's degree in "Management, Economics and Law of Territorial Governance in the European Union" by "LIUC" University of Castellanza, and University of Bari, "Aldo Moro", Italy. Currently, she is appointed to the position of Head of Business Department, and from 2016 to 2020 as Vice Dean of the Faculty of Economy, and from 2007 to 2009, as Head of Tourism Department, at the University of Vlora. She is also coordinating SMART CITY HUB, at the University of Vlora "Ismail Qemali". She is an accreditation external expert of the Agency for Quality Assurance in Higher Education, from July 2023. She is a participant and visiting professor of several academic and scientific exchange programs at international and national universities such as: University of Memphis,

School of Public Law and Policy in USA (2015), Academy of Arts and Science, Ljubljana in Slovenia (2007–2008), University of Durres (from 2020), Poland, Italy, Spain, Romania, Bulgaria, Portugal, etc. Participant and author of the studies presented and published in many periodical scientific conferences and scientific journals home and abroad, such as: USA, Germany, Romania, Turkey, France, Italy, Greece, Bulgaria, Bosnia and Herzegovina, Montenegro, Macedonia, Kosovo etc. She is an institutional and national project coordinator of some European Union funded and business consulting projects, such as: Erasmus KA2-CBHE "SMART" (2022–2025) project and leading the work package "Operational Set-up of Smart Innovation Centres" in Albania, Bosnia–Hercegovina, Montenegro partner universities, and also INTERREG IPA—South Adriatic Project "ILOFORDIGITAL", and also a member of different Cost Action Projects, such as ENIS COST, P-WILL COST, NEXUS Net, etc.

Alessandra Maria Pandolfi Arch. PhD, Contract Professor at DAStU/Department of Architecture and Urban Studies, School of Architecture at Politecnico di Milano. After graduating in Urban Planning at Politecnico di Milano and Torino (Alta Scuola Politecnica), she studied the dynamics and relations of economic values to development rights transfers in Italy and other countries. She is author of more than 60 international publications in Project Assessment and Economic Evaluation. Her main research interests span from Sustainability, Feasible Architecture and Sustainable Development to Spatial and Landscape Analysis, Urban Planning and Land Use Designing and Urban Sustainability.

Arxhenda Lipovica is Research Fellow at Politecnico di Milano (DASTU, Department of Architecture and Urban Studies) and teaching assistant in several courses. She holds a Master's Degree in Architecture and has several publications. In her approach to architecture, each project is the culmination of extensive exploration, analysis, and inquiry, laying the foundation for spaces that resonate with purpose and meaning. Her research seamlessly weaves together the threads of history, culture, technology, and sustainability, shaping environments that transcend mere functionality to become timeless reflections of human aspiration and ingenuity. The relationship between research and design, where every insight uncovered fuels the creation of experiences that transcend the ordinary, in a journey of discovery and inspiration.

Dicheng Yang is a PhD candidate of Politecnico di Milano with eight years study experience of Architecture Design and Urban Planning. Dicheng focus on Chinese rural studies. In collaboration with Alibaba Group and the International Forum of Urbanism (IFoU), he has conducted research on the challenges arising from E-commerce and the urban–rural relations in east China (2020). After graduating with honours from Politecnico di Milano in 2022, Dicheng research centres on urban studies, incorporating perspective from the neuroscience discipline. His research specializes in people well-being and perception related with urban complexities.

Hasibe Şimal Keskin is a researcher in urban studies and spatial analysis. She is pursuing a PhD in Critical Cartography and Spatial Analysis at the School of Architecture, KTH Royal Institute of Technology. She holds a Master of Science in Architecture and Urban Design from Politecnico di Milano, where her thesis explored the integration of circular agriculture, community collaboration, and green–blue infrastructure in peri-urban contexts. As an intern at UN-Habitat, she contributed to the assessment of the Nampula Strategic Plan and the development of the My Neighborhood Framework—Coastal and Delta Cities Edition. She also contributed to research activities on Metropolitan discipline, urban transformation and sustainable development in Europe, providing mapping and spatial analysis support through her affiliation with Politecnico di Milano. Her current work explores the role of critical cartography in advancing spatial justice, resilience, and sustainable territorial development.

Javad Kianisadr is an architect with a Master's degree in Sustainable Architecture and Landscape Design at Politecnico di Milano. Alongside his studies, he worked in a civil engineering firm, gaining cross-disciplinary design experience. He served as senior architect for a residential net-zero energy project and contributed to urban and infrastructure proposals. His research shaped his view of landscape architecture as a strategy for climate resilience and river–city reconnection. His work focuses on combining urban ecology, social inclusion, and adaptive landscape design. He aims to create environments where people and nature coexist in balance.

Tianduo Yan is an architect and music player. He graduated in Architecture with a thesis about "Benefits of resilient design approaches in urban waterfront spaces affected by flooding events: a possible proposal for Xiangjiang River, Xiangtan, China". His main interests of research span from landscape ecology to ecosystem services, passing from resilient design and urban space. His interests are also focusing on the value of landscape, which is often reflected in how it serves people, not only by ensuring a better quality of life and sensory experience to them, but also creating more significant values for the society. Due to climate change, natural disasters and flooding events are becoming more frequent and intense, and only by constantly reinventing and innovating can communities reduce crises and bring stability, health, wealth, and development to people.

Glen Carlo Canlas is a registered and licensed Filipino architect currently affiliated with Aidea, an architectural firm based in the Philippines. He holds a Master's Degree in Sustainable Architecture and Landscape Design from the Politecnico di Milano in Italy, and a Bachelor's Degree in Architecture from the University of Santo Tomas in the Philippines. His graduate thesis, "*Redefining Resiliency for Coastal Communities through Aquasilviculture in Navotas, Philippines*," reflects his research interest in climate-responsive and community-based design strategies. While in Italy, he worked as an intern at LAND Italia, a leading landscape architecture firm known for sustainable urban development. This experience deepened his understanding of landscape-led design and reinforced his commitment to ecological and socially responsive solutions. His academic and professional work focuses on the integration of sustainability principles in architecture and urban design, particularly in vulnerable coastal environments. At Aidea, he contributes to innovative and sustainability-driven projects in the built environment. Canlas has participated in international academic settings, enriching his cross-cultural approach to design. He currently resides in Makati, Philippines. His interests include urban resiliency, green infrastructure, and the role of architecture in advancing sustainable development goals.

Morteza Khajehahsani is an architect with extensive experience in professional practice and research, specializing in sustainable and community-focused design. In research, he collaborated on projects designing sustainable social housing in Toronto (Canada) to empower marginalized communities by using the sustainable governmental strategy. With a deep passion for sustainable social housing and community-focused design, he is committed to research that bridges the gap between sustainable architecture, social housing, and community empowerment. His work is driven by the belief that architecture should not only provide shelter but also foster inclusivity, resilience, and social well-being.

Nika Haghpanah is an architect with a Master's degree in Sustainable Architecture and Landscape Design from Politecnico di Milano. Her research focuses on design solutions for contemporary urban challenges, with an emphasis on social inclusion and sustainability. She explored homelessness and waste management in Toronto, proposing modular housing and community-based interventions. She is particularly interested in how architecture can contribute to shaping equitable and resilient cities. Her work aims to blend environmental responsibility with a strong commitment to social impact.

Eugenia Cicero is an architect and landscape designer specializing in ecological, urban, and environmental design. She completed her MSc. in Sustainable Architecture and Landscape Design at Politecnico di Milano in 2024. Her master's thesis focused on landscape interventions to mitigate the impact of climate change in the city Buenos Aires. To adress the physical and intangible challenges, a multiscalar approach was used, varying from regional analysis to site-specific strategies. The research seeked to reconnect people with the city's hidden history through natural and adaptive design solutions. During her master's studies, she was selected for the Alliance4tech program, completing her studies in TU Berlin and UP Madrid all while participating on academic initiatives for sustainable urban development. Driven by an interest in the intersection of research, urban design and landscape, she is currently working as a landscape designer at the Dutch architecture and urban design firm MVRDV.

Prarthana Ulaganathan is an architect based in Milan, currently working at Studio Novembre. A recent graduate of the MSc in Sustainable Architecture and Landscape Design at Politecnico di Milano, her work explores the intersection of art, science, and ecological thinking in architecture. With professional experience across India, Sri Lanka, and Italy, she brings a multicultural and interdisciplinary perspective to her work. Her interests span sustainable design, adaptive reuse, and climate-responsive strategies, with a strong focus on landscape-led approaches to urban transformation. She worked on a range of urban restoration and public realm projects, guided by a belief in placemaking as a tool for community resilience and cultural continuity. Her approach reflects a thoughtful balance of technical knowledge, creative inquiry, and environmental responsibility always rooted in the social and ecological context of place. Prarthana sees architecture and landscape as interconnected frameworks that can shape inclusive, regenerative environments for the future.

Aafreen Saleem is an architect and landscape urbanist with a focus on sustainable urbanism, ecological infrastructure, and the transformation of urban–rural territories. She holds a Master's degree in Sustainable Architecture and Landscape Design from Politecnico di Milano, where she currently works as a teaching assistant in landscape and urban design studios. Her research explores antifragility, water resilience, and circular economy strategies within desakota regions and metropolitan corridors. She has supported design critiques and collaborative fieldwork in the "Water Resilience and Water Diplomacy" studio and co-tutored international workshops including a recent urban–rural design lab in the Madrid Valley. Aafreen has also delivered guest lectures in India, sharing insights on placemaking, public space regeneration, and water-based design. Her approach integrates stakeholder engagement, cartographic storytelling, and spatial analysis, with a strong interest in reconnecting cultural knowledge with adaptive design solutions. She is committed to research that bridges academic theory with community-led practices to shape more resilient and inclusive landscapes.

International Collaborative Research "Indo-Pacific E-Gateways: Cases from Digital Economic Communities" and Conclusions

T. M. Vinod Kumar

Abstract This chapter has two parts. The first discusses the goals and organisational details of the international collaborative research project "Indo-Pacific E-Gateways: Cases from Digital Economic Communities." In the second part, the team leaders of the area studies, including the city study, present their general conclusions regarding the area study, consulting with one another.

Keywords Study Organisation of Study · Indo-Pacific E-Gateways: Cases from Digital Economic Communities · Conclusions

1 Smart City Research and Indo-Pacific Smart Economic Communities

The International Collaborative Research Project, "Indo-Pacific E-Gateways: Cases from Digital Economic Communities," published, is the seventeenth book in a series that Professor T.M. Vinod Kumar conceived, coordinated, implemented, and edited, which articulates the various aspects of smart cities and the roles of Smart People in Smart Cities. This book, the third in the series on the Indo-Pacific, is unique in its focus on the region. It discusses how smart people created the E-Gateway system for the Indo-Pacific as a digital economic community and utilised it as an entrepreneurial ecosystem with an international reach. There is an extended e-government role and a further role in developing the ecosystem to facilitate global economic transactions for entrepreneurs. Therefore, this book continues the sixteen books, emphasising each aspect of smart cities. The fifteenth book was The Indo-Pacific Megacity System: Emerging Architecture and Mega City Case Studies. The sixteenth book, "Indo-Pacific Digital Core and Peripheral Economic Communities," is an international collaborative research project documented in a selected chapter of a book and published by Springer in 2025. The forthcoming book, the Fourth on Indo-Pacific

T. M. Vinod Kumar (✉)
School of Planning and Architecture, New Delhi, India
e-mail: tmvinod@gmail.com

T. M. Vinod Kumar (ed.), *Indo-Pacific E-Gateways*, Advances in 21st Century Human Settlements, https://doi.org/10.1007/978-981-95-4824-8_4

and the eighteenth edited volume of T.M. Vinod Kumar, is expected to be published by Springer in 2026 and is entitled "AI-DIY Enabled Indo-Pacific Digital Economic Community Development."

While innumerable academic writers hailing from the East and the West postulated with their logic the rationale of Indo-Pacific, they are irrelevant when compared to a program development region, the conceptualisation of a political leader, the then Prime Minister Shinzo Abe of Japan who delivered the famous address to the Indian parliament famously remembered as the Indo Pacific the confluence of two great ocean Indian Ocean and the Pacific Ocean that covers an assortment of rich, middle income and low-income countries and tiny island nations to giant islands and small populated sovereign nations to giant populated Nations like India and China. It was in the aftermath of the Tsunami in 2004 that Japan pooled its efforts with that of the US, Australia, and India along with numerous countries in the region, jointly saving the Indo-Pacific devastation to the ecology and human life, the compassionate joint action of many like-minded countries moved Prime Minister Shinzo to conceptualise as an international program region for regional collective and cooperative developmental interventions.

The first book in this series, "Geographic Information System for Smart Cities," was published in 2014 [1]. It aimed to create a comprehensive and spatial self-awareness of Cities functioning every second and every day in real-time, which is the foundation of any Smart City or Smart Data City. Geospatial technologies, sensors, and analytics can enhance awareness and be utilised in real-time for various applications by Smart Individuals. The book discusses utilising GIS for various urban issues commonly seen globally in case studies.

These Smart People thereby progress towards their self-directed goals, which nobody must tell them, such as demanding a Smart City that connects more people for intelligent community action and smart economic development that changes their lives to prosperity. They aspire to the highest quality of life in "smart city living," which they can afford by expanding the economic development opportunities they identify without any help, thereby satisfying the urge for higher income and fulfilling employment needs to sustain Smart People. No intelligent person in a city is an island or isolated elite. Still, they share a common destiny and a common urban space—the urban realm—and shared social and physical infrastructure. As the regulator, the Government ensures that all citizens are provided with services, irrespective of their income level, social status, or whether they are above or below the poverty line. A hierarchy of Government exists in any city. Still, its governance needs to be for a smart city that is fully aware of itself every millisecond, not one that only becomes aware of an issue when a case is filed in court, which can take many decades to reach a final judgment. The existing governance systems need to be updated; a product of the sixteenth century or earlier, designed by colonial rulers for the intelligent inhabitants who would like to act quickly, rather than on the model of the East India Company's administration requirements in India or elsewhere, especially for the local natives. Local Governments shall refrain from using such approaches for Smart city economic development and intelligent global megacity creation and functioning. However, those who aspire to live in Smart Cities in the twenty-first century are no longer part of an exploitative colonial empire under the iron hand of a colonial administrator with a robust framework who cannot look beyond the interests of the colonial empire but effectively conceals them in administration. Therefore, twenty-first-century Smart Cities require a Smart City e-governance system, which is the subject of the second book, entitled "E-Governance for Smart Cities" [2]. The book is divided into three parts: State Art Surveys, Domain Studies, and Tools based on E-Governance in Smart Cities.

The third book in this series is "Smart Economy in Smart Cities" [3]. This book explores possibilities for rapid change in the income levels and employment opportunities of those individuals below or above the poverty level in a smart city, and for consistently achieving a higher NDP growth rate for the next many decades. Then, the current urban local economic development trend shall progress towards Smart City Economic Development. An ambitious 10% NDP growth rate envisaged for the next three decades in India, which was previously distant, can only be realised through Smart City Economic Development in many other countries. The projection that the Indian GDP in PPP or otherwise will be the third largest in the World after China can be recognised through an intelligent metropolitan or megacity economy. Smart Cities and the related conceptualisation boast of the Smart Economy, but more needs to be systematically researched and documented about it so far. This calls for a study of many cities worldwide to demonstrate what constitutes a Smart Economy. This book studies two groups of cities. Some have been designated as smart cities by learned societies, while others aspire to achieve this status. These call for different approaches to research design and studies. It was seen from case studies that both these cases in other countries emphasise different methods, proving that there are no cookbook solutions. The cities studied in this book spanned several major continents and regions, including North America, Europe, Africa, the Indian subcontinent, and East Asia. They are Ottawa in Canada; Stuttgart in Germany; Bologna in Italy; Dakar in Senegal; Lagos in Nigeria; Nairobi in Kenya; Cape Town in South Africa; New Delhi, Varanasi, Vijayawada, and Kozhikode in India; Hong Kong in China, Cape Town, Dakar, Nairobi, and Lagos in Africa.

The fourth book in this series is "E-Democracy for Smart Cities" [4]. The world over, participatory democracy is worshipped and preached, but what is practised is representative democracy at the city level and beyond, which is not participatory democracy. It is believed that in metacities, megacities, and metropolitan cities, representative democracy with elected representatives will work, and so will large democracies with a population of billions, such as India. However, democracy was practised in small cities like Athens in Greece and the Licchavi Republic in India in ancient times, much before Athens, which is well-documented in Buddhist scriptures and adopted by the Buddha in his Sangha. A face-to-face democracy existed in ancient times in many parts of the world. In these cities, everyone sat together and jointly decided on all aspects of the town or country, whether during war or peacetime. Today, it is prevalent in Indonesian village democracy and many places and countries in the Indo-Pacific, although it may be informal or have inadequate administrative, judicial, or financial powers. Citizens took part in decision-making, acted together as one government, and selected or elected a few among them to work, even as temporary administrators for a task or as regulators. There were no permanent administrators then, as colonial Governance and post-colonial republics propagated. With the advent of ICTs in Smart Cities of the twenty-first century, it is possible to go back to face-to-face democracy in meta city with a 20 million-plus population, a megacity with a ten million-plus population, and metropolises with one million-plus population that, by any measure, is much superior to representative democracy which is not participatory at all. We want intermediaries, the citizens' representatives, to refrain from making legislation in a diverse country like India. However, all citizens using electronic media demonstrate the creation of regulations based on their ability to think and act in accordance with the Constitution. The fourth book focuses on E-Democracy in Smart Cities, presenting case studies in action. It is divided into three parts: State Art Surveys, Domain Studies, and Tools that describe E-Democracy in Smart Cities.

The fifth book in the series was "Smart Metropolitan Regional Development: Economic and Spatial Design Strategies" [5]. However significant these cities may be, they need to be converted into smart metropolises by employing specific design strategies for economic and spatial development, rather than relying solely on off-the-shelf smart technologies. The city studies for the "Smart Metropolitan Regional Development" result in many insights on many smart spatial and economic strategies using the Internet of Things, Internet of Democracy, and Internet of Governance oriented to the specific issue of a town and its potential; taking into consideration that the Smart metropolitan City is an integrated six systems in which Smart Economy and smart people are dominant components. The smart Economy can relate to Smart Mobility or Smart Environment. Based on the elaboration of the Smart metropolitan city System, if one must develop any metro region, then country city-specific economic and spatial design strategies for a Smart metropolitan city must be designed based on the local ecological and cultural system of the town and not a type of universal design based not on a mechanical linear input–output model. Location-specific, exceptionally robust, ecologically and culturally acceptable economic and spatial strategies can be locally evolved by people, not their representatives, who may have other agendas, such as making money through corrupt means while in power, governing, and managing. This is the only way local culture will find expression in the Smart metropolitan City, utilising specific economic and spatial strategies, as well as the regional and creative talents of intelligent people, in various institutions within Smart Cities. Sixteen cities were studied in this project, namely Pittsburgh in the USA, Stuttgart in Germany, Naples in Italy, Dakar in Senegal, Conakry in Guinea, Abuja in Nigeria, Johannesburg in South Africa, and Nairobi in Kenya, Ahmedabad-Gandhi Nagar, Bengaluru, Chandigarh, Jaipur, Kozhikode, New Delhi, Surat in India and Hong Kong and greater Pearl River Delta Region from China.

The sixth book, published in mid-2019 by Springer Nature, is entitled "Smart Environment for Smart Cities" [6], a product of international collaborative research. This book aims to develop the Design Protocol and Practice of Smart Environmental Resources Management for Smart Cities. Environmental Resources are common properties where an active role of the Government and the People is needed. Hence, its management is a joint and synchronous effort of E-Democracy, E-Governance, and

IoT systems in a 24-h, 7-day framework on any resource in any smart city. Smart environmental resources management is a practice that uses information and communication technologies, the Internet of Things, Internet of Governance (E-Governance), and Internet of People (E-Democracy), along with conventional resource management tools to realise coordinated, effective, and efficient management, development, and conservation that equitably improves ecological and economic welfare without compromising the sustainability of development ecosystems and stakeholders. This book presents case studies of many cities, including Hong Kong in China, Ahmedabad, Gandhinagar, Chandigarh, Kozhikode, New Delhi, Patna, Surat, Yokohama in Japan, Nairobi in Kenya, and Dubai in the UAE, which are centred on one or all environmental resources within each city.

The Seventh and Eighth Books, published in two volumes by Springer Nature in 2020, focus on Smart Living for Smart Cities [7, 8]. The first volume focuses on city case studies, while the second explores community studies and their various approaches. Based on international collaborative research, this book aims to develop the state-of-the-art design of "Smart Living" for metropolises, megacities, and metacities at the community and neighbourhood levels. Smart living encompasses six key components of Smart Cities: Smart People, Smart Economy, Smart Environment, Smart Mobility, and Smart Governance. Smart living in any smart city can only be designed and executed through the active roles of Smart People and a Smart City Government, and is a joint and synchronous effort of E-Democracy, E-Governance, and ICT-IoT systems within a 24-h, 7-day framework for all activities. In addition to the uses of information and communication technologies, the Internet of Things, Internet of Governance (E-Governance), and Internet of People (E-Democracy), the design of intelligent living utilises domain-specific tools for many aspects of living by age cohorts to realise the coordinated, effective and efficient management, development, and conservation that improves ecological, social, biophysical, psychological and economic wellbeing equitably without compromising the sustainability of development ecosystems and stakeholders. This book presents numerous case studies covering cities such as Hong Kong, Dehradun, the Kozhikode Metropolitan area, and Shimla, focusing on one or two domain-specific innovative living components.

The ninth and tenth books are on Smart Global Megacities [9, 10]. The editor and coordinator of the book series, T.M. Vinod Kumar, and many authors who participated in the earlier eight books felt a gap in knowledge about the Smart Global Megacity. Like all earlier books, funding for such a collaborative research project was another issue, so international travel and consultations were ruled out. Universities and research centres collaborated on these smart megacity city research projects. We also found that, in addition to universities, some not-for-profit national and international networks and institutions, as well as city governments and regional governments in certain countries, came forward to participate in this collaborative research program. The editor and coordinator of the project again felt that this international project should not seek any external funding other than the internal resource mobilisation from within the participating universities. This book, the first volume, highlights

Tokyo, Mumbai, New York, Hong Kong-Shenzhen, Kolkata, Chennai, and Kochi-Kannur out of approximately 33 megacities in the World with ten million or more inhabitants.

The eleventh and twelfth books were entitled "Smarter Master Planning: Case Studies of Digital Innovations" and "Domain Innovations," respectively [11, 12]. Master Planning is practised all over the world, even though this well-tested instrument of Master Planning carries numerous defects that are apparent to the consumer citizen, planners, and other regulators of urban affairs. The two-volume books aim to address some of the shortcomings of Master Planning by exploring innovative approaches. These books research the Smarter Master Planning of Cities, discussing innovations using city case studies. Cities are urban agglomerations, accurately representing urbanised geographic space as per the census of India and UN-HABITAT, rather than municipal boundaries, arbitrary planning areas, or towns' administrative boundaries. This is where the not-so-satisfactory master planning of cities begins, which needs to be innovated since there is no replacement. The master planning approach differs among countries according to the governing system followed, whether it is a Communist, Capitalist, Secular, Theocratic, or democratic government, as well as based on city-specific urban issues and the city's long-range vision. The Constitution establishes the governing system and approaches to the City Master Plan through several regulatory legislations that encompass all aspects of city life. This varies from country to country and from state to state. The Constitution is a living document that enables the government system to function effectively.

Based on international collaborative research, this book presents a state-of-the-art design for "Smart Master Planning" for all metropolises, megacities, and sub-city zones, as well as communities and neighbourhoods. Smart Master Planning acknowledges that all cities are smart cities to some extent, as far as the six components of smart cities—namely, smart people, smart Economy, smart environment, smart mobility, and smart governance—are concerned. Smart Master Planning in any town can only be designed and executed through the active roles of competent individuals and a smart city government, and is a joint and synchronous effort of e-democracy, e-governance, and ICT-IoT systems within a 24-h, 7-day framework for all activities. In addition to the use of information and communication technologies and remote sensing, the design of Smart Master Planning utilises domain-specific tools for many aspects of a city to realise the coordinated, effective, and efficient planning, management, development, and conservation that improve ecological, social, biophysical, psychological, and economic wellbeing equitably without compromising the sustainability of development ecosystems and stakeholders. This book presents twelve case studies covering more than twelve cities, or cities centred on domain-specific smart planning components.

Case studies of Digital Innovation Master Planning include the Application of Artificial Neural Networks in Master Planning for cities, a 3-D GIS Planning support system, and Digital Spatial Planning Incorporating machine-to-machine Automation for Smart economic communities and M2M digital integration.

Case studies of Domain innovation include urban land management, Master Planning for Water Management, Comprehensive master Planning Innovations, Smart use of Master Plan Basics, Integrated master Planning, and citizen-centric Master Planning [11, 12].

The following thirteenth and fourteenth books focus on containment, life, work, and the restarting of cities and regions affected by COVID-19, utilising selected empirical case studies. This book presents the spatial and temporal spread of the Coronavirus, analyses containment strategies employed, and includes recommended strategies. Further, it analyses how life and work transform during the lockdown and gradual opening and presents the future of work and life in cities impacted by COVID-19. This book explores the concept of smart life and work in cities post-COVID-19, ensuring that they do not compromise the quality of work and life, and do not create adverse economic and living consequences, such as the restart of a town after COVID-19 [13, 14].

Using selected empirical studies, these two books (urban and regional case) attempt to model Coronavirus spread spatially and temporally by GIS Modelling in study areas, then analyse all containment strategies so far applied and come out with the authors' recommended strategies which may not be that was used by city Government or Union Government. Selected Cities of particular interest are studied. This is a special interest because the states of Kerala and Maharashtra are among the worst affected in India by the COVID-19 pandemic, and the book focuses on this aspect with numerous studies [13, 14]. These 13th and 14th books were COVID-19, Containment, Life, Work, and Restart: Urban and Regional Studies.

The fifteenth book is the outcome of one of the four projects envisaged in the Indo-Pacific. The first chapter of the fifteenth book, "Indo-Pacific Smart Megacity System: Emerging Architecture," utilises selected empirical regional studies. This book focuses on regional case studies through themes such as delineation and profiles. Regional Economy, Interconnections, International trade, mobility, Climate Change, Disaster Management, Security, Institutions, and Governance, with an introduction and conclusions. The book proposes to model the Indo-Pacific Megacity System spatially and temporally by modelling study areas around the World, then analysing all strategies applied and presenting the authors' recommended approach. Then, design smart cities for life and work post-COVID-19, ensuring that life and work do not compromise the quality of life and cannot create adverse economic and living consequences. The Indo-Pacific System of Smart Megacity explores the Indo-Pacific Region. In the second part, Megacity studies individual megacities, such as Chennai, Kolkata, Mumbai, and Osaka, as part of the megacity system. Although the project initially consisted of two books, the editorial board, in their wisdom, decided to consolidate them into one book entitled "Indo-Pacific System of Smart Megacity: Emerging Architecture and Megacity Case Studies" [15].

This book is an in-depth study of the Indo-Pacific region, focusing on effective interventions in the megacities system. First, based on several criteria, the area is identified as a diverse yet homogeneous country group, characterised by diversity within sub-regions. A multi-polar spatial design becomes a reality if one explores megacity systems and as program regions of QUAD and I2U2 for action programs and investment, transcending many nations but mostly the ocean space of the Indo-Pacific, connecting all megacities' sub-regions spatially and functionally. Then, the megacities with both problems and prospects for economic integration are examined from the perspectives of regional economics and international trade. Finally, the rural–urban interface is presented, along with case studies of selected countries. Prospects for systems of megacities and individual megacities in regional economies are examined. Existing interconnections through rail, air, and ocean of megacity systems, as well as their capacity, performance, and potential, are analysed for emerging issues. International trade among megacity systems and countries with emerging problems and barriers is presented. The mobility of money, goods, and services among megacities is analysed. Rule-based diplomacy and other emerging options are discussed to sustain the above calls for a study of the Security of the Indo-Pacific region. Finally, the emerging architecture for megacity system governance is also presented. Out of 21 megacities in the Indo-Pacific region, an in-depth study was conducted on a few in India and Japan to inform effective economic interventions in the megacities system at the city level. The city has people and communities, as well as smart individuals in select locations within the geography of individual megacities and their communities, for collective, shared action.

The sixteenth book, and the second on Indo-Pacific studies, examines the design of core and peripheral digital economic communities. In smart, community-oriented economic development, these community entrepreneurs can investigate urban planning imperatives to reduce the marginal cost of production and services to near zero, thereby achieving price competitiveness. These communities can demonstrate Industry-4 innovations on a broad scale, which is yet to spread in megacities. These smart communities can implement smart mobility to support logistics for an innovative economy, industrial production, and service provision. The emerging city-specific architecture can vary significantly across all megacities in the region, with no single design type prevailing. It can vary based on locational advantage, the strength of the gateway and country infrastructure, institutional development, and policy support such as participation in regional commitments. COVID-19 has affected most countries in the Indo-Pacific. The number below the poverty level increased with a GDP contraction and a negative GDP growth rate. Foreign Direct Investment is not forthcoming in any of these countries. Job creation becomes a priority, alongside addressing public health concerns related to COVID-19. This opens up the possibility of investigating low-cost options of intelligent economic development possibilities of smart communities in selected locations in the megacity who act as entrepreneurs that reap the opportunities for knowledge-based products and services that meet the need of supply chains of Indo-Pacific region with a focus on megacity economy, income, and employment—essential and non-basic, multiplier, the current level of interconnectedness in the Indo-Pacific area, and infrastructure for megacity systems for integration of existing gaps [16].

The study teams from respective countries studied individual megacities. In contrast, the region is studied by teams located within the Pacific or outside, such as those from Italy and Slovakia within the European Union. Some of these countries, including France, Germany, and the Netherlands, have individual country-specific engagement policies, while others, such as Italy and Slovakia, do not. This editor aimed to present a balanced view of sectoral studies, free from the influence of individual countries' Indo-Pacific policies.

2 Design of the Collaborative Research Programme

Research Collaborations worked out are purely voluntary and without any financial support that appears to bind a project together. We could achieve this without money. Since collaborators include universities, governments, research institutions, professional networks, and not-for-profit associations, as is the general pattern in several countries, this approach provides complete independence for pursuing research, free from the ideological baggage of the granting organisation or its government. They need to refrain from accepting existing smart city policies of study cities in their research, as well as their governments' policies for engagement with the Indo-Pacific region. This was never an issue in the project's execution or timely completion. The project coordinator editor has no financial or administrative control over any institution participating in the project, as he did not receive any grants or distribute them. Typologies of the institutions involved in this international project: All these autonomous institutions are guided by the highest standards of scholarship and strive for the prompt completion of research and publication within the milestone dates.

3 General Research Questions on Smart City Research and Indo-Pacific Projects

Collaboration in this international research project requires that all participating institutions formulate research questions and methodologies appropriate to their country, federal state, or the Indo-Pacific region, where these study cities or areas are located. The approaches must differ depending on the type of City; some lead to Smart cities, but others decline and do not. However, the lack of empirical evidence on smart cities opens a new area of research: What strategy interventions bring about a Smart city? This is the central focus of the book series.

Do smart cities awaken social, cultural development, and ecological (environmental) management through smart city development? This question lies at the heart of the proposed international collaborative research program on the Pacific, and unpacking it gives us four interrelated research questions, as follows:

I. What constitutes a Smart city beyond its official definition by UN-HABITAT? This will need to identify the key ingredients and their role in creating smart cities and Driving Spatial changes in cities.
II. What changes does the Smart City bring to social development, cultural preservation, heritage conservation, and ecological management? Understanding the

interlinkages between a Smart city and social development, cultural protection, heritage conservation, and environmental management is crucial.

III. How and what processes help the changes to the Smart City? Do cities contribute to social development, cultural preservation, heritage conservation, and ecological management? These may include:

   (a) innovation–diffusion (by ICTs and other modes),
   (b) spatial planning,
   (c) sectoral planning (including economic, social development, cultural preservation, and ecological management),
   (d) Heritage conservation and management plan and
   (e) institutional and Governance processes, among others.

IV. How and what changes can be brought to improve the processes to achieve enhanced/best results? These changes relate to the various methods, as Research Question III mentions.

V. The study aims to gain a deeper understanding of the changes in the social, cultural, and ecological systems of the Smart City, driven by the advent of Smart Cities and Smart People, for innovative city development. This research program, in collaboration with the selected institutions as academic partners, aims to address this research gap.

4 Scope of Research

The following outlines the areas that may be covered when researching the "Smart City" program. This is a sign only; it is left to the team to decide what is appropriate.

I. A time-series study of changes in the urban parameters and finding distinctive features of the evolution of the Smart City.
II. Study of theories of the smart city interventions at the smart city level and modelling for the study city.
III. The concept of accessibility in the Master Plan and its changes to the increasing use of ICT in Smart Cities for smart city development.
IV. The increasing use of ICT to make a city bright influences the role of the hierarchy of service areas or watersheds in a Smart city.
V. Evolving Structure of Megacity Urban Agglomeration and Changes Needed in a Smart City When Transformed.
VI. The increasing use of ICT requires changes in the evolving structure of cities in urban agglomerations.
VII. Changes in Spatial Standards in a Smart Metropolitan City.
VIII. Changes required in zonal policies and plans.
IX. Study Town and Country Planning legislation and suggest changes to meet the Smart City requirements.
X. Change of role of community-based organisations (for example, Residential Associations in India) in a Smart city with an increase in the use of ICT.

XI. Change of role of the Ward Committee in a Smart city with an increase in the use of ICT.
XII. The role of the Municipal Council in a smart city has changed with an increase in the use of ICT.
XIII. Change the role of the Planning system in a Smart city with an increase in the use of ICT.

Note: The collaborating institutions can further elaborate on the scope of research, but it does not need to be uniform for all study cities. Each university department participating in this research program shall incorporate relevant Smart City Development features tailored to the goals of each department. The coordinator of this project does not intend to dictate the direction of the research and has a diverse group of collaborating universities. They should orient their study strictly based on the academic goals of their department.

5 Study Cities

The study city will be selected independently by each collaborating university as the study area, where a one-year and two-semester combined effort will be conducted to conduct this research. Universities participating in this program adopted various types of collaboration. Some universities utilised their doctoral and postdoctoral students, while others employed students at the master's and first professional degree levels. A post-doctoral student in the department can work on a narrow subject area in the study as individual work. While graduate and undergraduate students can work on designing solutions for Smart City, research institutions can charter their strategic research areas.

6 Project Details

Each collaborating university will independently select one city as the study area, where the one-year and two-semester combined effort will be conducted. Universities participating in this program adopted various types of collaboration.

The project details of the study city are given in Table 1.

Table 1 Project details TOC

Chapters	Title	Author
1	The Indo-Pacific E-Gateways in Digital Economic Communities	T M Vinod Kumar
2	Okun's law and digitalisation in ASEAN-10 economies	Martin Boďa, Mariana Považanová, Katarína Vitálišová
3	E-Global Gateways of Digital Economic Communities as Metropolitan Strategy for Anti-Fragility Experiences from Intelligent Megacities spatially and institutionally connected by international economic linkages	Antonella Contin, Fioralba Vela, Alessandra Pandolfi, Arxhenda Lipovica, Dicheng Yang, Simal Keskin, Javad Kianisadr, Tianduo Yan, Glen Carlo Canlas, Morteza Khajehahsani, Nika Haghpanah, Eugenia Cicero, Prarthana Ulaganathan, Aafreen Saleem
4	International Collaborative Research "Indo-Pacific E-Gateways: Cases from Digital Economic Communities" and Conclusions	T.M. Vinod Kumar

7 Ways of Working the Programme

7.1 Integrating Smart City Research with Academic Programmes

As tabulated above, this international collaborative research program and study were conducted by multiple diverse university departments, research institutions, and other organisations, as shown in the table above.

7.2 Role of Students

This international collaborative research program is designed for students, as they are the future and are part of an internal academic program at the university. We consider them to be the leading actors and shall give them a vital role in this program. Many in that age group will live in Smart Cities rather than their older faculty. Under the direction of the faculty, new concepts were introduced in the studio, and empirical studies were conducted around these concepts.

7.3 Role of Faculty

The faculty is the program's Designer within the framework of existing curricula in each participating university's design studios and theory courses.

I. The project duration is one academic year or two semesters.
II. They guide and monitor student work as usual in the academic program.
III. They check students' input to the monthly progress report.
IV. They rewrite the output of the project for a book to be published by an international publishers give due credit to their work.

7.4 Co-design and Co-production of Knowledge

This international collaborative research program is founded on the principles of co-design and co-production of knowledge. In today's interconnected world, such collaboration is physically and intellectually possible, thanks to the Internet and ICTs. The collaborative aspect of the research program will be actualised in the form of:

I. Co-design the program with the partner academic institutions.
II. Co-production of knowledge through an interactive process of sharing, reviewing, and completing research findings.
III. Within each partnering institution, the co-design and co-production of knowledge can be implemented through design studio or laboratory work between faculty and students.

7.5 Research Output

The key output of the "Smart City" research program will be these books, edited by the coordinator, Professor T.M. Vinod Kumar, to be published by Springer Nature, an internationally renowned publisher.

8 Summary of Conclusions

8.1 The Significance of Indo-Pacific E-Gateways in Digital Economic Communities

Below is a comprehensive summary and the conclusions of the chapter "The Significance of Indo-Pacific E-Gateways in Digital Economic Communities." The summary concerns seven major themes that capture the document's arguments and findings.

1. **Introduction and Overview**

This chapter sets the stage by introducing the concept of "E-Gateways" as an evolution of historical city gates that once controlled access, trade, security, and communication. In ancient times, physical gateways served multifaceted roles—from regulating the flow of goods and people to displaying public and governmental information. Today, those functions have been transposed into the digital domain. The chapter argues that as the world's economies shift toward digital modes of operation, these digital gatekeepers become crucial in connecting global digital economic communities.

The primary purpose is to illustrate how Indo-Pacific E-Gateways can serve as vital infrastructure to spur economic development and cross-border collaboration. These gateways are positioned as the new "front doors" to international commerce, e-governance, and even smart city constructs by enabling rapid connectivity and secure digital transactions. This initial section frames the discourse on digital economic ecosystems, emphasising digital trust, transparency, and the need for protocols governing data, transactions, and taxation across nations. It underlines the historical significance and the forward-looking imperative of integrating e-governance with economic strategies in an increasingly interconnected geopolitical space.

2. **Historical and Conceptual Foundations**

Building on history, the chapter revisits the evolution of the "gate" concept. Traditionally, city gates were not only defensive structures but also symbols of authority and sites of public information dissemination. Over time, as wars, technologies, and urban design evolved, the physical gate gave way to more complex networks of controlled entry, culminating in today's digital interfaces. The text draws on historical analogies to emphasise that even though the physical form has changed, the core notion of "gatekeeping" remains.

This transformation is essential to understanding how digital platforms mediate interactions among households, industries, and transnational trade corridors. The chapter details that digital economic communities are formed when individuals, entrepreneurs, and even regional governments coalesce into networks that operate on shared computing platforms. In this way, local economic dynamics—once managed by municipal or military authorities at physical gates—are now governed by integrated digital protocols. The fusion of entrepreneurship with digital infrastructure forms the underpinning argument: well-designed "E-Gateways" are fundamental to achieving the last-mile connectivity required in today's economic ecosystem, ensuring that even peripheral communities can participate meaningfully in global trade and e-governance.

3. **Technological Enablers and Emerging Innovations**

A core driver behind the shift to digital gateways is the rapid advancement of technology. The chapter details how the evolution from 2 to 5G—and the emerging promise of 6G—has transformed connectivity, enabling high-stakes applications such as telesurgery and remote industrial automation. Fundamental to the digital economic ecosystem are several enabling technologies:

- Internet and Broadband: Affordable and widespread access to high-speed broadband now underpins the digital transformation of communities.
- Mobile Technology and IoT: Devices such as smartphones, tablets, and networked sensors provide the physical interface through which digital communities interact.

  Artificial Intelligence, Machine Learning, and data Analytics offer the computational glue for securely processing transactions, automating routine tasks, and delivering personalised user experiences.

  Blockchain and quantum Computing promise enhanced security and transparency, ensuring that transactions, from digital signatures to cross-border payments, remain tamper-proof.
- Digital Public Goods: Innovations like the Unified Payment Interface (UPI) and initiatives like Digi Locker exemplify how public services can be digitised to reduce costs and improve trust.

The text argues that these digital innovations streamline trade and governance and are essential in enforcing transparency, accountability, and security. Such technologies create an ecosystem where the traditional roles of intermediaries are minimised, and entrepreneurial activities are amplified through secure, efficient digital processes.

4. **Regional Focus—The Indo-Pacific and Case Studies**

The Indo-Pacific region is singled out as a hotbed for leveraging digital gateways. This section describes how nations bordering the vast Indian and Pacific Oceans are uniquely positioned to benefit from e-governance and digital economic integration. Despite enormous cultural and political differences, these countries share common goals: accelerated economic development, enhanced regional trade, and coordinated security.

Two detailed case studies further illustrate these concepts:

1. Thrissur, Kerala (India):

The district of Thrissur is presented as a microcosm of a smart global economic community. The case study describes Thrissur's cultural heritage, infrastructural growth, urbanisation trends, and labour participation challenges. It highlights how robust digital infrastructure, embedded within local governance (via municipal projects and connectivity initiatives like fibre-optic networks), can support transforming local industries (e.g., traditional handicrafts, agro-based enterprises) into globally competitive digital economic entities.

2. Estonia's E-Governance and E-Residency:

Estonia emerges as a world leader in digital public services. By offering an innovative e-residency program, Estonia enables non-citizens to remotely access its state-of-the-art digital governance services—from digital banking and taxation to secure online signatures. This model demonstrates how trust, transparency, and efficiency in public service delivery can serve as a blueprint for other regions wishing to integrate digital economic communities into global supply chains.

Together, these case studies underscore that the practical implementation of E-Gateways can vary according to local strengths and challenges. Yet, the underlying objective remains the same: to create a seamless, secure bridge between local and global economic activities.

Global Digital Governance and Regulatory Frameworks

An effective digital economy relies on more than technology—it needs robust governance. This section delves into the role of international legal and regulatory frameworks that underpin digital transactions. Global digital governance entails establishing standardised protocols and guidelines that ensure data protection, facilitate secure cross-border transactions, and promote fair market practices.

The text highlights several initiatives and regulations:

- European Union's Acts: Proposals such as the Digital Markets Act (DMA), the Digital Services Act (DSA), and the Artificial Intelligence Act (AIA) are discussed as frameworks intended to control the power of major digital platforms ("gatekeepers") while ensuring fair competition and data portability.
- Data Protection Frameworks: Instruments such as the GDPR, the ePrivacy Directive, and complementary regional regulations create the legal basis for privacy, accountability, and secure digital identification.

Global Initiatives: The discussion extends to multilateral efforts by organisations such as the OECD and the World Trade Organisation, as well as initiatives within the Indo-Pacific region (e.g., QUAD, I2U2, and the India-Middle East-Europe Economic Corridor). These collaborations aim to harmonise standards and stimulate economic cooperation despite geopolitical uncertainties.

The chapter argues that establishing standardised e-governance stacks—protocols that allow interoperability between national systems—is essential for building trust. Without such harmonisation, digital communities would remain fragmented, hampering global trade and political cooperation.

5. **Future Trends and the Evolving Digital Economic Ecosystem**

Looking forward, the chapter identifies several emerging trends that promise to reshape both digital economic communities and the way work is organised globally:

Artificial Intelligence and automation: AI will improve data processing and decision-making efficiency and redefine the workforce. Routine tasks may be automated, while high-value roles in analysis, innovation, and creative problem-solving will gain prominence.

- Digital Transformation of Work and Education: With technology-driven changes, remote work and online learning will become mainstays. Adaptive learning platforms and digital tutors will help build skills and digital literacy, ensuring citizens can thrive in a digital economy.
- Cybersecurity and Data Integrity: As the volume and sensitivity of digital transactions increase, cybersecurity will become even more central. The evolution of blockchain and quantum-resistant protocols is expected to mitigate risks.
- Global Supply Chain Resilience: Digital ecosystems are predicted to allow a more agile response to geopolitical disruptions. With integrated digital networks,

supply chains can be rapidly reconfigured to avoid dependency on any single region, thereby cushioning economies against sanctions or political instability.
- Human-Centric Digital Transformation: Amid rapid technological change, the human element remains paramount. Building digital trust, ensuring inclusivity, and promoting AI ethics are critical to sustaining a vibrant digital economy.

These trends are not without challenges: digital divides, resistance to change, and the need for ongoing investment in infrastructure and skills development must be acknowledged and addressed. As global economic power shifts, digital sovereignty becomes a strategic imperative for nations seeking to secure their financial future.

6. Conclusions and Recommendations

Conclusions

The chapter presents a compelling argument that E-Gateways are not merely digital interfaces, but pivotal elements in creating and managing digital economic communities. Tracing the evolution from physical city gates to virtual platforms shows how historical functions—security, regulation, trade facilitation, and public communication—are now being reinterpreted within digital ecosystems. With its dynamic geopolitical landscape and rapid technological adoption, the Indo-Pacific region is an ideal setting for these innovations.

Key case studies from Thrissur and Estonia demonstrate that digital economic empowerment is achievable locally and internationally. Thrissur's integration of innovative urban planning with digital commerce illustrates the transformative potential for local economies. Estonia's pioneering e-residency model stands as a global exemplar of how government can re-engineer citizen services to operate seamlessly across borders. These cases underscore the vital interplay between technology, governance, and economic strategy.

Moreover, the chapter emphasises that sustainable digital transformation depends on robust technological infrastructure, progressive regulatory frameworks, and a human-centric approach. Successful digital economies must ensure interoperability between national systems, foster fair competition via international standards, and constantly update cybersecurity measures. Equally important is cultivating digital skills and literacy to support a workforce that can thrive in this evolving ecosystem.

7. Recommendations

1. Strengthen Digital Infrastructure: Governments in the Indo-Pacific and beyond should invest in robust broadband, 5G/6G networks, and next-generation digital platforms to ensure seamless connectivity.
2. Harmonise Regulatory Frameworks: Concerted international efforts, through initiatives like the EU's Digital Acts, QUAD, and I2U2, are necessary to standardise digital governance protocols across borders.
3. Promote Public–Private Partnerships: Successful E-Gateway projects require collaboration between governments, technology providers, and local communities to co-create secure, inclusive, and scalable digital ecosystems.

4. Invest in Digital Literacy and Education: To mitigate job displacement due to automation, policies must focus on skill development, lifelong learning, and adaptive education systems that prepare citizens for emerging digital roles.
5. Foster Innovation in Digital Governance: Countries should emulate best practices such as Estonia's e-residency and continuously innovate to ensure that public services are user-friendly, transparent, and accessible 24/7.
6. Build Resilient Supply Chains: Integrating digital platforms to reconfigure supply chains quickly in response to geopolitical disruptions will establish a resilient economic framework.
7. Adopt a Human-Centric Approach: As digital transformation accelerates, ensuring that technology serves citizens and addresses ethical, privacy, and equity concerns remains paramount.

The chapter clarifies that the successful design and activation of digital economic ecosystems hinge on a balanced interplay between historical insights, emerging technologies, and forward-looking governance frameworks. By embracing digital transformation at all levels—from local communities to global regulatory bodies—stakeholders can build an inclusive, resilient, and dynamic digital economy that benefits everyone.

This structured summary traces the historical evolution, technological underpinnings, regional deployment, global governance challenges, future trends, and final recommendations. It encapsulates the key themes and conclusions drawn from the chapter on Indo-Pacific E-Gateways in Digital Economic Communities.

8.2 *Okun's Law and Digitalisation in ASEAN-10 Economies*

The integration of digital technologies is a characteristic feature of contemporary economic and social life. Among the countries or groups of countries that significantly benefit from the rapid development of digitalisation is the Association of Southeast Asian Nations (ASEAN), comprising Brunei, Cambodia, Indonesia, Laos, Malaysia, Myanmar, the Philippines, Singapore, Thailand, and Viet Nam.

Digitalisation as a process has manifold impacts on society and the economy. It changes labour market conditions by affecting labour demand, the organisation of labour requirements placed upon employees and even the availability of jobs and their offerings. It also affects total economic production by enforcing sectoral changes and introducing new technical and organisational modes of enterprise operations. To better understand the impact of digitalisation on the economy, it is necessary to examine its effect on the relationship between labour markets and the real economy. These components of economic life are isolated here into unemployment and total output, and their relationship (the so-called Okun's law) is studied and matched against digitalisation and its trends. Despite considerable attention being paid to ASEAN countries, research on the relationship between unemployment and output in this region of the world, and how it is affected by digitalisation, remains limited.

This study addresses the aforementioned research gap by focusing on Okun's law across the ten ASEAN countries from 1991 to 2022 and explaining it, inter alia, in terms of digitalisation tendencies. Although Okun's law is merely an empirical relationship between unemployment and output changes, it is, in fact, a standard and globally recognised analytical framework for examining the relationship between economic performance and unemployment. A two-stage procedure is employed for this purpose. First, the time-varying framework of Schlicht's VC method is adopted, and Okun coefficients are allowed to update realistically in response to changes in the economic environment. The examination of the unemployment-output relationship is further extended by a second regression stage, in the hope of explaining the cross-country heterogeneity in the estimated Okun coefficients.

Although the motivation was to prove that digitalisation affects the intensity with which unemployment and output co-move across the business cycle, this ambition was only partially successful, at least for the 10 ASEAN economies over 1991–2022.

The results indicate that only in three ASEAN countries did Okun's law show regular patterns for the entire period, with both the difference and gap versions; that is, in Cambodia, Okun's law was found not to hold, while for Malaysia and the Philippines, it was found to be constant. For the other seven ASEAN economies, it displayed less stable patterns and frequent insignificance for most of the period, interrupted by consecutive or intermittent years of significant Okun coefficients. At any rate, the estimated values of Okun coefficients are not only unstable in terms of significance, but they are also small in comparison to typical estimates for advanced economies, which implies that unemployment and output variations or fluctuations are not sensitive for ASEAN economies, or this sensitivity is insubstantial. It means that economic growth and output fluctuations are not virtually reflected in unemployment, which is driven by other forces.

On the one hand, the results of the regression analysis reveal that digitalisation of society—proxied through metrics of access to, and use of, elements of the infrastructure transmitting digital services—was not ascertained to be relevant for the explanation of the differences in the estimated Okun coefficients. On the other hand, two factors associated with the digital transition were found to be appropriate, namely the share of medium- and high-tech manufacturing in GDP and the share of agricultural land. On average, across the ASEAN region, with a higher contribution of medium- and high-tech manufacturing to value added, Okun's law becomes somewhat stronger, and unemployment commoves with output more significantly. In all ASEAN countries except Singapore, the share of agricultural land increased during the period, while the share of agricultural production in GDP trended downward, with a high proportion of employment still concentrated in the farming sector. On average, in the ASEAN region, this increase in agricultural land was found to be associated with the strengthening of unemployment and output fluctuations across the business cycle, albeit year-on-year mutual changes were found to be weaker. That said, even though the magnitude of these effects is low, digitalisation of the economy, manifested through its several channels, was found to have an impact on the unemployment-output relationship in ASEAN economies.

8.3 E-Global Gateways of Digital Economic Communities as Metropolitan Strategy for Anti-fragility Experiences from Intelligent Megacities Spatially and Institutionally Connected by International Economic Linkages

The chapter proposes a Green and Digital Transition, a shift towards relational urban living that integrates communal existence, regenerative knowledge, and digital tools. The Twin Transition policy emphasises digital technology in metropolitan construction and the growing role of intangible components, such as knowledge, local culture, and labour. The research aims to develop guidelines for sustainable habitats in vulnerable territories in metropolitan regions, focusing on data-driven innovation and strategic partnerships. A "E-Gateway for digitisation in the metropolitan economy" strategy is proposed to enhance economic potential in Italian urban systems, including the development of digital infrastructure, skills enhancement, supportive policies, and collaborative partnerships. E-Global Gateways are digital platforms that facilitate global economic integration, enhancing local digital economy participation and strengthening the competitiveness of metropolitan systems in the worldwide marketplace.

The proposed vision for Europe's future focuses on integrating multi-level governance, civic innovation, sustainability, and digital sovereignty at the metropolitan scale. The E-Gateway will serve as a cognitive infrastructure, promoting economic, cultural, institutional, and anthropological exchanges. The Twin Transition will significantly impact the future, necessitating policymakers to address existing inequalities and socioeconomic effects. Strategies that embrace de-risking and anti-fragility principles are being implemented to enhance community well-being, preserve local identity, and safeguard natural heritage. The study will investigate the intersection of the Twin Transition with disparities in the Far East, Mediterranean, and Latino-American metropolitan areas. Anti-fragility, a concept introduced by Nassim Nicholas Taleb, refers to systems that thrive in the face of stress and uncertainty, promoting growth and improvement. By applying anti-fragility to metropolitan areas, cities can transform potential challenges into opportunities for growth, ensuring their vibrant, resilient, and sustainable future.

The concept of developing an AI-powered cartographic tool model to map territorial assets, with a focus on regional cohesion and sustainable development, is emerging from case studies. This multidisciplinary approach integrates historical, political, economic, environmental, and social data, providing a place-based approach for regional planning and policy-making. The approach utilises Metropolitan Cartography, Development, and E-gateway Strategies, as well as the Quintuple Helix Model, to generate synthetic maps, providing a holistic view of territorial interactions.

The multidisciplinary nature of the approach fosters cross-sectoral insights, benefiting public authorities, urban and regional planners, researchers, academic institutions, and local communities. Potential beneficiaries include public authorities and policymakers, urban and regional planners, researchers, and local communities, who benefit from improved governance and sustainability planning.

This approach not only enhances territorial knowledge and decision-making but also serves as a model for future applications in regional policies, addressing a complex, multidisciplinary challenge with high innovation potential. It could deliver innovative, interdisciplinary results that align with the objectives of EU territorial cohesion, sustainable development, and data-driven policymaking.

The multi-scale approach, focused on sustainability and the enhancement of cultural and environmental capital, represents a replicable model for other metropolitan contexts, as it extends beyond mapping and is a proactive system that integrates the physical dimension with the virtual one. This becomes even more important in the light of climate change, which affects economic communities and landscape architects are tasked with proposing interventions to improve city conditions.

The European Digital Public Goods represent shared, scalable, and democratically compliant infrastructures that can drive digitisation in partner countries inclusively and sustainably. The role of SMEs, support from public institutions, and integration of local communities are indispensable elements in making the EU not only an exporter of technologies but also a builder of collaborative and resilient platforms.

The European Union faces the strategic challenge of building a single, inclusive, and interoperable digital identity, with metropolitan cities emerging as central hubs to implement a federated vision of the European digital infrastructure. A federated metropolitan network would promote a culture of interoperability and facilitate the secure circulation of digital identities between territories, avoiding duplication, enhancing transparency, and reducing administrative costs.

Multi-level and participatory metropolitan governance, based on the quintuple helix of innovation (environment, citizens, research, industry, and institutions), finds the space to be practiced in metropolitan cities, becoming a laboratory and platform for a polycentric digital ecosystem capable of integrating public and private resources, technological infrastructures, and social capital in the construction of an equitable, open, and resilient European digital identity. In this scenario, digital diplomacy, a distinctive feature of Europe, emerges as a fundamental pillar for science diplomacy.

References

1. Vinod Kumar TM et al (eds) (2014) Geographic information systems for smart cities. Copal Publishing Group, New Delhi
2. Vinod Kumar TM (ed) (2015) E-governance for smart cities. Springer, Singapore
3. Vinod Kumar TM (ed) (2016) "Smart Economy in Smart Cities" International Collaborative Research: Ottawa, St Louis, Stuttgart, Bologna, Cape Town, Nairobi, Dakar, Lagos, New Delhi, Varanasi, Vijayawada, Kozhikode, Hong Kong. Springer, Singapore
4. Vinod Kumar TM (ed) (2017) E-democracy for smart cities. Springer, Singapore
5. Vinod Kumar TM (ed) (2018) Smart metropolitan regional development: economic and spatial design strategies. Springer-Nature, Singapore

6. Vinod Kumar TM (ed) (2019) Smart environment for smart cities. Springer-Nature, Singapore
7. Vinod Kumar TM (ed) (2020) Smart living for smart cities: case studies. Springer-Nature, Singapore
8. Vinod Kumar TM (ed) (2020) Smart living for smart cities: community study, ways, and means. Springer-Nature, Singapore
9. Vinod Kumar TM (ed) (2021) Smart global megacities: Tokyo, Mumbai, New York, Hong Kong-Shenzhen, Kolkata. Springer-Nature, Singapore
10. Vinod Kumar TM (ed) (2021) Smart global megacities: Chennai, Kochi-Kannur. Springer-Nature, Singapore
11. Vinod Kumar TM (ed) (2022) Smarter master planning: case studies of digital innovations. Springer-Nature, Singapore
12. Vinod Kumar TM (ed) (2022) Smarter master planning: case studies of domain innovations. Springer-Nature, Singapore
13. Vinod Kumar TM (ed) (2022) COVID-19, containment, life, work, and restart: city studies. Springer-Nature, Singapore
14. Vinod Kumar TM (ed) (2022) COVID-19, containment, life, work, and restart: regional studies. Springer-Nature, Singapore
15. Vinod Kumar TM (ed) (2023) Indo-Pacific system of smart megacity: emerging architecture and megacity case studies. Springer-Nature, Singapore
16. Vinod Kumar TM (ed) (2025) Indo-Pacific core and peripheral digital economic communities. Springer-Nature, Singapore

Professor T. M. Vinod Kumar has 55 years of experience in Urban Planning as an author, editor, teacher, researcher, and adviser/consultant. He has worked in many Indo-Pacific countries such as India, China, Bhutan, Nepal, Malaysia, Indonesia, and Hawaii, USA. He was Dean of Studies, Head of the Department of Urban Planning, Head of the Centre for Systems Studies and Analysis, Head of the Centre for GIS and Remote Sensing, and Centre for Urban Studies of the School of Planning and Architecture, New Delhi; Visiting Professor National Institute of Technology, Calicut, and Institute of Technology Bandung, Indonesia and Professional Associates, East-West Resources Systems Institute Honolulu, Hawaii, Fellow Centre for the Study of Developing Societies, Delhi, Project Manager in Council for Social Development, New Delhi, Regional Program Coordinator at the International Centre for Integrated Mountain Development (ICIMOD) and Planner-Engineer at the Ford Foundation. He is the author of many books and journal articles. "Innovation" (ed) and edited "Geographic Information System for Smart Cities" (Copal:2014), "E-Governance for Smart Cities" (Springer:2015), "Smart Economy in Smart Cities" (Springer:2016) and "E-Democracy for Smart Cities" (Springer:2017), "Smart Metropolitan Regional Development: Economic and Spatial Design Strategies" (Springer:2018), "Smart Environment for Smart Cities" (Springer:2019), "Smart Living for Smart Cities Case studies" (Springer:2020), "Smart Living for Smart Cities Community study and Ways and Means" (Springer:2020), "Smart Global Megacities Collaborative Research-Tokyo, Mumbai, New York, Hong Kong-Shenzhen, Kolkata" (Springer 2021) and Smart Global Megacities Collaborative Research-Chennai and Kochi-Kannur"

(Springer:2021); "Smart Master Planning: Case Studies of Digital Innovation" (Springer: 2022), "Smart Master Planning: Case Studies on Domain Innovations" (Springer:2022); "COVID-19, Containment, Life, Work, and Restart, Urban Studies" (Springer: 2022); "COVID-19, Containment, Life, Work, Restart, Regional Studies" (Springer: 2022); "The Smart Megacity System of Indo-Pacific: Emerging Architecture and Megacities Studies" (Springer: 2023); "Indo-Pacific Core and Peripheral Digital Economic Communities" (Springer: 2025).

MIX
Papier aus verantwortungsvollen Quellen
Paper from responsible sources
FSC® C105338

If you have any concerns about our products,
you can contact us on
ProductSafety@springernature.com

In case Publisher is established outside the EU,
the EU authorized representative is:
Springer Nature Customer Service Center GmbH
Europaplatz 3, 69115 Heidelberg, Germany

Printed by Libri Plureos GmbH
in Hamburg, Germany